Population-Environment Dynamics

Population-Environment Dynamics

Ideas and Observations

Editors:
Gayl D. Ness
William D. Drake
Steven R. Brechin

Ann Arbor
THE UNIVERSITY OF MICHIGAN PRESS

Published in the United States of America by
The University of Michigan Press
Manufactured in the United States of America

1996 1995 1994 1993 4 3 2 1

Population-environment dynamics : ideas and observations / editors,
Gayl D. Ness, William D. Drake, Steven R. Brechin.
p. cm.
Includes bibliographical references and index.
ISBN 0-472-10395-4
1. Environmental policy—Congresses. 2. Population—Environmental aspects—Congresses. I. Ness, Gayl D. II. Brechin, Steven R., 1953- . III. Drake, William D., 1936- .
HC79.E5P666 1993
304.2-dc20 92-41581
CIP

A CIP catalogue record for this book is available from the British Library.

In Memory of

George S. Simmons
1940-1990

George Simmons was an esteemed colleague and friend. He chaired the Department of Population Planning and International Health in the last years of his life. He was one of the founders of our Population-Environment Dynamics Project, enriching it immensely with his vision and wisdom. One of his students remarked wisely at his memorial that much of what we are now we have gained from George; what we do from now we give back to George. We offer this volume to George.

Royalties from this volume will be contributed to the George S. Simmons Memorial Fund, Department of Population Planning and International Health, School of Public Health, University of Michigan, Ann Arbor, MI 48109-2029.

Preface

This volume is the product of efforts of the University of Michigan faculty to grapple with theoretical and practical issues of population and environment interaction. The real world is one in which population and environment are closely interrelated. Yet it is precisely this interaction that has been neglected in teaching and research. These are truly interdisciplinary problems which demand interdisciplinary approaches and the close interaction of theory and action. The approach at Michigan reflects the objective of keeping theory building close to empirical reality and practical problems. It has also been largely inductive and eclectic. We begin with neither grand theory, nor quantitative models from which specific research questions can be derived.

There are, to be sure, some general models and equations that have addressed the population environment issue. The *Limits To Growth* work of Meadows (1972)[1], for example, represents an early attempt to produce a global model for future extrapolation from the current industrial system. A more recent and very general formulation is found in the work of the Ehrlichs (1990). They propose a simple model, $I=PAT$, where environmental impact (I) is a function of population (P) times per capita consumption (A) times the environmentally degrading technology that provides the consumption (T). A slightly more precise and specified set of equations is also found in the work of Paul Harrison (1990 and 1992).

Our approach differs from these models not because we think they are flawed or ineffective. Rather, it is dictated by what can be called our institutional character, which determines what we can do best. A brief history of our Population-Environment Dynamics Project makes clear how and why we adopted this inductive and eclectic strategy.

Faculty from the University of Michigan Schools of Natural Resources and Public Health (Department of Population Planning and International Health) began to meet together in the fall of 1987 to discuss possible collaboration. We knew that Natural Resources was doing effective research and training in environmental issues and management. Similarly, in Public Health, we felt we had effective training and good research in population planning. We recognized, however, that no population analysis was included in environmental research, and we were rather blind to environmental issues in our population work. It was also apparent that we were poorer for this lack of collaboration. Our discussions were aimed at enriching both sides.

---

1. See also Meadows and Meadows 1973, and for critiques see Cole 1973, and Roberts 1975.

The problem was one of organizing faculty and students from traditionally separate areas into research and training programs that would promote interdisciplinary integration. The University of Michigan is characterized by what organizational analysts call a flat structure. Collaboration and cooperation are not built by directive, but through generating personal interest. Thus a strategy that would attract the hearts and minds of the faculty was needed. This called for the creation of an interdisciplinary research seminar. Initiated in January 1989, the biweekly seminar marked the beginning of the collaborative effort that eventually led to this volume. The seminar was designed to provide faculty and students with a forum to present their research to people with backgrounds in environment, policy, health, global change, population, and engineering.

In January, 1992, the seminar enters its fourth year, with a regular attendance of fifty to seventy members. In the seminars, discussions move from examinations of detailed research within particular areas to broad theoretical questions raised by individuals from other disciplines. The seminars illustrate that a true understanding of population or environment cannot be achieved without an examination of their dynamic interaction. This cannot be achieved within the intellectual and cognitive confines of a single discipline. Through these sessions, participants also found that generating an interdisciplinary approach is a rich and rewarding intellectual challenge.

Interaction between seminar participants led to a series of interdisciplinary pilot projects. Students and faculty from Natural Resources, Public Health, Engineering, and Biology, among others, have been involved in pilot projects in twenty-five countries. Ten research projects in Population-Environment Dynamics were funded in the summer of 1989, fourteen more were launched in 1990, and grants for another seventeen projects were awarded for 1991-1992. They have explored everything from the relationship between family size and wildlife resources in southern Africa to the impact of war on environment and health. The pilot projects focus on practical situations where these issues must be confronted in all their complexities and provide the raw empirical data necessary to develop and test theories of population-environment dynamics.

In October 1990, the University of Michigan hosted an International Symposium on Population-Environment Dynamics to bring together recognized leaders in this area. The meetings, which represented the culmination of two years of consideration of population-environment dynamics, focused on identifying the underlying theoretical issues and developing a research agenda for the field. Experts from the University, other research and teaching institutions, and a number of government agencies throughout the world participated in these discussions. The papers presented in this volume are the combined product of the international symposium and the interdisciplinary pilot projects generated from the seminar series.

The symposium and its resulting papers represent only one step in a journey. It was recognized from the start that the field is broad and the issues cover a wide range of disciplines and subject areas. One cost of our inductive strategy was the omission of important substantive issues and geographic areas. At Michigan, much work has been done in Asia and Africa, less in Latin America. A great deal of research has focused on population (especially fertility decline), on natural resources, agriculture, and deforestation. Yet, it is clear that a whole range of issues such as urban problems, the metabolism of the industrialized world, pollution and waste management, oceans, and energy still remain outside the purview of the activities reflected in this volume. A number of steps are being taken to expand our coverage. The specific set of relationships involved in population and environment are now set in the broader framework of global change, where faculty at Michigan feel they belong. Since 1990, the Population-Environment Dynamics Project has been an integral element of a newly formed Project for the Integrated Study of Global Change. Plans are now underway at the University to develop a Institute for the Study of Global Change which will promote a broader series of specific projects in relation to population, environment, and global change. These areas leave an ample agenda for future activities.

Gayl D. Ness
William D. Drake
Steven R. Brechin
University of Michigan
Ann Arbor, Michigan
October, 1992

Acknowledgments

Throughout this project, we have been assisted by the generosity of major funding sources. The MacArthur Foundation has provided financial support for the Population-Environment Dynamics Project as well as valuable insights and excellent substantive and technical suggestions along the way. We are especially indebted to Dr. Daniel Martin for his early encouragement and continuing support. The United Nations Population Fund (UNFPA) provided an initial grant to support our first summer of research projects and a portion of the International Symposium. Our own university contributed early seed monies as well as substantial support for the larger Project for the Integrated Study of Global Change. The University's Center for South and Southeast Asian Studies acted as cosponsor for the Symposium. The Principal Financial Group also provided a generous grant to assist in the publication of this volume.

We would also like to thank, for their invaluable assistance in formatting and editing, Sandra Thomas, Clara Haggerty, Danielle Gordon, Kurt Slawson and, for their added contribution in constructing the graphics, Mita Sengupta and Meera Sripathy.

Contents

CHAPTER

I. Introduction 1

SECTION I

Global Perspectives: History, Ideas, Sectoral Changes, and Theories 11

II. The Missing Links: The Population-Environment Debate in Historical Perspective 17
Michael Teitelbaum, Jay Winter

III. The Long View: Population-Environment Dynamics in Historical Perspective 33
Gayl D. Ness

IV. Constraints on Sustainable Growth in Agricultural Production 57
Vernon Ruttan

V. Population as Concept and Parameter in the Modelling of Deforestation 71
Alan Grainger

SECTION II

The State as Actor: Population-Environment Dynamics in Large Collectivities 103

VI. The Powers and Limits of State and Technology: Rice and Population in Southeast Asia 109
Gayl D. Ness

VII. Population Policy and Environmental Impacts in Rural Zimbabwe 133
Alison McIntosh

VIII. Indonesia: Stresses and Reactions 153
Kartomo Wirosuhardjo

IX. The Phases of Agricultural Modernization in Brazil 167
George Martine

SECTION III

The State as Environment: Population-Environment Dynamics in Small Communities 187

X. Resource Control, Fertility and Migration 195
Bobbi S. Low, Alice Clarke

XI. Protected Area Deforestation in South Sumatra, Indonesia 225
Steven R. Brechin, Surya Chandra Surapaty, Laurel Heydir, Eddy Roflin

XII. An Ecosystem Approach to the Study of Coastal Areas: A Case Study from the Dominican Republic 253
Richard W. Stoffle, David B. Halmo, Brent W. Stoffle, Andrew L. Williams, C. Gaye Burpee

XIII. Population-Environment Dynamics in a Constrained Ecosystem in Northern Benin 283
Valentin Agbo, Nestor Sokpon, John Hough, Patrick C. West

SECTION IV

Emergent Ideas: Theory and Method 301

XIV. Towards Building a Theory of Population-Environment Dynamics: A Family of Transitions 305
William D. Drake

XV. Perceiving Population-Environment Dynamics: Toward an Applied Local-Level Population-Environment Monitoring System 357
Frank D. Zinn, Steven R. Brechin, Gayl D. Ness

SECTION V

Summary, Conclusion, and Next Steps 377
Gayl D. Ness, William D. Drake, Steven R. Brechin

References 407

Contributing Authors 439

Index.. 441

Chapter I

Introduction

Population and Environment: An Unnatural Separation

Population and environment are topics of intense current discussion and international action. Population growth rates have given rise to the idiom of a bomb exploding in the world. Great hoards of famine refugees fill television screens, alongside masses of street urchins produced and abandoned by poverty-stricken parents. In a lighter vein, tee-shirts sport images of the earth with people stacked up and falling off. The international community has organized to address the problem of population growth with substantial amounts of international assistance, and the creation of hundreds of specialized organizations to reduce human fertility. Environmental issues also crowd the human agenda. Oil spills, droughts and famines, rain forest destruction, and the spread of toxic materials appear daily in the news, rendered dramatic and immediate by television. Scientists portray the earth as a greenhouse whose recent atmospheric changes carry the prospects of rising world temperatures, and a future that reads like science fiction. In response, the international community is mobilized around issues of environmental protection, generating funds, organizations and projects in an attempt to mediate or arrest some of the massive destruction the human species is wreaking on the planet.

Despite the crowded agenda and the intense discussion, population and environment relationships still defy clear and comprehensive understanding. While they are closely intertwined in real life, they are separate in much of our thought and action. Public and private agencies and scientific disciplines define the two topics as specialized and consequently address them separately. A recent colorful presentation of the United States Agency for International Development (USAID) environmental work displayed great activity in a variety of settings, but contained not a word on population. The vast population establishment[1] has given extensive attention to growth rates and fertility decline, but until recently was virtually blind to environmental conditions. Why should this be so? Why are these two areas of thought and action so divided when their subjects are in reality so closely intertwined? Further, is there any way to bring the two closer together, to increase the dialogue between those who carry the interests of each?

These two questions--why the separation and how to integrate--provide the driving force for this volume. The second question, how to integrate, will

---

1. The term was coined by J. Nicholas Demerath (1976) in *Birth Control and Foreign Policy: The Alternatives to Family Planning*. New York: Harper and Row.

underlie much of the discussion in the chapters and will be the specific focus of the final chapter. First, however, we must try to explain the separation.

The separation of the two areas in our thoughts and actions can be attributed in large part to three types of imbalances: conceptual, organizational, and power. We can identify them briefly here and note that they are elaborated explicitly by Teitelbaum and Winter in their discussion of the continuities in our thinking about population and environment (as well as by a number of other authors in this volume).

Conceptual Imbalances

Conceptually, population is simple, narrowly conceived, and easily confined. Demography's six variables[2] and stable population theory make population conceptually quite self sufficient. It can be easily grasped, observed and measured, and the measures permit projections and interpolations that have a high degree of validity. The environment, on the other hand, is broad, ambiguous, conceptually messy, apparently boundless, and anything but self sufficient. Rowe (1989) called it "an obscuration, a grab bag of elements so hazy in their relationships that attempts at structured thought about them face certain frustration." Note that it is almost impossible to conceive of the environment without thinking about such things as technology and organization, which themselves lack powerful taxonomies and clear boundaries.

Organizational Imbalances

Organizationally, population is dominated by one academic discipline, demography, which is located more in sociology than in any other discipline. There are, of course, economists, geographers, and a few political scientists who deal with population, but sociology is the dominant home and within that home, demography is a highly cohesive discipline. The environment, on the other hand, is the subject of many specialized disciplines, from atmospheric science to zoology and almost everything in between. All these disciplines, including demography, have specialized academic curricula, distinctive research technologies, professional organizations and journals. These are essentially political structures, which carry the power to define boundaries and certification processes. Such structures may well strengthen the detailed pursuit of a specific type of research, but they also inhibit communication across the boundaries they carefully erect and jealously protect. The same can be said of government agencies. We organize action around specialized

---

2. Births, deaths, and migration constitute the dynamic variables of population; size, age-sex composition and geographic distributions are its most commonly used comparative static variables.

sectors, such as health, agriculture, population, environment or natural resources. We know that specialized action does work to promote specific ends, such as vaccination against infectious diseases or protecting forests, but we also know that such actions often cannot take account of the larger aims (public health, sustainable development) for which these are in some sense only means. Further, we know that the ability to coordinate interorganizational activities, in the public or the private sector, remains one of the most critical stumbling blocks to organizational performance in general, and to the promotion of sustainable development more specifically.

Power Imbalances

Finally, in power, the imbalances are reversed. Here population becomes the difficult partner, especially since today the central population issue is that of the rate of growth. Although population policies to limit growth often emerge from rational considerations of public health or national economic and demographic goals, they tend to be far more controversial than environmental policies. Population issues, after all, touch deeply held, primordial values and sentiments. In all societies, ethnic or national identity, gender roles, sexuality, and reproduction imply core values that are highly elaborated, well articulated, and reinforced by our most powerful national institutions, those codified in what we call religions. These are the values that affect population character and dynamics. As deeply held primordial values, they are essentially nonnegotiable. One cannot easily make compromises on these fundamental issues of morality, or on what is essentially the identity and the viability of the group. Teitelbaum and Winter point out how ideological the population arguments are, and how sides are taken and held with great intolerance for other positions. The current world-wide struggles over abortion were recently and aptly termed "the Civil War."[3] Many environmental issues, on the other hand, tend to be more economic than moral.[4] They involve calculable costs and benefits, and are thus much more capable of being negotiated. Policies can produce prices that either destroy or protect the environment, and those policies can usually be changed readily, especially through a calculation of costs and benefits. To be sure, all economic interests attempt to enlist or generate political power that will subsidize their interests. They are sometimes quite successful, often at the expense of portions of the environment. But even

---

3. This was the subject on a National Broadcasting System television debate led by Peter Jennings in November, 1990.

4. To be sure, some environmentalists attempt to make the issues moral ones, for example, with the argument for species protection on the moral grounds of the intrinsic value of all life. But there are also environmentalists who argue primarily on the technical grounds of the failure of economics to account properly for the value of overall ecosystems and for a future of sustainable activity.

conflicts between political interests can be negotiated when it is possible to assess the costs and benefits of those interests.

With these profound imbalances, it is not difficult to understand why population and environment are more noted for their separation than integration in our thought and action. These significant imbalances raise three fundamental questions:

The Fundamental Questions

To begin to understand the dynamic relationship between population and environment, we will ask the following questions of the papers in this volume.

What is population?

What is the environment?

What is the character of their relationship?

These three questions will also be used to introduce each of the book's three sections, thus providing a framework that draws the disparate papers together into a more coherent whole. The three sections of the book reflect a major division of the *scale* of observations of population-environment interactions. Especially today, many observations are at the *global* scale, where natural or social processes are examined not necessarily with respect to a specific location, but with the entire world. Other observations commonly use the nation state as the unit of observation and analysis. We designate this the *macro* level. These observations reflect the historical rise of the nation-state and its importance today as a form of social organization. Finally, many observations are concerned with small scale societies, local communities or groups, below the level of the nation-state. We designate this the *micro* level, noting that the state is a major part of the environment for these small societies. These three levels of observation and analysis define the three sections of the book.

The volume is an attempt to use the papers as a basis for a fuller consideration of population-environment dynamics. This implies the addition of chapters that were not presented at the first Population-Environment Dynamics Symposium in October 1990. The first section of the book deals broadly with the historical change of the population-environment relationship on a global scale, and with the human attempt to make sense of the relationship. The second chapter, reviewing the past thousand years of population was added to set the other papers in the specific and very important historical context of the rapid population growth of the past three centuries.

Without this recent demographic change, we would not now be addressing the issue of population and environment dynamics. Sections II and III began as simple divisions of the empirical papers by scale, distinguishing studies of whole states from those of smaller communities. This led to an important observation of the role of the state in the population-environment relationship, which is now reflected in the titles of these two sections. In Section II, we find that the state is a major *actor* in the population-environment relationship. In Section III, the papers deal with smaller social systems, where the state becomes part of the *environment* of the small societies within which the population-environment relationship is worked out. Section IV presents innovations in both theory and methods. Drake's theoretical chapter attempts to deal more systematically with the *dynamics* of the interaction, by conceiving of them as a family of transitions. Zinn, Brechin, and Ness outline an emerging technology that is designed to permit the actors in local settings to observe their own population-environment relationships in a new integrative way. Our concluding chapter is our attempt to draw some of these disparate strands together and to suggest a research agenda for the future.

Although each of the three sections is introduced by asking the three basic questions, some general observations about these questions are provided here to help us identify some of the underlying themes that will emerge from the chapters.

What is Population?

Population is, in most respects, the easiest concept with which to deal. Some of the chapters define population simply as total numbers, others identify many of the critical dimensions of variance in the human population. Especially important are the forms of organization that place individuals in different socially defined positions. Nation, ethnicity, race, religion, language, class, family and occupation all emerge as critical dimensions of variance that affect the population-environment relationship. There are also qualitative differences in populations that indicate something of the outcome of the relationship. These vital signs of a population, birth and death rates, life expectancy, infant mortality, and educational attainment, mark the difference between more and less *healthy* populations. As we move from global to micro levels of analysis, we see that the view of the population becomes more and more complex as more internal dimensions of variance are added.

It is in the population dynamics that we have seen one of the most remarkable transitions of the modern period. As Ness points out, the demographic transition represents a major transformation of the character of human life. For thousands of years, the human population grew only slowly, and was marked by high birth and death rates. The radical, sustained decline of death rates has contributed substantially to the world's current problems of population pressure and environmental degradation. The more recent decline of fertility draws lines of conflict over basic values, and draws increasingly complex observations of *human groups* into the center of the population-environment dynamics. Teitelbaum and Winter examine the deep ideological conflicts that have arisen out of this transition, as it portends a potential decline of a population, or more importantly of a specific sub-group of the population.

What is the Environment?

Environment is far more difficult to define. There are at least two major problems presented by the concept: vagueness and the conflicts between various intellectual traditions. In any intellectual tradition, the term is very messy. Section One and the Teitelbaum and Winter chapter, identify the different meanings of the term and show how they have changed with time. Adding to the messiness is the fact that we have at least two very different intellectual traditions, the natural and social sciences, which give us two very different environments: the natural or biogeochemical environment and the social systems that are also part of the environment.

Ecologists have given extensive consideration to conceptualizing the environment. Rowe (1984) reviews the history of these attempts and makes a series of important proposals. He proposes first, that we see the entire planet as an integrated, coherent *ecosphere*. This "thin life-filled skin of air-water-earth that girdles the globe" is a complete system. He argues further that this *ecosphere* is the only complete ecosystem that science knows. This ecosphere can be divided into *ecosystems*, structured segments of earth-air-organisms. Ecosystems are properly viewed as three-dimensional spaces, or *landscapes*. Their vertical dynamics-leaf fall, biogeochemical cycles, infiltration, transpiration, etc.--constitute the physiology of the system. Horizontal dimensions are the ecological anatomical character of the ecosystem.

Much of Rowe's purpose in this review is to argue against viewing *communities* as distinct, organic entities. He finds that ecologists have too often defined ecosystems in terms of biotic communities, likening them to living organisms. Countering this argument, Rowe contends that communities are simply aggregations of individuals. Ecosystems, on the other hand, are the conceptual equivalent of organisms. Both have a three-dimensional integrity, which communities lack. Communities are quite different things. The proper units of analysis, therefore, are *landscape ecosystems*, three-dimensional pieces of the ecosphere. Referring to Schreiber (1977), Rowe argues that the focus should be neither on vegetation nor soils, but on the landscapes of which these are a part. These can be called, with Jenny (1980), *ecotesseras*, "whose vert space above and soil space below are coupled components."

This focus on three dimensional spatial elements provides useful insights for our attempts at understanding population and environment, especially with our new technology of remote electronic sensing. This gives us a new way to view the landscapes that are the basic elements of the environment. In the chapter on Population-Environment Monitoring, Zinn, Brechin, and Ness use this technological innovation to suggest strategies for promoting local participation in addressing current population-environment problems.

When we deal with the human population, however, the rejection of *communities* is quite inappropriate. It may well be true that animal and plant communities are simply aggregates of individuals, but for humans that aggregation is always more than simply a sum of the individuals. Consider for example, the difference between antelope populations and human populations on the Great Plains of the United States over the past five hundred years. The human populations certainly have been marked by far greater variety in *communities* than the antelope. The capacity for symbolic communication has always meant that human communities can *create* structures of organization and technology that are far from being genetically determined. They are therefore more variable, more flexible and less environmentally constrained than are the structures of organization of other life species.[5] Human structures of organization can change much more rapidly than those of other life structures, as we have seen in the recent radical change to urban industrial organization that gives us the current problem of population-environment interactions. This, of course, further confounds the concept of environment, for in the human population, we cannot speak of the environment without drawing in technology and social organization. Humans create rules for interaction and they create technologies, both of which change the character of the resources in the landscapes. Humans also create ecologically arbitrary boundaries around parts of the landscape, which have a profound impact on how that landscape is used.

We shall see the import of this distinctive character of the human community repeatedly in the papers in this volume. It emerges clearly in the political boundaries that surround specific landscapes. The impact of the political element in action, such as the modern state, penetrates local communities that previously had some control over a specific landscape. We can also see the import as we look ahead and note the need for different human institutions that recognize the limitations of those political boundaries and call for specific actions to protect the entire ecosphere. The ecological referents to the environment are important, and should be drawn more fully into our discussions, but it will also be necessary to deal with the distinctive character of the environment that is created by the human species.

---

5. See Ness and Ando (1984) for a more extended treatment of this condition of human organization, and its application to the study of population planning in Asia.

What is the Population-Environment Relationship?

The most important question is clearly, what is the character of the population-environment relationship. As all ecological perspectives would have it, the population- environment relationship is dynamic. Things are never still, they are always changing. While there may be movement toward some abstractly contrived equilibrium state, this state is never achieved. But although there is constant change in the ecosystem, not all changes are alike. Some changes are gradual and almost routine. Rivers change course, ponds fill in, species succeed one another in natural processes that can be gradual or cataclysmic. For the human species there are also constant changes. Generations come and go, each one both relying on past behavior and innovating to respond to the new problems arising from those "solved" by the past generation. The stability of tradition, it should be noted, is often more a mythical dream than an historical reality.

Some of the natural and human transitions can be seen as gradual ebbs and flows of the process of life. Others are more dramatic *transformations*, changing forever the character of the population-environment relationship. In human history, the advent of agriculture and cities represent such radical transformations in the distant past. In just the past three centuries, rapid population growth and the associated transformation from rural agrarian society to urban industrial society are examples of radical transformations. It is, in fact, the latter that has led to the current problem. Without that change, the world's population probably would not be more than one or two billion, and we would not, perhaps, be discussing issues of population-environment dynamics with quite the same urgency as we are today.

Drake observes that the chapters of the volume deal with various types of *transitions*, or rapid change from one relatively homeostatic condition to another. The period of rapid change provides a time of great potential and a point of special vulnerability for a society. A proposition to emerge from his analysis is that the speed of the transition increases the vulnerability of a society. Drake also notes that our modern social transformation is marked by a number of transitions, which can be conceived of as a *family of transitions*. This permits him to ask how the different parts or members of the family fit together. Is there a regular pattern of transitions, or are they chaotically strewn over the historical landscape? If there is a pattern to the transitions, what shapes that pattern? In the final chapter, we are led to ask how human institutions affect the speed and character of these transitions. Finally, we can ask whether how the different transitions are linked together matters to the ultimate quality of human life.

All authors see the population-environment interaction as a complex and dynamic set of relationships, and all see some aspect of human technology and institutions as mediating the relationship. Perhaps most important is that all see the relationship as a qualitative variable: it can be *better or worse*, more or less healthy, especially for the human population. Here again a reference to the ecological literature is useful, especially its attempt to conceive of the *health of the ecosystem* (Rapport 1989). We can find similar attempts in the way both the natural and social sciences deal with this qualitative outcome of the interaction. Common approaches in ecology have included the use of vital signs to identify healthy from sick ecosystems. These have clear analogues in epidemiology and public health. There are also attempts to conceive of ecosystem health through an observation of a system's capacity to recover from major stresses or perturbations. These have their human analogue in economic analysis. Finally, there are approaches through the focus on stresses, or risk factors, which we find currently at the forefront of much medical sociology and public health. All of these approaches are useful, and will have to be fully incorporated into the future planning that will be required if the human population is to survive. As we shall see, although the papers in this volume may deal only partially or indirectly with this dimension of the population-environment dynamic, behind every discussion there is some idea of the *varying quality* of the outcome.

Finally, many of the chapters at least indirectly refer to the development of new *monitoring technologies* as critical for mediating the population-environment dynamic. From earlier population censuses, through the recent monitoring of atmospheric changes, to remote electronic sensing, human populations have developed immense capacities to observe their relationship with their environment, and this observation has always had a substantial impact on the population-environment relationship. It might not be too much to suggest that it is precisely this monitoring capacity that may permit the human population to create a more sustainable productive organization and thus to save itself from possible extinction. In the penultimate chapter, simple new monitoring technologies, *Population-Environment Monitoring Systems* (PEMS), are discussed as a tool to examine political, institutional, and environmental conditions simultaneously, by making it possible for small local groups to monitor, and thus to control, their own interaction with the environment.

The three questions which guide this volume emerge as a logical sequence from the simpler to the more complex. Further, they lead us to consider both the character and outcome of the population-environment interaction, and what can be done about the outcome through premeditated social intervention. That is, our considerations run from ideas and concepts to empirical observations to policy considerations. We have some opportunity here, therefore, to sustain the link between theory and action, between scientific observations and practical problems, which has been an enduring aim of The University of Michigan's Population-Environment Dynamics Project (PEDP).

Section I

Global Perspectives:
History, Ideas, Sectoral Changes, and Theories

Introduction

This section presents observations and analyses of the population-environment dynamic at the global level. Here we have conditions not necessarily tied to a specific country or community, but those that can be found throughout the world. Teitelbaum and Winter begin the section with an examination of two centuries of Western thought on the population-environment relationship. They show how ideological and conflictual human thought has been on this issue, and how intolerant adherents are of one another's position. Ness follows with a thousand year history of population growth and economic development, showing how the human population has been so successful in creating a niche for itself in the world that it threatens to destroy that niche and perhaps the world itself, as a living organism. Ruttan and Grainger deal with modern developments in the specific sectors of agriculture and forests, respectively. They identify the successes as well as the destructive nature of that success, and look ahead to the future with a guarded optimism.

Asking how each of these authors deals with populations, environments, and the population-environment dynamic can provide an introduction that draws them together into a coherent whole.

Population

Teitelbaum and Winter begin with a simple view of population as a total aggregate. Its changing numbers, either growth or decline, form the subject of much debate. This leads, however, to a discussion of population *quality*, specifically focussing on the variables of racial, national, and class differences in sizes and growth rates. The fear of population decline in the nineteenth century arose with the observation of fertility decline, and was often couched in terms of race or nationalism. As scientists searched for explanations of fertility decline, they identified them first in biological terms. This line of thought persisted until it was finally rejected in reaction to the Nazis' genetic theories of superiority. The new science of demography proposed a purely social and economic theory of fertility decline, in the demographic transition theory.[1] The observation of fertility differentials by class, income, or education, led to a fear of the decline of population quality that could be couched in social rather than racial terms.

Ness deals with population only as total aggregates. Ness sees the total aggregate engaged in a major transition over the past three centuries as both absolute numbers and growth rates have increased to unprecedented levels.

Grainger views the populations in the deforestation process largely in terms of numbers. The size of the population is a positive determinant of deforestation. Ruttan introduces a variance in the quality of the population, with a new interaction between health and age. For Ruttan the health of the population becomes important as a constraint on agricultural production. He also makes an important observation on the character of the new impact of AIDS. Past epidemics have generally attacked the very young and the very old. It is, demographically, relatively easy to replace the loss of the young through high fertility, supported by a larger productive age population. But AIDS tends to attack people in their most productive years, leaving the young and the old to carry the burden of increasing agricultural output.

Environment

Teitelbaum and Winter begin by telling us that the term "environment" is vague and highly ambiguous. Environment can mean anything, and has had at least four very common meanings in the historical discussions they review: the entire global setting, natural resources, the rural as opposed to the urban setting, or anything that is not hereditary. They also note that it is often defined by the scientific discipline that is observing the environment, and that discussion of it is marked by the antinomian character they found in discussions of Western population theory. The environment is viewed as *opposed* to population and either highly supportive or highly constraining on population growth. Finally, they note important changes in the assessment of the environment. Malthus represents early classical economists who viewed the environment largely in terms of agricultural land, whose growth was quite problematic. With the rise of modern industry and western imperialism, the environment came to be seen as a stock of cheap natural resources.[2] Ness treats the environment as an undifferentiated set of constraints and resources that the human population can employ to a greater or lesser degree depending on its organization and technology. In his view there is something akin to a

---

1. The authors tell us that this is less a theory than a grand generalization. We would argue that it is a general empirical observation of the trajectories of mortality and fertility rates over the past century (chapter 1).

2. Note that until very recently economic textbooks referred to air as a free good, and thus *not* within the purview of economic analysis.

natural environment whose basic life sustaining attributes may be changing due to human activity.

Following the specialized disciplines of their topics, Ruttan and Grainger define the environment in quite unambiguous, disciplinary terms. For Ruttan it is largely the land and water, and to a lesser extent the air, that sustain agricultural production. Grainger considers tropical forests, but in the process, provides a richer taxonomy of other uses for which those forests are cleared. This includes: four types of shifting agriculture; four of permanent agriculture; and three others--mining, hydroelectric power, and narcotic plant production. These are not incomplete treatments of the environment. Rather, they represent well a point made by Teitelbaum and Winter. The specialization of the sciences that gives us many different environments provides for a powerful capacity to observe very distinctive conditions. The strength of the sciences, as well as their limitations, lies in specialization.

It is important to note that at this global stage, the environment is not defined by social boundaries. If it is limited at all, as it is for Grainger's consideration of *tropical forests*, it is limited by natural conditions. We have yet to address those human institutions that create social boundaries which intermix with natural boundaries.

The Relationship

All authors see the population-environment interaction as a complex and dynamic set of relationships, and some aspect of human technology and institutions as mediating these relationships. There is a consensus that the modern period is one of an unprecedented *transformation.* Perhaps most important, however, is that all see the relationship as a qualitative variable: it can be better or worse, more or less healthy, especially for the human population. Here again a reference to the ecological literature is useful, for its attempt to conceive of the *health of the ecosystem* (Rapport 1989), and we have already seen that the common approaches in ecology have their analogues in human social analysis. As we shall see, the chapters in this volume deal only partially, and sometimes only indirectly, with this dimension of the population-environment dynamic, but at least behind every discussion is some idea of the *varying quality* of the outcome.

Teitelbaum and Winter remind us of the continuities in the basic qualitative judgment made about the population-environment relationship. For more than two centuries, at least in Western thought, there have been two major camps: pessimists and optimists. For *pessimists*, from Malthus to many modern environmentalists, the environment cannot sustain the kind of rapid population growth and economic development the earth has been experiencing over the past two centuries or more. For *optimists*, from Godwin, Marx or Mao to the modern conservatives like Julian Simon, there are really no limits to growth.[3] All that is needed is better human institutions. They also note the non-rational commitments that tend to mark many of the followers of both camps. The invective that Marx heaped on Malthus was characteristic of much of the passion and intolerance found in population debates for two centuries, and which continue unabated today. And, of course, these debates have often had an impact on institutional practices that, in turn, have a strong impact on the population-environment dynamic.

Ness views the population-environment relationship as relatively simple and straightforward, even if its causal connections and mechanisms are not fully understood. It is a relationship dominated by human institutions and technology. He points out that changes in energy technology have produced the current unprecedented rise of population and growth rates. These technological changes are, however, inseparable from the character of the human institutions in which they are born and utilized.

Ruttan makes most explicit the technological and institutional conditions mediating the population-food-environment relationship. He notes the great successes in making food cheaper and more abundant, at least over the past half century. This was the result of considerable technological progress which, he argues, is determined very much by human institutions. Current population projections, however, imply that food production will have to grow at annual rates of two to four percent over the next generation or two. This will require new institutions and new technologies, and will come at greater costs than the past increases. Health will require the same kind of location specific research and extension organizations that we have for agriculture. And agriculture will require a larger network of location and crop specific research and development centers than we now have.

Grainger focuses on a relatively simple relationship. Population growth, together with increased physical infrastructure, causes tropical deforestation, and government policy affects both of these causal conditions, though the specific connections to policy are not spelled out in his model. Grainger also suggests, however, that there may be a natural limit, a point, at about 0.1 hectare of tropical forest per capita, at which deforestation slows and fundamental economic incentives for reforestation begin to operate. The issue of the health of the ecosystem can be inferred from this discussion, with a proposition that market forces will provide some protection against full scale devastation.

This section provides some guidance in dealing with the population-environment interaction as a set of transitions. Although all authors note that human institutions and technology are important mediators of the interaction,

3. The authors note well the irony of modern Washington D.C. conservatives sharing positions with Marx and Mao.

there is much yet to discover about the character of those institutions and how this affects the character of the transitions. The most explicit treatment of this issue comes from Ruttan, who suggests that the success in increasing agricultural output lies in a unique blend of research and extension activities, focused on increasing future output. These institutions also show a distinctive level of decentralization. Agricultural research must be location specific, carried out on the specific soils, topography, and climate on which the agricultural production will occur. Further, research findings, even from centralized laboratories or experiment stations, must be *extended*, or presented to farmers in their own settings. Ruttan suggests that in contrast with agriculture our health infrastructure is woefully inadequate. It is oriented to the past rather than to the future, is curative rather than preventive, and lacks the extension character that has been important for agricultural successes. These are powerful insights, which will be complemented by materials in the next two sections to provide more empirical observations from which we can attempt to draw generalizations in the final chapter.

Chapter II

The Missing Links:
The Population-Environment Debate in Historical Perspective

Michael Teitelbaum
Jay Winter

It is impossible to disentangle contemporary discussions of population-environment issues from a discourse nearly two centuries old on growth and resources. The language we use on these matters today reflects a complex mixture of philosophical and scientific debates which emerged in the eighteenth century and which have taken on the color of the political and ideological conflicts of subsequent generations. Before dealing with particular facets of this literature, it may be useful to consider two problems, one related to styles of thought, the other to language.

Styles of Thought: Priorities

A striking and enduring feature of this discourse is the propensity for participants to adopt a mode of analysis in which one central issue comes to dominate, eclipse, or eliminate virtually all other relevant elements. Some scholars whose work is marked by this tendency towards unduly restrictive analysis are (strictly speaking) one-factor determinists; others select a dominant issue or subject, while admitting the possibility of multiple causation. Thus, some writers have selected biological factors as dominant or determining ones. They have clashed with those emphasizing economic elements, at times to the exclusion of all others. Both have been criticized by ecologists, whose order of priorities is at times profoundly different. What unites these observers is a tendency to adopt an analytical approach and a language of interpretation which may or may not claim (or have) predictive force, but which are defined in such a way as to marginalize or exclude other modes of thinking. Such restrictive selection of evidence and emphases, whether or not they entail simple cause and effect statements, tends towards reductionism. This paper highlights the historical pedigree of this style of writing on population and environment issues within Western intellectual and scientific tradition, and suggests ways of using the more liberal versions of several schools of thought to advance our understanding of what are inevitably profoundly complicated issues.

Over time, there have been many different sources and inspirations for the vast literature which has appeared in European and American environmental issues. Before the twentieth century, the deep strain of pastoralism in Western cultural life helped create a binary vision of social problems, in which a prelapsarian myth of a benign rural past was constructed to highlight the miseries of modernity. Thus, the key element in such thinking was the fall from a stylized rural environment to some urban nightmare (Williams 1973; Hulin 1978; Hobsbawm and Ranger 1983).

To explain the fall, one central theme or motif was chosen and explored to find the key to contemporary ills. In the hands of Carlyle, the harmony of medieval village life was ruined by the 'cash nexus'; for Marx, by enclosures and early capital accumulation; for William Morris, by factory production; for Emile Zola, by the pull of migration to the seductions of urban life, and so on. Each writer had a story to tell of a vanished and luminous past, lost largely through the corrosive effects of one primary and powerful process. Thus, the return of a noble aristocracy of manufacture was Carlyle's hope. For Marx and William Morris, proletarian insurrection was the answer. To Zola, a return to the land was the only way to retrieve individual and national dignity. Despite their differences, they and many other subsequent writers idealized preindustrial agrarian society, and thus reinforced the tendency to see the ills of urban life through a single lens (Carlyle 1834; Morris 1890; Disraeli 1845; Zola 1899).

This fundamental anti-urban vision was certainly not universally shared. The celebration of the vitality of the city, its excitement and potential for creative minds, was a common theme in European culture and political life.[1] (Dyos and Woltt 1973) But a darker, more pessimistic vision haunted most of those who wrote about population-environment issues in the nineteenth century.[2]

We should not underestimate the force or longevity of this pastoral concept. Its Western European variant has appeared in various forms of Utopianism, from Owenism to the Israeli kibbutzim, from Tolstoyan images of honest toil to Gandhian celebrations of the peasant at his loom. They all have in common the tendency to posit a mythical past of harmonious labor and individual

---

1. The literature on Dickens is vast. A good place to begin is Grahame Smith, *Dickens, Money, and Society* (Berkeley: University of California Press, 1968).

2. See generally, F. Stern, "Capitalism and the Cultural Historian," in Dora B. Wiener and William R. Keylor (eds.), *From Parnassus. Essays in Honor of Jacques Barzun* (New York: Harper & Row, 1976). What Stern has termed 'the politics of cultural despair' captures the brooding character of much of this writing.

dignity, utterly remote from a era of industrial conflict and social degradation.[3]

Such a vision of a vanished world gave primarily urban elites the means of highlighting the evils of the metropolitan era against the backdrop of a peaceful and harmonious (and imaginary) agrarian past. Once constructed, this model of history served as the basis for discussions of present predicaments and for predictions of the future. This religion of the rural landscape can be found in painting, poetry, and in the creation of what one English philanthropist calls 'gardens for the gardenless,' green public spaces in the heart of the metropolis.[4]

This rural/urban motif is the most powerful cultural element underlying the debate on population and environment issues over the past two centuries. But it may be helpful to introduce three other salient features which characterize much of this literature. The first is what may be termed 'negative reference' thinking. That is, one mode of mono-causal explanation has bred another, which reacts in an excessively narrow manner to the restrictive reasoning of its predecessor. Thus, as we shall see, Malthus's celebrated essay on the principle of population elicited excessive denunciations by Utopian socialists and Marxists, who denied (at times) that there were any limits on population growth. Similarly, demographers in the West found some facets of eugenics so distasteful that they eliminated biological elements entirely from their discussion of the 'demographic transition'. What they have in common is the adversary mentality, so common to lawyers and among political and religious sects, whose adherents do not notice how similar their style and attitudes are to those they condemn as heretical or misinformed.

Secondly, restrictive thinking has been common among those obsessed with the idea of 'scientific laws' of human behavior, which supposedly enable highly complex issues to be understood in a straightforward, objective way. The prestige of natural science was prized by many nineteenth-century commentators on these issues.

This was as true among political writers as in the academy. What mattered fundamentally to Marx and Engels was that their approach to social questions was scientific, not utopian. In their view, utopians saw the world not as it was but as they wished it to be. Ironically, this echoed Malthus's exasperation with Godwin and other writers, who saw what they needed to see, not what was. Malthus deliberately adopted an inductive approach to many demographic issues, and did so despite a set of moral and religious convictions which framed his work. It is debatable whether his observations produced the

3. The film industry has done its bit to foster this idyll. A recent, and moving example, is Akira Kurosawa's *Dreams*, the last segment of which extols a natural life, untrammeled by any urban malaise.

4. The phrase is on the statue at the entrance to Waterloo Park, in north London. In general see P.D. Klingender, *Art and the Industrial Revolution* (London: Noel Darrington, 1947).

theory, or the theory informed his observations, but what is clear is that his distaste for utopianism was as deep as that of Marx and Engels (Malthus 1987; Pullen 1981, 13: 1981; Spengler 1945). The fact that Malthus's distinction between geometric and arithmetic progressions of population and food was contradicted within a few years of its publication, or that Marx's law of emisseration under conditions of capital accumulation was unable to account for social trends and conditions within his own lifetime did not undermine the scientific faith they both held. The passage from Darwin's writings to those of Social Darwinism and from genetics to eugenics is well known, and similarly describes the pervasiveness of the mystique of science in late nineteenth-century political arguments (Kingsland 1988, 167-98). Where the masters lead, the epigone followed, making up in conviction what they lacked in originality.

There is a third, more narrowly sociological, phenomenon which may help to account for the restrictiveness of such theorizing. Over the past century, the rigors of academic specialization have made 'true believers' of many new scholars and professionals. They demonstrate their adherence to the craft and the cause in characteristically zealous ways, and on occasion adopt an exclusive language and mode of interpretation in presenting their research findings. Socialization into a branch of scholarly activity thus has also had its price: the narrow concentration on one kind of evidence or one branch of knowledge to the exclusion of others. Thus some economists or ecologists give sole pride of place to variables which find little room in the others' academic arsenal. There is no intrinsic reason why this must be so; but the nature of academic training may help account for the fact that so many have succumbed to the temptation of defining population-environment questions in terms which display the specialists' knowledge -- and only that knowledge -- to best effect.

Styles of Language: Four 'Environments'

So far, we have examined the stance of many Western writers on population-environment questions in terms of their preconceptions and styles of thought. But there is the additional, linguistic, level of difficulty in interpreting this abundant literature. Writers have tended to use the concept of the 'environment' in entirely different, though overlapping, ways.

Let us consider four here. The first is the global environment, or the 'ecosystem.' Such all-inclusive usages tend to produce explanations on a level so general as to defy contradiction or verification. This is as true of those who speak of world systems as of the ozone layer. Both may be right, but the subject of discussion is so vast as to make critical analysis difficult, to say the least.

The second usage of 'environment' is as a shorthand phrase for all natural resources, and more particularly, all sources of raw materials, energy and food. This is more restrictive than the first usage, but tends to group together very different industries and milieux under one rubric, such as the 'frontier,' 'peasant farming' or the 'tropical rain forest.'

The third sense in which the 'environment' has been understood in Western debates over the past century is in terms of the urban/rural dichotomy, to which we have already referred. William Blake's poetic incarnation of 'dark satanic mills' captured an image of a landscape, besieged by impersonal and mechanical structures and condemned to an ugliness created in the service of greed. Once more, the 'environment' is not a neutral term, but one defined narrowly, in this case, as newly urbanized space, beset by all the problems of city life.

Lastly, the 'environment' is a term widely used on a still smaller scale, as an antonym of 'heredity.' Thus, 'environmental' effects are those which are socially, not biologically, determined, though feedback possibilities are usually accepted. Here we come to the lowest level of generalization, since such 'environments' have been defined in small-scale terms, as the local community, the neighborhood, or the family. Socialization certainly involves inputs from larger social groups--classes, gender and ethnic groups, or nations, for example--but frequently the language of the 'environment' has conveyed an intimate, domestic configuration.

Population and Environment in Western Political Thought

Given these problems of definition and discourse, it may be helpful to isolate, for heuristic purposes only, two fundamental phases of the developing debate on population and environment in the West. The first extends from 1750 to 1880 and straddles the period of the industrialization of Western Europe and North America. The second covers the period 1880 to 1940, when rates of industrial growth in Western Europe both peaked and reached a plateau, imperial power reached its apogee and began to wane, and political instability and collective violence reached unprecedented levels. The discourse on population and environment in these two periods reflected first the ascending European political and economic power and then its slow but irreversible decline.

1750 to 1880

In the first phase, the problem of the balance between demographic growth and the environment, defined either as the fertility of the soil, or the urban-rural balance, produced fundamentally antagonistic arguments.

The optimists, from Condorcet to Marx, posited a virtually unlimited potential for the expansion of the material resources necessary to support a growing population. The sole constraint on the environment was the shackling effect of social relations of production geared to the needs either of a rentier or a capitalist class. Once these social relations were changed to take account of the needs of the majority, then the plasticity of the environment became unlimited.

As the social environment changed, so did human character and behavior. Here we encounter the fully promethean form of Marxian and utopian thought. If Marx and the Marxists sought the transformation of urban, industrial society, many utopians sought to escape from it and create a new environment within which human potential could be realized. In Britain, France and America, utopian communities were envisioned as an alternative environment, which would nurture a new kind of society, remote physically and morally from the corruptions of the old (Hecht 1988, 49-73; Harrison 1969).

Here too is the crucial point of entry for those either unconvinced or horrified by this vision. For the key error in this outlook, for pessimists like Malthus and de Maistre to Spengler and Freud, was its inattention to the unchangeable nature of human character and the propensity of men and women to destroy in the act of creation.

Malthus saw no truth in the argument of human perfectibility which he found in Godwin and other optimists. Instead, he explored principles of human behavior whereby the sexual instinct remained constant, and the resultant population growth presented a constant threat of disaster. Through limits governed by positive checks, catastrophe was put off, but its specter was never completely eliminated.

Attempts to raise the income of the poor through transfer payments were self-defeating, he argued, since higher incomes would produce higher fertility, recreating the initial poverty trap. By putting additional pressure on the food supply through encouraging high fertility, social action to relieve poverty thus worsened the environment. The only hope, Malthus wrote, was for individuals to restrict their fertility through the 'prudential restraint,' or the postponement of marriage until the couple had sufficient resources to resist the descent into poverty. If they did not act in this way, the positive checks of famine, plague and war would reassert the older balance between numbers and resources (Wrigley 1988, 30-48).

Although we now know that Malthus underestimated the potential for the expansion of food production and economic activity as a whole in the society in which he lived, and understated the sophistication of fertility regulation then practiced, his prime role was in stimulating a debate on population and the environment which has lasted to this day (Wrigley and Schofield 1981). Marx's railings against Malthus certainly were caricatures, but the distance between their positions was related less to their views on population than to their alternative visions of the possibility of the improvement of the human lot.

The same pessimism about the impossibility of fundamental social transformation governed other aspects of conservative thought in the nineteenth century. For our purposes, the crucial elements in the first phase of this debate between 1800 and 1880 are the emphasis on the plundering of the land and on the aesthetic disaster of urbanization. Marx held out the prospect that such deleterious developments were transitional; conservatives remained unconvinced. They held that the false prophets of progress made the fundamental error of building an urban desert and calling it progress. This is the cry of a host of nineteenth-century writers horrified at the physical and moral environment of urban capitalism (Karlyle 1843; Engles 1892).

1880 to 1940

Decadence is a term which is crucial to an understanding of the second phase of the population-environment debate. From the 1880s, a scientific component was injected into the debate about human perfectibility (Pick 1989). Much of this literature was ill-informed about the scientific arguments in question; nonetheless, its authors advanced many propositions derived initially from Darwinian biology and genetics. The 'science' of eugenics, or the study of the means to improve the human gene pool, was primarily devoted to demonstrating the extent to which disability and decadence were hereditable and proliferating. Negative eugenics aimed at the elimination of undesirable traits in individuals or populations; positive eugenics, at the cultivation of traits deemed socially beneficial.

Most eugenists believed in the existence of laws of human development, in this case laws of inheritance, which had a powerful, even decisive, effect on social organization. The inheritance of social characteristics, in their view, constituted a strong check on social improvement. Many believed that urbanization was genetically damaging because it drew into the metropolis people who produced stunted, unhealthy offspring, who in turn raised large families of equally unhealthy children. This enfeeblement of the many was made more dangerous by the relative infertility of social elites. The combination of the two processes presented profound dangers of reduced military and international power, under conditions of conscriptions (as in France and Germany) or of voluntary enlistment of recruits (as in Britain). Differential fertility between social classes presented the prosperous with a vision of being swamped in a morass of urban disability.

This set of worries was common to people of different political persuasions. What they had in common was fear of decadence and an anxiety about instability, both internal and international, in a period of imperial rivalries. Some eugenists were liberals or socialists; most were conservatives.

Relatively few were fascists, though the movement later was fatally compromised by the overlap between its notion of decadence and those of the reactionary or revolutionary right which, in its extreme form, informed Nazi race laws and ultimately genocide.

In both phases of what may be termed the classical period of modern political thought in the West, writers succumbed to the temptations of monocausal explanations of complex social processes. In much of this literature, we find visible in abundance the search for 'laws' of development, the concentration on one unifying theme to the exclusion of much other relevant information, the pronounced polemics of denunciation, and the pastoral idyll. In reaction to the social philosophy of some Enlightenment thinkers, Malthus created an opposite but equally general vision. In turn, Marx wrote in a way which made totality the key to social thought, again provoking in his adversaries a one-dimensional reaction positing biological limits to human progress. Biologism of eugenicists was a reaction to the hopes of reformers, both moderates and revolutionaries, that environmental or political change could ameliorate the lot of the poor. In turn, eugenics spawned an exaggerated response among writers in the 1930s and later, who shunned all attempts to discuss population trends in biological terms. Here is one of the origins of the theory of the demographic transition.

It is also the context in which to place the appearance of an alternative school of Soviet heredity, associated with Lysenko, which in the interwar years created an entirely fraudulent science of environmental determinism. If hereditary characteristics could be acquired in a relatively short period, and immediately transmitted to a new generation, then there were limitless possibilities of changing the gene pool. Its flaws as science were brushed aside given its utility as a Soviet 'answer' to eugenics (Graham 1968).

The classical paradigms of the population-environment debate are, therefore, a polemicists' dream: full of emotional and exaggerated rhetoric, of sweeping generalizations, and ringing accusations. Those in the Marxian or socialist traditions, as well as many liberal Christians, clung stubbornly to the belief that once political power changed hands and the majority ruled, a more humane environment could be constructed, whatever the nature of demographic change. Those following Malthus and later the eugenists, countered by arguing that human progress was bounded rigidly, if not hermetically, by demographic and biological laws. The echoes of present discussions are there for all who wish to hear (Kevies 1985).

Economic and Demographic Approaches

Unduly narrow thinking was by no means restricted to those in biology. Despite many developments in neoclassical and Keynesian economic analysis, little serious attention was paid to environmental and natural resource questions. Nor, indeed, were demographic issues high on the agenda of most economists. Meanwhile, the parallel growth of demography was also marked by the tendency of scholars to work within overly constrained models.

The essential absence of attention to resource and environmental issues among leading economic thinkers of the late nineteenth and early twentieth centuries reflects the nature of the economic system of the day. Those writing in the neoclassical tradition did so at a time when economic growth was so rapid as to divert attention from natural resource and environmental issues. During this period too, the leading economy of Great Britain experienced a spectacular decline in rural employment--from 50 percent in 1851 to 10 percent in 1911. Unsurprisingly, economists seeking to understand the nature of economic change focussed their energies not on the rural economics of land and agriculture, but upon the urban industrial base that had come to account for the bulk of both employment and economic activity. Furthermore, the expansion of formal and informal imperial systems rendered access to essential natural resources essentially unlimited, with declining real prices prevailing. Meanwhile, the technological advances and investments in transportation--steam replacing sail, refrigeration, the extension of railways--reduced sharply the costs of bulk transport of food and natural resources from world regions (e.g., North America, Australia, and Russia) where earlier classical economists' interests in limited land resources seemed utterly irrelevant.

Meanwhile, especially in Britain and America, demographic analysis and measurement flowered in a setting somewhere in-between economics and biology. The professional training in economics of figures such as Frank Notestein was combined with the powerful mathematics emerging from biological traditions (e.g., the regression analyses of Pearson, the intrinsic rates of Lotka), the improving official data on fertility and mortality, and the emerging epidemiology of public health, to produce a dynamic multi-disciplinary approach directed toward improved understanding of the profound changes then underway in European and American fertility and mortality.

As this effort proceeded, however, it came to be influenced deeply by the growing tendencies among some biologists and lay commentators to explain Western fertility declines in exclusively biological terms. A long tradition of analysis and prescription had emerged in this vein, portraying such fertility declines in often deeply pessimistic language as resulting from profound biological degeneration of the species, race, class, or nation. Such arguments, originally driven by serious efforts at scientific analysis, can be linked with, and ultimately corrupted by, conservative political ideas and movements (Teitelbaum and Winter 1985).

The still-small but creative cadre of demographers was confident that such biologistic explanations of Western fertility decline were profoundly wrong in scientific terms. In addition, many of them found the constellation of political views surrounding such explanations to be personally offensive. With the rise of German fascism, there came a parting of the ways. The International Union for the Scientific Study of Population (IUSSP), founded in 1930 by the Italian demographer, Corrado Gini and the American, Raymond Pearl, decided to relocate its 1936 meeting from Rome to London to avoid any chance that it would be exploited for Italian fascist propaganda purposes. American demographers boycotted another IUSSP meeting in Berlin for similar reasons (Notestein 1982, 462; Pogiliano 1984, 61-79).

One result of this scientific and ideological opposition to biologistic explanations of fertility decline was the elaboration of an alternative set of explanations that coalesced into the "theory of the demographic transition." The main idea of this "theory" was to show how the dramatic Western fertility declines could be explained on the basis of economic and social forces intrinsic to the industrial revolution that had begun in England two centuries earlier.

This is hardly the place to recapitulate the arguments of demographic transition theory. Suffice it to say that the theory was really not a theory at all in the sense of offering testable predictive hypotheses. Instead, it represented a rather grand historical generalization as to the forces underlying the preceding decades of fertility decline in Europe and America. The important point for our purposes is that demographic transition theory was almost entirely economic and social in its explanations of fertility decline, virtually excluding biological elements at either the human or ecosystem levels. So convinced were the proponents of the theory of the deterministic power of the socioeconomic forces of the industrial revolution, that the "theory" gave short shrift even to non-biological elements such as culture, language, religion, etc., except to the extent these could be seen as driven by socioeconomic change (e.g., the weakening of the extended family consequent upon urban-industrial life).

It is perhaps surprising that the buoyant field of demography, with its long-standing ties to biological ways of thinking and analysis alluded to earlier, should have produced a cosmic generalization that essentially excluded biological factors. The solution of this puzzle may lie in the realm more of political ideology than of scientific analysis. In important respects, demographic transition theory was a reactive response to the overconfident and exclusionary explanations of fertility decline as purely biological. Many of those troubled by this perspective were motivated by nationalist concern about fertility differentials vis-a-vis other competing nations (especially true in France) or by social anxieties about fertility differentials as between social classes (especially in England). The excesses of such arguments, and their underlying political agendas, may well have led demographers to elaborate an alternative explanation minimizing biological elements. Thus did the reactive and rhetorical elements of demographic transition theory lead it into its own kind of parochialism.

The pall of the Great Depression of the 1930s also affected the thinking of both economists and demographers. For economists, the fundamental issue was underutilization of capital and human resources. In the 1930s and 1940s, Keynes and his disciples developed an approach for coping with this problem. For some demographers, the fact that the Great Depression coincided with all-time low rates of fertility (often below the notional "replacement" rate of 2.1 children per woman) evoked parallel concerns about the underproduction of children. Gunnar and Alva Myrdal in Sweden and Enid Charles in England produced gloomy, sometimes apocalyptic writings anticipating the terrible consequences of continuing low fertility. In the Myrdals' case, the analysis of "the population crisis" produced powerful arguments for the adoption of social legislation that became the foundation of the Swedish welfare system. Demographers played a similar if less explicitly political role in other Western European countries (Winter 1988).

The onset of World War II ended one phase of the debate about the problem of underproduction, in either its economic or demographic forms. The end of the war opened an unparalleled period of economic prosperity, and fertility levels in the 1940s and 1950s were higher, sometimes sharply so, than the lows of the 1930s. Gloom about overcapacity and deficient demand was supplanted by expectations of unlimited growth, under the powerful guiding force of American investment and Keynesian thought. The dramatic and sustained baby booms experienced during the 1950s in the United States, Canada, Australia and New Zealand, and the smaller and shorter-term booms in Western Europe, focussed population concerns more in the direction of possible overcrowding rather than the earlier concerns about a supposedly dwindling "race."

By the early 1960s, the shadow of the political and economic upheavals of the 1930s and 1940s had faded. New worries began to emerge in political and academic circles about the negative environmental consequences of economic growth. Such anxieties, combined with recognition of the dramatic acceleration in demographic increase due both to the Western baby booms and the rapid mortality declines in Third World countries, resulted in a new level of awareness and activism about environmental problems.

By the early 1970s, the postwar optimism regarding unlimited growth was shaken by the Vietnam War, the 1973-74 Middle Eastern War and attendant oil crisis, and by the phenomenon of economic "stagflation" presenting still-unresolved questions as to the adequacy of Keynesian economics. Later, in 1979, came the public opinion trauma of the American nuclear accident at Three Mile Island (anticipating by a decade the far worse Soviet disaster at Chernobyl), challenging profoundly the technological optimism prevailing during the postwar years. Together these somber developments suggested that there might indeed be limits to natural resources, technological advance, and economic management capabilities, limits that could either cause irreversible damage to the global environment or make states powerless to prevent it. A new political environmentalism emerged both in America and Western Europe, with the latter taking on a more direct electoral form in the "Green" party in Germany and the "Green movement" elsewhere.

Political and intellectual reactions to environmentalism followed rapidly. In a mixture (at times garbled) of economic analysis, technological enthusiasm, and libertarian ideology, a new form of "cornucopianism" emerged during the 1980s. These arguments represented an unlikely amalgam of earlier utopian, Marxian and laissez-faire thought, coupled with the use of apparently scientific data and economic analysis.

In an extreme form, this view held that there were absolutely no natural limits to economic or to population growth, and indeed that natural resources were in far greater supply than ever before. Indeed, some went so far as to argue that since the ultimate resource was the human intellectual power to innovate technologically and create new resources, the more people, the more resources there would be, *ad infinitum*.[5]

The polyglot (and especially the Marxian) pedigree of such arguments do not appear to have been understood by many who embraced them. Ironically, they were (and still are) strongly promoted by the public relations and political advocacy apparatus of American new right groups such as the Heritage Foundation, the Cato Institute, and the editorial board of the *Wall Street Journal*, all of which during the Reagan years became highly influential "inside the [Washington] Beltway," or among what the French call "the political class." It remains to be seen how such groups will react to the new political conditions of the 1990s.

Environmental and Ecological Approaches

Environmental scientists have not been immune from the tendency toward overly narrow thinking displayed by many political thinkers, biologists, economists, and demographers. Frequently (though by no means universally) ecological analyses--from the macro-level of the global to the micro-level of the local ecosystem--have emphasized the complexities of the ecological system, and often its vulnerability to the external insults of human technologies, while at the same time tending to minimize or ignore the potentially positive environmental impacts of political, medical, or economic action.

Many ecologists, for example, do not consider there to be any important distinctions between animal and human populations. The assumption that humans are a biological species that performs according to ecological rules is

5. See, for example, Julian Simon, *The Ultimate Resource* (Princeton: Princeton University Press, 1981).

fundamental and unquestioned. Thus, ecologists reason by analogy from their experimental and field studies of animal populations to humans' (Golley 1988, 200). This perspective has led some in this tradition to describe the rise in human crime rates and similar phenomena as analogous to the increases in other species of social disorder, violence, aggression, stress disorders, and disease that have been observed in experimental studies of overpopulation in confined mammalian populations (Golley 1988, 204).

A second example of overly narrow thinking is offered by some ecologists' views on the "epidemiologic transition," the dramatic improvement in health conditions that has transformed the level and structure of human mortality over the last two centuries. This triumph of human knowledge over many endemic infectious disease organisms--perhaps the major cause of rapid human demographic growth since 1750--can also be described as a significant transformation of complex and long-standing ecological systems, brought about by the effective application of human science and technology. While such consequences for aggregate human welfare have been viewed positively by many environmentalists, a more typical emphasis is upon more negative effects of science and technology, such as those surrounding energy consumption and production, modern agricultural practices, etc. To the extent that the epidemiologic transition has been considered in many environmental discussions, it has been seen in terms of the resulting rapid population growth as yet another corrosive force acting upon the environment. While few environmental writers portray improved public health as a negative factor, equally few emphasize its positive elements.

Some environmental analyses have also tended to minimize the potential of new technologies for expanding the natural resource base, or for eliminating or redressing the environmental damage done by other human activities. In the same way that some economists on the disciplinary/ideological extremes take as an article of faith that human ingenuity can (or will) overcome any limits imposed by the natural environment, some early environmentalist analyses of the "limits to growth" failed to anticipate the potential of such innovation to overcome resource scarcity.[6]

While there is persuasive evidence of environmental degradation in many settings--persuasive to all but those who for ideological reasons reject such arguments--ecologists have found it much harder to handle problems of causation. In part, this is because environmental experimentation is difficult to implement, and partly because the very high levels of complexity in ecological systems present profound analytic difficulties of multiple causation and feedback in any ecological system.

In some ways, ecology represents a polar extreme from economics in its treatment of the role of human technology. Indeed, Shepard and McKinley

6. See the journal, associated with the work of Paul Ehrlich, *ZPG Reports*.

call ecology "the subversive science" because of its fundamentally divergent attitude to the impact and potential of technological innovation. To the ecologist, the technological optimism of much modern economics, and of Marxists and the new right, ignores the laws of nature by which all growth must ultimately approach unbreachable limits (Worster 1987, 87-103).

Once again we have returned to older terrain, and will miss much if we ignore the echoes of Malthus, Fourier and Marx. It is certainly true that ecology has developed tools of analysis unavailable to earlier generations, but the utility of the best methodology is a function of the content and character of the questions it is meant to solve. Those questions are hardly new, and present the same conceptual and philosophical difficulties as they did two centuries ago.

When environmentalists decry the decay of the city, and the rape of the rain forest, they are reiterating older arguments, no less eloquent than D.H. Lawrence's lament at the ruination of the English countryside, or Dickens' devastating portrayal of Coketown in *Hard Times*. What happened to England, some environmentalists claim, is now endangering the world as a whole. Whether or not this is true, it is a *cri de coeur* profoundly rooted in Western intellectual traditions.

Conclusion

One-dimensional approaches have been common to most of those who have addressed population-environment questions over the past two centuries. How can we account for the salience of this mode of thinking, and are there ways of preventing similar narrowing of political and analytical perspectives in the future?

The most striking feature of this debate is its intensely ideological character. Most of the major political thinkers of the early to mid-nineteenth century were drawn to the problem of population change and its interaction with the environment, which was defined either as cultivable land or growing urban concentrations. They could hardly avoid it since they were following the natural propensity of social thinkers to try to account for the dynamic features of the world in which they lived. Among the most visible and most significant of these changes was an increase in population, producing unprecedented aggregate growth in Britain and later on the Continent, unparalleled urbanization, and a concomitant shift of labor from agriculture to manufacture and extractive industry.

In Britain and elsewhere on the continent, these developments were accompanied by profound shifts in economic and political power, from a regime dominated by rural elites whose wealth and influence were based on land and (to a lesser extent) commerce, to one in which urban elites--manufacturers and traders to burgeoning international markets--predominated politically. While demographic change anticipated and made industrialization possible, the industrial revolution created its own labor force. This new mass of factory-based workers were neither docile nor inarticulate. Their demands grew in the course of the century, and they came to constitute a profound threat to the industrial order they helped to create.

Observers trying to understand these monumental changes were unsatisfied with an ecumenical approach to them. They sought instead the one unifying principal which would give form to the transformations of their world and provide clues to its future destination. Their approaches were cyclical rather than progressive. Instead of one generation of writers building on the work of their predecessors, they all tended to return to first principles in order to refute, root and branch, their adversaries' error. In the first two generations, Malthus, the utopians, Marx and their allies clashed over the potential for growth at the moment of the emergence of industrial capitalism. From the 1880s, the debate centered on the potential for decadence and decline, defined as falling fertility rates, declining rural populations, and increasingly 'enfeebled' urban concentrations.

Since 1914, the argument over the balance between population and environment has changed, in line with more general political and economic developments. First, the political and economic instability of post-1914 Europe made environmental questions recede into the background. The pervasiveness of mass unemployment, the consolidation of Soviet power, the rise of fascism, and the experience of total war eclipsed issues of the limits to growth or the trajectory of urban decadence.

After 1945, affluence rather than instability further eclipsed environmental and demographic questions. It was only in the 1960s and later that there emerged the three fundamental problems dominating Western debate in this field: What can be done about declining fertility? What are the consequences of in-migration by non-European populations? What can be done to stem the tide of environmental pollution and degradation?

The above survey of earlier discussions on these matters suggests that there is a good chance that current debate will simply reiterate the ideological preoccupations of the past. But there is also an opportunity to avoid the trap of selective causality. The way forward entails the rigid adherence to a multi-disciplinary approach, in which pride of place is denied to any one school of thought. Resource economists have shown one way in which this is possible, so have demographers who have escaped from economic or biological reductionism (Kneese 1988, 281-309). It is not that they have abjured ideology: no one can; rather, they are committed to an investigation of these phenomena from more than one point of view. Tolerance is rare enough in this field for us to claim that its virtues may be greater than those of certainty, and perhaps even greater than truth.

Chapter III

The Long View: Population-Environment Dynamics in Historical Perspective

Gayl D. Ness

Introduction

The human population is most clearly related to global environmental change through its historical pattern of growth in numbers and productivity. This pattern is now fairly well known, if not fully understood. It consists of thousands of years of variation producing exceedingly slow net increases in total population. Only the last three centuries show exponential growth. The recent growth is closely associated with the significant increase in human productivity that has accompanied the rise of urban industrial society. The recent exponential population growth is associated with two major energy transformations that extend back for about five centuries. One was the shift to sails, starting in the fifteenth century. More important for modern global change, however, was the shift to fossil fuels, beginning in the late eighteenth century.

There is an association between this rapid population growth and the full range of environmental changes that is coming to be known as *Global Change.* This includes atmospheric chemical changes related to global warming and ozone depletion; degradation of the environment, including deforestation and the release of toxins into the earth, water and air; and the destruction of species.

This chapter provides a brief summary of the long historical trends that link population growth to environmental change. It begins with a review of the past millennium of population growth on a global scale. It then presents the underlying population dynamics that mark our modern period, the *demographic transition,* which helps to explain differential growth rates in major regions of the world. Finally, it examines broad patterns of population and economic growth over the past four decades, with projections to 2025.

One Thousand Years of Population Growth and Economic Development

Population Growth

The population-environment relationship has always been a reciprocal one. A brief review of the history of the human population makes this abundantly clear. From its probable origins in East Africa, the human species took perhaps up to 500,000 years to spread throughout the world. Growth rates of the total population were usually very close to nonexistent, and for many local populations the growth must often have been negative.

Even this slow population growth, however, was accompanied by substantial environmental change. McNeill (1976) notes the elimination of large animals (mammoths etc.) from much of the territory invaded by man the hunter. The domestication of animals and plants, beginning around 9000 B.C., had a substantial impact on the environment in relatively small, sparsely settled and disconnected societies. The population environment relationship was, however, certainly a two-way street. The emergence of agriculture in MesoAmerica and the Middle East at roughly similar periods suggests the importance of environmental change resulting from the recession of the glaciers. Over the next few thousand years, agriculture spread to many parts of the globe, often implying radical alterations in the environment. The most dramatic of these are seen in irrigation and terracing, which molded the land to produce great increases in plant yields. While these transformations produced pockets of relatively high population density and some periods of substantial growth, the overall process was still very slow.

By the year 1000 B.C. the human species had come to number about a quarter of a billion. For the next 700 years the historic pattern of slow growth continued. Growth rates were kept low by high human mortality, usually in the range of thirty to forty per thousand population. A variety of mechanisms had evolved, however, to offset these high death rates, such that fertility rates in the best of times usually hovered around five or ten points above mortality rates.[1] Under these conditions, population growth was predominantly governed by mortality, which rose and fell with both social and environmental changes.

This pattern of high birth and death rates began to change about 300 years ago, when the world entered into a series of *demographic transitions*. Death rates began to decline as some populations began to experience a transition from infectious to degenerative diseases as the leading causes of death.[2] Falling death rates brought a period of rapid population growth, as birth rates

1. Note this is *in the best of times*. Then, growth might have been as high as 0.5 per year though usually only for relatively short periods. Even those growth rates implied a doubling time of 140 years, or about four generations (Coale and Watkins, 1986, Cipolla 1974, Deevy 1960).
2. This is often known as the *epidemiological transition*.

remained high. This was then followed by declining fertility and lower population growth rates. This transition was completed in most of the more developed countries by the early part of the twentieth century and is now being experienced in much of the less developed world. The expectation today is that the demographic transition will be accomplished throughout the world by the middle of the next century.

For the world as a whole, the demographic transition has given us rising numbers and rising rates of growth for the three centuries from 1700 through 2000. The population of roughly 250 million in the year 1000 rose to about 610 million by 1700, to 2.5 billion by 1950, and is expected to reach 6.2 billion at the end of this century. The growth rates rose steadily from less than 0.1 percent five centuries ago to a peak of 2.06 percent in the period 1960 to 1965, when the population reached 3.5 billion. Since then the growth rate has declined to about 1.7 percent today, though it will continue to grow in absolute numbers for some time.[3] Not all of the world's regions have experienced these changes at the same time, however.

The different timing of these transitions in different world regions can be seen in detail from an examination of average annual growth rates, shown in table 3.1.[4] The low growth rates that were universal up through 1700 rose in the eighteenth century to .4 percent in Europe and Asia, and to .6 percent in the Americas. Asia retained that rate of growth and was joined by Africa in the nineteenth century, while Europe's growth doubled to near .8 percent, and the Americas reached levels twice that. Through the twentieth century, European growth rates declined slightly and those in North America rose through 1950 then declined. Africa, Asia and Latin America showed high and rising growth rates throughout the century, and are only expected to decline in the next century.

Current (1988) United Nations projections place the world's total population at about 8.5 billion in 2025. Beyond that, projections become very uncertain, and much depends on what happens to human fertility in the near future. For example, if we were to reach replacement level (2.1 total fertility

---

3. This results from what is known as *population momentum*. When fertility and growth rates are high, many new babies will be born. Even if these new additions experience reduced fertility when they reach reproductive ages, their sheer numbers will keep population growing in total numbers.

4. McEvedy and Jones (1978) is used for these global figures to 1950, and the United Nations (1988) for the years 1950-2025. There is some dispute over the population of the Americas on the eve of the European conquest, which McEvedy and Jones place at 14 million. William McNeill (1976) places the number at 100 million. I am inclined to accept McNeill's figure, but have retained the McEvedy & Jones figure for consistency. There is more agreement after 1700 on the total numbers, but the implication of the difference concerns the extent of the demographic collapse of the American Indian population resulting from the European contact. It is not unreasonable to accept McNeill's judgment of a full decimation of the Amerindian population as a result of the impact of European diseases.

TABLE 3.1 Population of Major Regions 1000-2025 Total Population in Millions and Average Growth Rates

Year	1000	1500	1700	1800	1900	1950	2000	2025
Region								
World	265	423	610	902	1622	2515	6248	8466
(a.a %)		.09	.18	.39	.58	.87	1.84	.62
Europe	36	81	120	180	390	572	816	863
(a.a %)		.16	.20	.41	.77	.72	.71	.22
Asia	185	280	415	625	970	1375	3698	4890
(a.a %)		.08	.20	.41	.44	.70	2.0	.56
Africa	33	46	61	70	110	224	872	1581
(a.a %)		.11	.14	.14	.45	1.25	2.75	1.19
Americas	9	14	13	24	145	332	831	1093
(a.a %)		.09	-.01	.61	1.8	1.6	1.9	.55
L. America	8.6	13.2	11.8	18	66	164	537	760
(a.a %)		.09	-.05	.04	1.3	1.8	2.4	1.4
N. America	0.4	0.8	1.2	6	79	168	294	333
(a.a %)		.15	.20	1.6	2.6	1.5	1.1	.5
Oceania	2	2	2.25	2.5	7	14	30	39
(a.a %)		.09	.06	.10	.58	.87	1.54	.53

Source: McEvedy and Jones 1978. The average annual growth rates (a.a%) given are for the full period from the prior date shown.

rate) by the year 2000, which is possible though highly unlikely, the world's population would rise to just over eight billion by the year 2100 and then level off or decline. If we do not reach replacement level fertility until 2080, in 2100 we shall have almost fourteen billion people and still be growing rapidly.

This general pattern of centuries of slow growth and recently rising growth rates is illustrated in figure 3.1. It also shows the two energy transformations and the rise of urban society, to which we now turn.

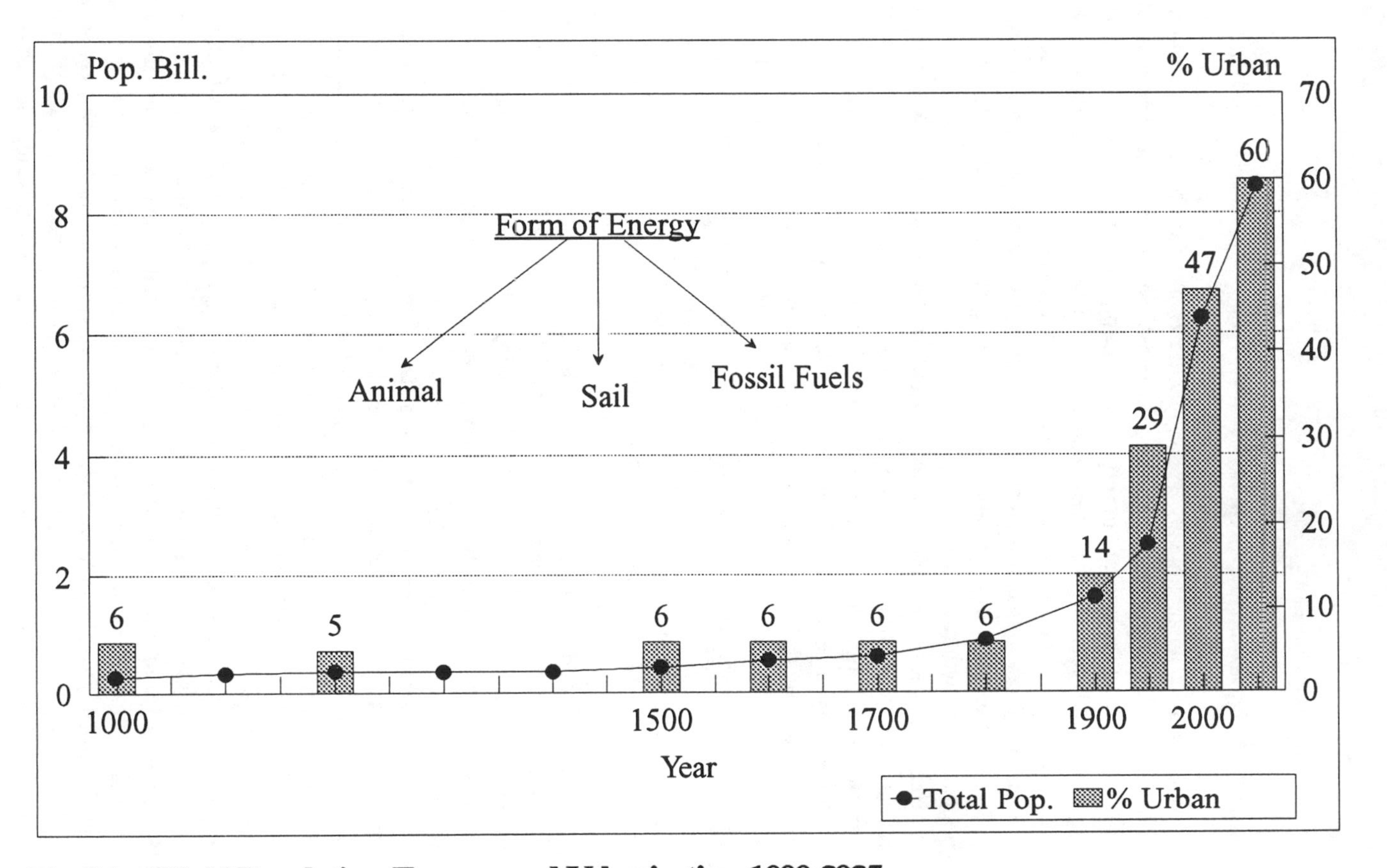

Fig. 3.1 . World Population, Energy, and Urbanization, 1000-2025

Growth of Population and Output

What is most remarkable about this recent exponential growth in population is its association with continued rising per capita output. Both agricultural and industrial output have risen more rapidly than population throughout this past three hundred years. The phenomenon is especially remarkable in the past fifty years, when the population growth rates have been so high. For at least the past fifty years annual increases in world cereal output, for example, have averaged over three percent, about one to two percentage points above world population growth rates.

The increase in output associated with rapid population growth is quite contrary to the gloomy predictions of Thomas Malthus. It is the result of the emergence of a new type of society: urban industrial society. The emergence of this new type of society was a consequence of two sequential revolutions in energy use: first the use of sails and then the use of fossil fuels.

Energy Transformation I: Sails

The change from the oars of the Mediterranean galley to the sails of the Portuguese caravel occurred in the first half of the fifteenth century and represented a major development in naval architecture. The galley's one mast amidships was replaced with three--after, amidships, and forward. Oars were eliminated, the hull was raised farther out of the water and ultimately protected by guns rather than by swords (Cipolla 1965). The transition from oars and swords to sails and guns gave the west the technological capacity to "discover the seas".[5]

The transition began in 1415, when the Portuguese Prince Henry led a successful attack on Ceuta, giving them a permanent base in North Africa. From there the great Portuguese oceanic explorations pushed down the coast of Africa, turning the Bight of Benin in 1472, and rounding the Cape of Good Hope in 1488. Finally in 1498, Vasco DaGama reached the coast of India, linking for the first time Asia and Europe by sea.[6] Just over a decade later, in 1511, the Portuguese captured Malacca, a major port and seat of Islamic learning for all Southeast Asia. A decade later Magellan reached the Philippines by sailing across the Atlantic and Pacific, then returned through the Indian Ocean for the first circumnavigation of the globe. Thus in this one brief century, from 1415 to 1521, the world was encompassed by ships at sea.

This technological advance transformed world trade and transportation routes, linking all the continents by the seas, in effect making the world a

---

5. The term is J.H. Parry's (1974).
6. Boxer (1961) provides a succinct survey of the Portuguese expansion. Parry (1959) considers the broader European involvement.

single integrated environment for the human population. Henceforth human transportation would permit the spread of all localized flora and fauna to other parts of the globe. It took the human population half a million years to spread throughout the world, finding ecological niches in which it could adapt to the environment and survive. In the last 500 years, the human species turned the entire globe into one environment, in which human activities would become paramount in changing that environment.

The transformation of trade routes between 1500 and 1600 is shown in figures 3.2 and 3.3. These new routes brought a series of highly productive crops from the Americas to many part of Asia, Africa and Europe. Everywhere these new crops increased the carrying capacity of the land, thus permitting the human population to greatly increase its numbers.

If the impact of the discovery of the seas was positive for Europe and Asia, it was anything but that for the Americas. There a population of perhaps as many as 100 million had emerged, isolated from the micro-organisms existing with people throughout Asia, Africa and Europe. Because of an absence of autoimmunity, the Americas suffered disastrously from the external contact. Their populations were reduced to one tenth their pre-Columbian populations in less than a century. In some cases, as in the Caribbean, the entire native population was wiped out.

It is important to note that this discovery of the seas was only partly a technological transformation. Equally important were its political and social dimensions. Half a century before Vasco DaGama reached the coast of India, and a century before Magellan's circumnavigation, the Chinese launched five major naval expeditions into the Indian Ocean, reaching the East Coast of Africa. In 1400, Chinese naval technology was far more advanced than that of the West. The great nine masted flagship of Admiral Zhang He, which made the last of the five African expeditions between 1403 and 1433, was five times the size of the tiny vessels that carried DaGama, Magellan or Columbus. It had water tight compartments, double hulls, a stern rudder and was navigated by complex and accurate astronomical calculations. Thus the Chinese had the naval technology to discover the seas. They would also have had the capacity to people the West Coast of the Americas, and halt the western advance into Asia. In effect they had the technological capacity to make the world a Chinese world, rather than the Western world it became. The Chinese decision not to use its technological advantage for conquest of the seas was thus of momentous importance.

The full explanation of the Chinese refusal and the Iberian rush to oceanic conquest is quite beyond the scope of this review,[7] but an instructive contrast can be drawn between the two in the history of the interaction between human institutions and ecological forces. China has been a land-based empire since

7. Paul Kennedy (1987) has a good summary contrasting the Chinese and Western systems.

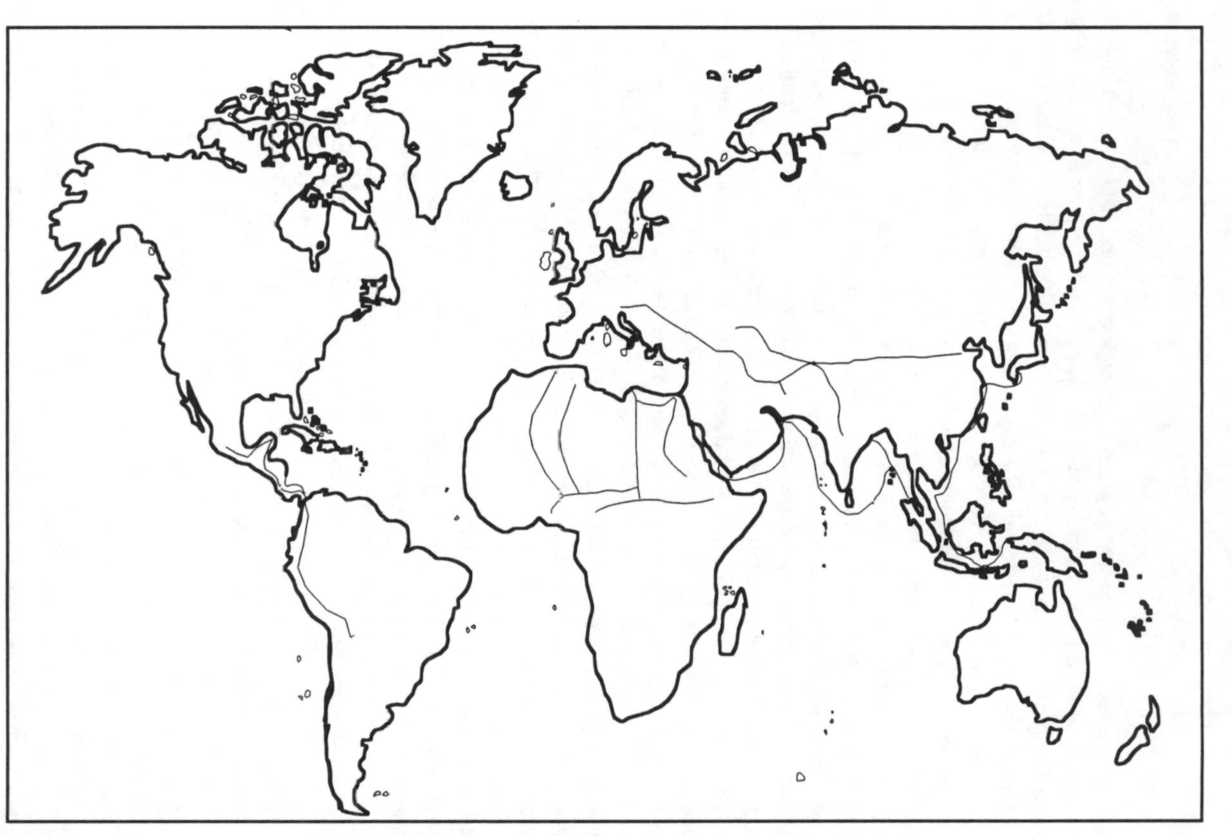

Fig. 3.2. Major World Trade Routes to 1500

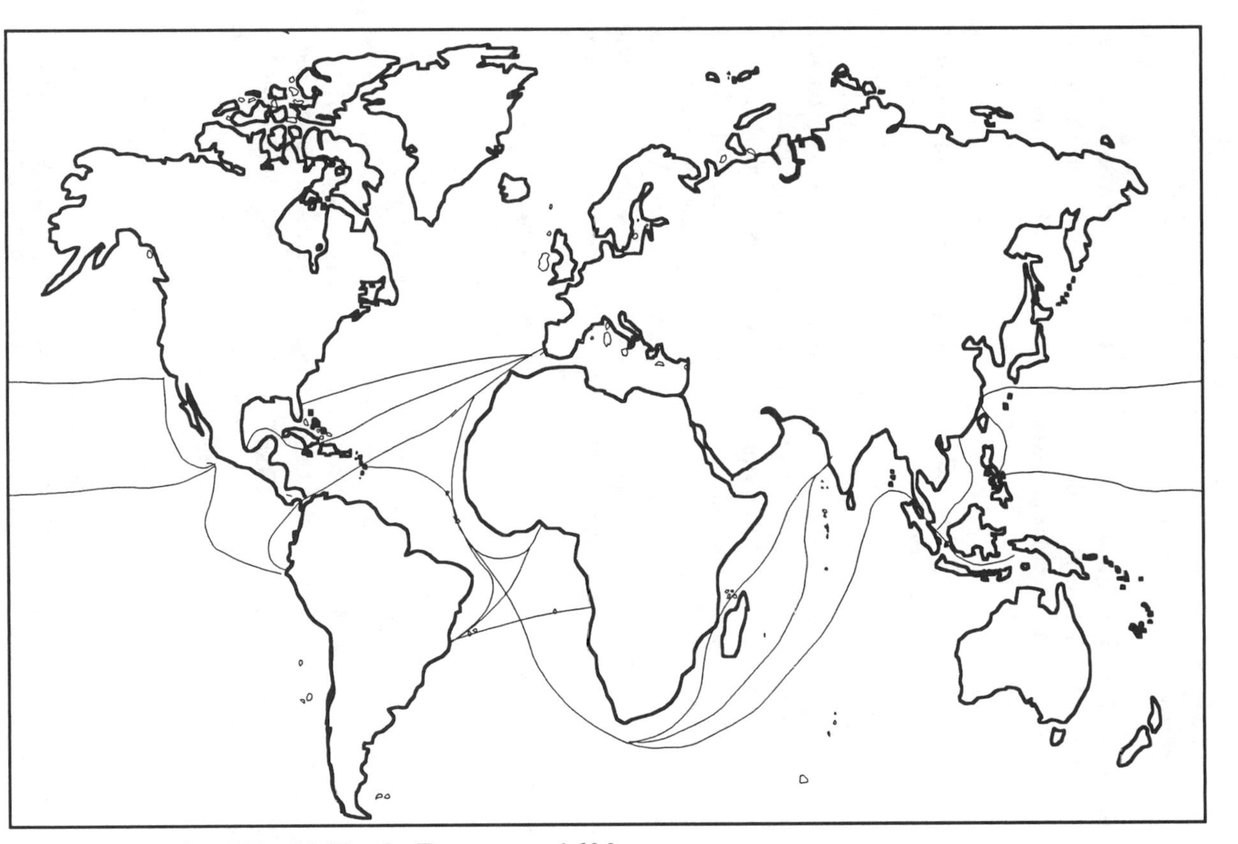

Fig. 3.3. Major World Trade Routes to 1600

its inception. Powerful groups arose by using the state to mold the land, digging canals for transportation and managing water for irrigation, drainage and flood control. The result was greatly increased agricultural product, which brought immense wealth to the empire. Given the great land mass of China and its openness to Mongol populations, much of the empire's wealth was used for *defense of the land.* Three major power groups emerged: the court and the Emperor, the Bureaucracy, and the merchants. The first two were constantly in conflict over the control of the state, and both were allied against the merchants, who were seen as a lowly but lucrative tax base for the empire. It is instructive to note that Admiral Zhang He was a Muslim and a eunuch, marking him a loyal personal servant of the Emperor and an outcast from the bureaucracy and the Chinese gentry. By contrast, Portugal, thrust out into the western seas and with a long coast line, made fishing and sea transport well established activities. Further, in contrast to Admiral Zhang He, Prince Henry the Navigator, who promoted the Portuguese explorations, was a son of the Emperor, a protector of the Church, and was financed by Lisbon merchants. In Portugal and Spain, crown, church, and merchants had built an alliance, first to wrest the peninsula from Islam, and then to continue that conquest to the seas. These institutional and geographical differences, far more than the technological differences, explain why the world is now a Western rather than a Chinese world.

For the next two centuries, the seas were used to begin to tie the world environment together into a single unit. A major product of this new integration was that the new crops from the Americas increased the carrying capacity of the earth in Asia, Africa and Europe. Overall population growth rates rose to .18 percent in the sixteenth and seventeenth centuries, and to .39 percent in the eighteenth century. By 1800 the world's population was just under one billion. To raise growth rates to the higher modern levels, however, would require another form of energy.

Energy Transformation II: Fossil Fuels

It is the second energy transformation, to fossil fuels, that lies behind the rise of modern industrial society. This began slowly with the invention of the steam engine and the expansion of coal production at the end of the eighteenth century. It grew more rapidly with the invention of the internal combustion engine and the exploitation of oil and natural gas in the nineteenth century. It has now exploded into exponential growth of fossil fuel consumption in the twentieth century. Without coal and oil, steam and internal combustion engines, modern urban society as we know it would be quite impossible.

Even as late as 1700, the world's urban population accounted for less than 10 percent of the total population. There were some large cities, mostly in Asia, but the social organization of the human species was primarily rural and agrarian. Fossil fuels in transportation permitted high concentrations of populations in urban centers to be provided with food produced by others. Fossil fuels also permitted those urban populations to produce a surplus of goods that could be traded for food produced elsewhere. The use of fossil fuels would, however, increase the level of carbon dioxide and other greenhouse gases in the atmosphere, producing a marked human impact on the entire global environment.

Both of these energy revolutions stimulated increased population growth. The second has been especially important. It is considered doubtful that the world's population could have grown beyond one or two billion without the transformation to fossil fuels. If this is true, the link between population growth and environmental change is especially salient for today. The fossil fuel revolution produced massive increases in both human numbers and human production, and it is precisely through these numbers and productivity that the human population is having its remarkable, and destructive, impact on the environment.

Modern Urban Society and Population Growth

The Western World experienced rapid urbanization and industrialization in the nineteenth century. This was already evident in 1800, when London had a population of 865,000 and fully 10 percent of the population of England and Wales lived in cities of 100,000 or more: By 1900 this increased to 35 percent (Davis 1955, 1965). For the world as a whole, the proportions in cities of 100,000 or more in 1800 was only 1.7 percent, rising to only 5.5 percent in 1900. England and Wales experienced the greatest spurt of urbanization between 1811 and 1851, with the U.S. following between 1820 and 1890. Thus by 1900 Europe and North America had become substantially urbanized, displacing Asia as the continent with the largest cities. Until 1800, for example, fifteen of the world's twenty-five largest cities were in Asia, with two more in the Middle East. In 1900, fourteen of the twenty-five largest were in Europe and North America (Chandler and Fox 1974).[8]

In the first half of the twentieth century, the economic growth and urbanization of the western world were spreading to Asia and Africa. By 1950 almost 30 percent of the world's 2.5 billion people lived in urban areas. In the more developed regions the level was 53 percent, and in the less developed world it was only 17 percent, or not much above the 10 percent that many societies had reached throughout history. By 2025, it is projected that over 60 percent of the world's population will live in urban areas, 80 percent in the more developed regions and almost 60 percent in the less developed.

---

8. The five largest in 1900 were London, New York, Paris, Berlin, and Chicago.

While this massive transformation of the world community, the rise of urban industrial society, has been associated with the increase of population growth rates, it also contains conditions that lead to the slowing of population growth rates. To understand this phenomenon, we must turn to the major dynamic underlying modern population changes, the *demographic transition.*

The Demographic Transition: Past and Present

Figures 3.4 and 3.5 show the two variants of the *demographic transition*, or the move from high to low birth and death rates, that distinguish the more developed from the less developed countries. This can help us to understand some of the dynamics behind past and present population movements and can also serve as an introduction to the modern population policies that represent a revolutionary change from the past.

Figure 3.4 uses the experience of England and Wales to illustrate the demographic transition, which has now been completed in all of the industrialized countries. The transition began with high levels of mortality and fertility, which can be called the traditional condition of the human species throughout most of its earthly existence. Often death rates rose above birth rates to bring a period of absolute population decline. In the best of times, mortality declined and population increased, but this did not last long and overall the growth rates over long periods must have been only a little above zero.

In the early 1700s the death rate in England and Wales began a gradual and persistent decline, leveling out around ten per thousand in the early twentieth century. Fertility remained high until near the end of the nineteenth century, and then dropped rather rapidly to about fifteen, coming into line with low mortality. This transition, from high to low mortality and fertility, with an intervening period of rapid population growth, marked the transition from rural agrarian society to urban industrial society. While many details are obscured, the general pattern seems clear. Mortality declined through the combination of an epidemiological transition (McNeill 1976), a gradual rise in the earth's temperature (LeRoy and LaDurie 1988), trade expansion through the seas, agricultural and industrial revolutions, and, to a far lesser extent, through improvements in medical technology. Only at the end of the nineteenth century did medical advances play much of a role in mortality declines.[9] Together these changes implied a slowly rising standard of living, or an increase in the carrying capacity of the land.

9. The one major medical advance that precedes this is the discovery of a vaccination against small pox that is dated about 1740. Until about 1800, however, the vaccination used small pox itself, which often resulted in death, and was not widely accepted, especially in urban England and on the continent. At the turn of the century cowpox was used to vaccinate humans, with few serious side effects. This new form of vaccination then spread rather widely throughout the world during the 19th century.

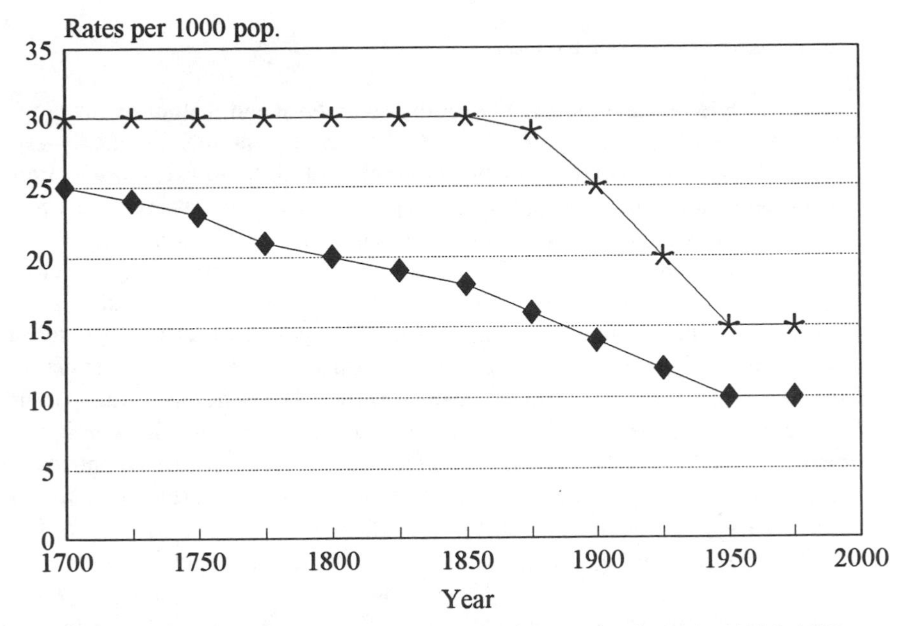

Fig. 3.4. Past Demographic Transition England & Wales 1700-1980

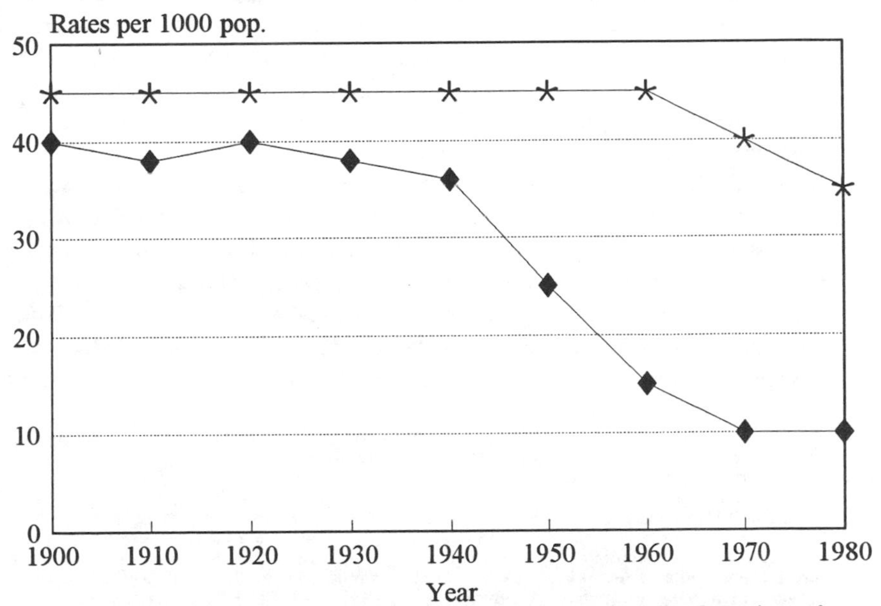

Fig. 3.5. Present Demographic Transition Africa, Asia & L. America

CBR CDR

The decline in fertility came with the emergence and maturation of the new type of society, urban industrial society.[10] Perhaps the most important aspect of this transition involved the changing value of children and reproduction in what John Caldwell (1976) has called the transition from upward to downward net intergenerational capital flows. Children were transformed from economic assets to economic liabilities. This was, however, far from a simple change in the economic calculations made by individual families. It was, rather, a broad social and cultural change that moved rather quickly, after it began, through groups identified by language, culture or ethnicity. In the broadest sense, it was a change that we tend to call modernization.[11] With this change, fertility declined and came into line with low mortality rates. The demographic transition was completed. Every industrial society has now completed that transition, though there have been substantial differences in the timing and trajectories of the declines in birth and death rates.

That same transition appears to be taking place in the less developed countries of Asia, Africa and Latin America (figure 3.5), though there are considerable differences in the character of the change. First, the birth and death rates at the beginning of the transition were higher than those in Europe and North America.[12] Second, the recent decline of mortality has been much more rapid, and has been due to major advances in medical and public health technologies, which arose largely out of World War II. The development of antibiotic drugs, vaccines, pesticides, and fungicides all permitted new widespread health networks to reduce mortality from infectious diseases. Mortality declines that required a century or more in the past, now took place in decades. Because of this rapid mortality decline, the population increase associated with this transition has been more rapid and of greater magnitude than that experienced in the past. Past transitions brought rates of 1 to 2 percent per year; current transitions have brought rates of growth of 3 percent and more.[13] Finally, the fertility declines that we now witness in the developing world are, in part, associated with a remarkable set of policy

10. See Coale and Watkins (1986) for a full review of the decline of fertility in Europe. This provides an exposition of the variety of economic, health, social and cultural conditions that played a role in the fertility decline.

11. Cleland and Wilson (1987) make this point clearly for both past and present fertility declines. In both cases fertility declines tend to run along broad cultural, language or ethnic lines rather than along clear class or income lines. The latter is more a part of the common perception and is also well articulated theoretically in what is called the new household economics of fertility. As Cleland and Wilson show, however, empirical support for this well developed theory is weak or lacking altogether. Caldwell (1977, 1987a, 1987b, 1988) also makes this point for Africa and for comparisons with Asia and Africa.

12. This is probably related to differences in kinship patterns, but that discussion must be left to another time and place.

13. It may well be that the speed of mortality decline and the magnitude of population growth are the most powerful deleterious effects of the current transition. Ogawa and Suits (1985), for example, have performed a simulation exercise for Japan's past 100 years of demographic transition. They simply assigned recent Asian mortality declines to Japan after 1870 and concluded that with the more rapid mortality decline and consequent higher population growth rates, Japan would not have been able to achieve the savings and investment rates necessary for its own economic take-off.

revolutions, from pro- to anti-nationalism. In the industrialized world, past fertility declines often came *despite* the wishes and policy of government. Today, many governments in the less developed world are *leading* the drive to fertility decline.

It is difficult to overemphasize the revolutionary character of these policy changes. Until 1952 virtually all governments throughout human history were pro-natalist. People were a resource, which translated into power. Governments taxed people, worked them, and sent them off to war. Thus governments have always tended to want more rather than fewer people. This led virtually all governments to be pro-natalist.[14]

This changed dramatically in 1952 when Japan and India became the first of the modern governments to announce official policies to limit population growth through limiting fertility. Since 1952 almost all developing country governments have followed. Throughout the world today, governments have launched, supported or permitted the formation of national family planning programs, to distribute the new contraceptives and to disseminate the message of fertility control. There is no doubt that this combined technological and policy change is having a major impact on fertility reduction and population growth.[15]

Like the declines in mortality, both the policy changes and the current declines in fertility in the developing world, have been greatly assisted by the development of a new contraceptive technology. It is highly unlikely that governments would have adopted wide ranging fertility limitation programs, or that they could have been as successful as they have been, had there not been this technological innovation (Ness and Ando 1984). It is also important to note that the new fertility limiting technology, like the mortality limiting technology is what can be called bureaucratically portable. It can be set in the specialized hierarchic organizations that governments throughout the world have developed to administer to their populations. This ties population environment interactions of the present closely to technology and human organization, especially to the rise and spread of modern bureaucratic organizations.

It seems quite likely that current fertility declines will continue, and will be driven by two major forces. One is the same urban-industrialization that in the

---

14. The logic of this statement requires a connecting argument. Governments that want more people can get them through conquest, encouraging immigration, and encouraging high fertility. All three have been extensively used as government policies. Conquest has always been risky, however, and in the modern world promoting immigration also raises problems. In all societies, probably the safest and surest method of increasing, or sustaining population, is simply to encourage people to do what comes naturally -- reproduce. Hence the near universality of pronatalism as official government policy

15. It is also necessary, however, to note that the policy change is neither a sufficient or necessary condition for fertility decline. Policy changes have been made in Egypt, Kenya and the Philippines, for example, with little apparent impact on fertility. On the other hand, Brazil has had no real policy change, yet fertility has declined, largely because contraceptives and abortions have become much more widely available in that rapidly urbanizing society.

past transformed children from assets to liabilities. This is often referred to as the "demand" side (i.e., demand for children, or for fertility limitation) forces in fertility decline. In addition, the expansion of national family planning programs, bringing a greater supply of the new contraceptive knowledge and methods, also works to depress fertility. There is considerable controversy over the relative impact of these demand and supply sides of the forces, but there is also agreement that the two together work more powerfully than either does alone.[16] The issue of policy implementation, specifically of the organization and management of modern family planning programs, has been dealt with extensively in the literature. It is evident that governments vary considerably in their willingness and ability to promote fertility limitation. This is related to something that can be called a political culture that greatly affects what a government can and cannot do (Ness and Ando 1984; Finkle and Ness 1985). Finally, there are powerful cultural forces that work directly through individual and family reproductive orientations, and these, too, profoundly affect both what governments can do and what individuals themselves will do to limit fertility.

---

16. As in most discussions of current fertility decline, even this broad generalization requires some qualification. Bangladesh illustrates a case, especially in its experimental *Matlab* district, where fertility decline appears to come solely from contraceptive distribution, with no appreciable economic improvement. On the other hand, Brazil illustrates a case where economic development, and especially urbanization have brought rapid declines in fertility in the near complete absence of any government efforts to distribute contraceptives. Burma also shows some fertility decline without economic development, and even with government attempts to limit the distribution of contraceptives. In this case, however, abortion appears to be the most generally used method of fertility limitation, usually with very high costs for women.

The Future

Population and Development Policy Change

In addition to new population policies, governments throughout the world have also adopted policies to promote economic growth. This major policy transformation followed the Second World War and the dissolution of the extensive Western overseas colonial system.[17] In fact it preceded and in many respects produced the transformation of population policy (Ness and Ando 1984). These policies have also had a massive impact on economic organization and human productivity in the Third World. But this has resulted in an increasing assault on the environment through increasing energy consumption, greater use of fertilizers and pesticides and the discharge of a wide variety of industrial toxins.

Both these population and development policies portend major changes in the future of the population environment dynamic. We can suggest some of the dimensions of this dynamic, which will vary by major world region, even though the condition in any region will also have a major impact on the world as a whole.

Varying Scenarios

Population and economic growth rates and their varying impact on global change can be seen by examining separately four major world regions. Simple aggregate measures of recent and projected population growth can be used along with cereal output per capita as an indicator of one important element of economic growth. Figures 3.6 through 3.9 provide a visual portrayal of the changes we shall discuss.

Asia: Progress and Questions (Figure 3.6)

The development in Asia is generally hopeful, at least as far as population and food are concerned. The population has grown from 1.4 billion in 1950 to over three billion in 1990, and will most certainly reach near five billion by 2025. Asia's projected rate of population growth is expected to fall drastically to about .5 percent, from the high of over 2 percent in the second half of the twentieth century. Most countries in Asia have entered into the long term irreversible fertility decline that marks the completion of the demographic transition. A substantial portion of the Asian population has already reached or will very shortly reach replacement level or below through a combination of

---

17. See Gunnar Myrdal's *Asian Drama* (1968) for a good discussion of the spread of the ideology of national planning. See also Nigel Harris (1986) for a review of recent changes in the concept of the state and its role in development promotion.

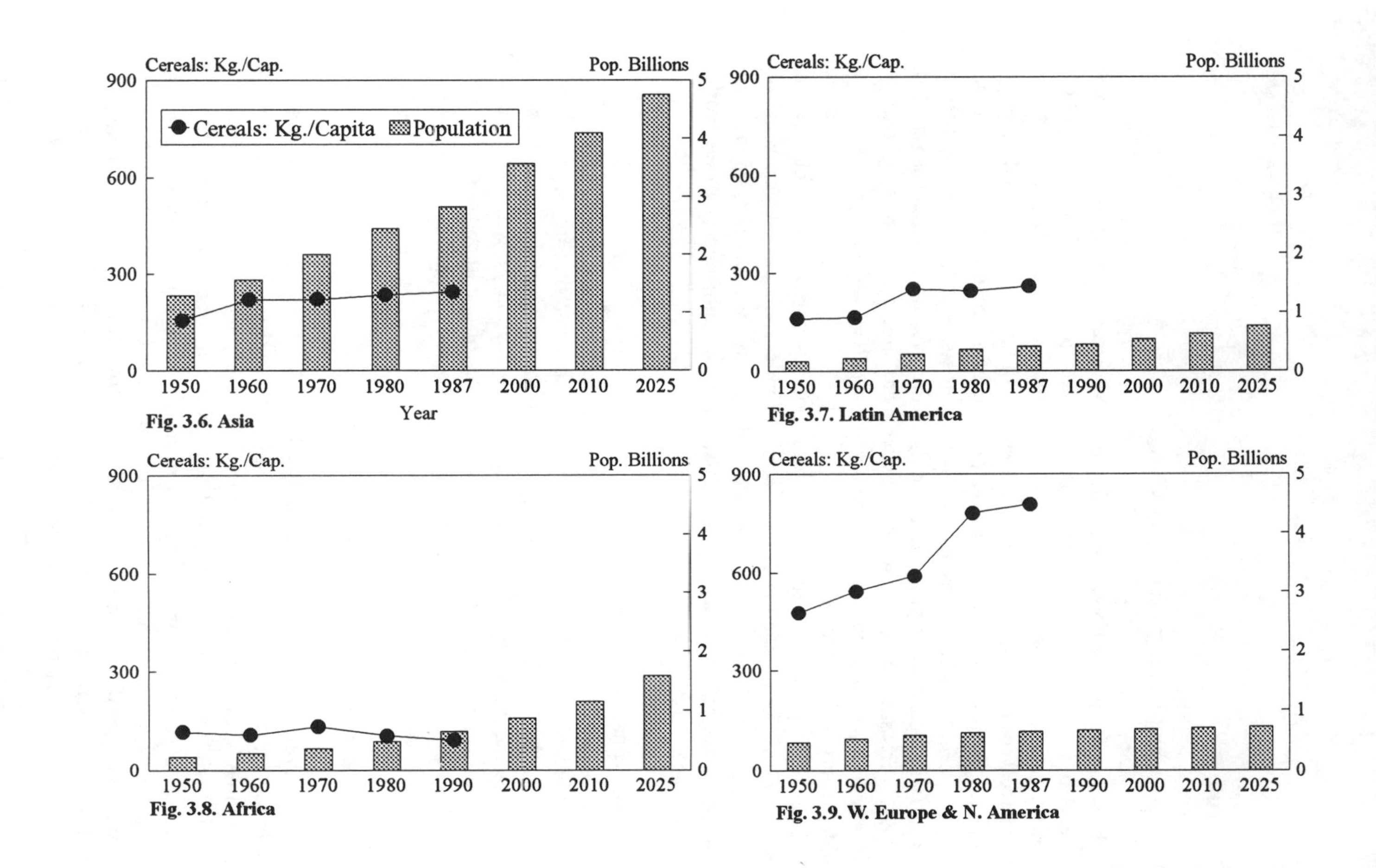

Fig. 3.6. Asia

Fig. 3.7. Latin America

Fig. 3.8. Africa

Fig. 3.9. W. Europe & N. America

economic development and family planning program extension.[18] China, Hong Kong, Japan, Singapore, South Korea, and Taiwan, are currently at or below replacement level, with Indonesia, Sri Lanka, Thailand, and some of the states of India coming close to this level.

Cereal output per capita since 1950 has also risen steadily, however, with fairly good prospects for continued increases.[19] The rise in cereal output over the past few decades has come from the technological change known as the Green Revolution. Modern science and technology have been applied to rice and wheat production with striking results in output and yields. Even with the decline of population growth rates, however, Asia will experience increasing industrial development. Asia is planning and mobilizing resources for national economic development and there is little doubt that the region will become highly industrialized. But if Asia makes this transition with current energy technology, as appears likely, the impact on the world environment will be striking. For example, China's current per capita energy consumption stands at about one-thirteenth that of the U.S., or one-fifth that of Japan. Much of China's industrialization will be fueled by high sulfur content coal.[20] If China attained Japan's current relatively efficient standard of per capita consumption by 2025, it would reach almost three times the current total U.S. energy consumption (World Resources Institute 1988).[21]

Further, there is reason to doubt that Asia's past increases in agricultural output can continue at the recent rate. There is even more reason to ask what the price will be for that increase, both in capital and in environmental degradation. Fertilizer consumption has already begun to show reduced marginal returns. Water tables are dropping as increased irrigation strains the underground supplies, and pesticide use has reached prohibitive costs and seriously threatens human health.

Finally, much of Asia's development is paid for by the export of primary products. This implies increasing deforestation as tropical forests are logged for export. Deforestation and agricultural development result in species destruction with, as yet, unassessed impacts. Oil and mineral deposits are also extensively exploited with various environmental degradation results that are also, not yet, well assessed. Thus, as hopeful as Asia's population and food

18. See ESCAP (1988) and Richard Leete (1987) for reviews.

19. Asian population densities have always been near those of Europe, and much higher than Africa or the Americas, indicating a strong social ability to raise the carrying capacity of the land.

20. For example, from 1971 to 1987, world coal production increased 50 percent, from 2.1 to 3.3 billion metric tons, an increase of 1.2 billion metric tons. China alone accounted for almost one half of that increase, rising from 375 to 879 million metric tons. India increased its output by about 108 m.m.t. Thus together India and China alone accounted for about half of the total increase in coal output over roughly the past two decades.

21. Total U.S. energy consumption in 1986 was estimated at 66,766 petajoules, or 278 gigajoules per capita. Japan's per capita consumption stood at 106 gigajoules. With a population near 1.5 billion in 2025, at Japan's per capita rate, China would consume over 158,000 petajoules of energy, the overwhelming majority of which would come from high sulfur content coal (World Resources Institute 1988).

scenario appears, it also has a more ominous side when we consider the potential environmental impact of its future industrialization and agricultural development.

Latin America: More Hope With More Questions (Figure 3.7)

As in Asia, the population of Latin American has also increased greatly, and is projected to continue growing rapidly through the first quarter of the next century. From below 200 million in 1950, it rose to over 400 million in 1985, and is projected to reach 760 million by 2025. Like Asia, its population growth rate is now slowing, as most nations have entered well into the sustained fertility decline that will mark the end of the demographic transition (Merrick 1986). Also like Asia, Latin America's cereal output in the past few decades has kept well ahead of its rapid population growth. The large spurt in agricultural growth came after 1955 with the generation of high yielding varieties of wheat and corn. Prospects for continued increase are generally good.

At the same time, the region has experienced rapid urbanization and industrialization. It is now almost as urbanized as North America and Europe and it is industrializing rapidly. If it reaches the Japanese per capita energy consumption level by 2025, its 750 million people will consume almost 20 percent more energy than the U.S. consumes today. That level of consumption will certainly have a major impact on greenhouse gas emissions. Further, while much of the agricultural increases have come from yield increases, much has also come from the extension of agricultural land. Most of the new agricultural land comes from the destruction of the tropical rain forest. In the 1980s Latin America lost an average of seven million hectares of forests per year, balanced by less than half a million hectares of reforestation. There are also the same questions found in Asia about the capital, environmental, and human health costs of future agricultural development. Thus, as in Asia, future increases in food output and industrialization raise serious questions about costs and environmental impacts.

Africa: The Nightmare Scenario (Figure 3.8)

Africa's population grew from about 200 million in 1950 to over 600 million today. It is projected to more than double to about 1.6 billion by the year 2025. Africa today has the world's highest levels of fertility and the most rapid population growth. Further, with the possible exceptions of Botswana, Kenya and Zimbabwe, African nations have not yet entered into the sustained fertility declines that have marked the Asian and Latin American experiences (Weeks 1988; Goliber 1989; van De Walle and Foster 1990). Thus the projections of population doubling in thirty-five years seem frighteningly probable.

During the past period of population growth, Africa's cereal output per capita has shown a general downward trend, from initial levels that were already considerably below those of Asia or Latin America. The recent development of a network of international agricultural research stations in Africa provides some promise of increasing the yields of a variety of indigenous and new crops. At the same time, the hydrological picture shows much of Africa deficient in basic water resources even for its current population (Falkenmark 1990). It is today a food deficit region, and there seems scant prospect of reversing that condition in the near future. It is also deficient in the capacity to pay for needed food imports. How Africa's population can more than double in the next generation, or what implications that will have for either population or environment, are at best difficult to discern. Even the best of the World Bank's scenarios for Africa today are termed nightmare scenarios.

Europe and North America: The Breadbasket (Figure 3.9)

The populations of Europe and North America will grow more slowly over the next forty years, since they have completed the demographic transition and reached replacement fertility levels or below. Nonetheless even these populations will continue to grow for the near future. From a population of 470 million in 1950, this highly industrialized region grew to over 600 million today, and will reach 722 million by 2025.

During this period, cereal output per capita has grown at impressive rates,[22] and stands today at very high levels. Although North America can produce a great deal of food, it is produced with very high energy costs, and perhaps unsustainable costs in land and water. There are serious questions about whether future increases in food can keep up with world population growth. The increased population will also consume more energy, but here there seems to be considerable room for improvement of efficiency.

Europe's high standard of living is achieved with much greater energy efficiency than the United States'. European per capita energy consumption levels are less than half (130 vs 278 gigajoules) those in the U.S. (World Resources Institute 1988). Thus if the increased population consumed energy at the more efficient European level, total consumption for the two together would fall from about 131,000 to 94,000 petajoules by 2025. On the other hand, if the consumption followed that in the U.S., the total would rise to over 200,000 petajoules. It is not difficult to imagine what that higher level of

---

22. North American cereal output has been about 2.5 times that of Europe throughout this period, and its population has grown from about 55 percent to about 70 percent of the European population (FAO 1988).

energy consumption would mean for atmospheric conditions. Here, however, there is cause for some optimism. Over the past two decades, the United States has reduced its energy consumption per capita by 12 percent, and its consumption per dollar of GNP by 33 percent. Most European countries and Japan have also substantially reduced their energy consumption per GNP levels. Obviously there are both economic and policy forces supporting the development of cleaner and more efficient energy.

The environmental degradation associated with this region's mature industrialization poses more serious threats both to individuals and to the entire world ecosystem. It is, in fact, this massively productive industrial system that has given the human species such dominance over the environment, and also threatens the very survival of the species.

All of these scenarios, taken together, indicate the need for new institutions on a world wide level. The human species will have to find ways to increase its output sufficiently to feed and cloth the population that is all too likely to double before the next century is over. It will also have to find ways to do that without expanding its current destructive impact on its environment.

Conclusion

In considering the impact of population and development on global environmental change, it is necessary to keep in mind the full range of global changes that are affected by human action. Population growth and economic development have resulted in massive increases in energy use, especially in fossil fuel consumption, and thus the increasing release of greenhouse gases into the atmosphere. They are also associated with massive deforestation and the destruction of many plant and animal species. This species destruction continues today, perhaps at increasing rates. The impact of this aspect of global change may be less readily apparent than the threat of global warming, and thus may have received less attention. In addition, urbanization and industrialization have the led to increased emission of toxic wastes in land, water and air. This aspect of environmental degradation has received more attention recently, but it has yet to become fully integrated into our views of global change. In effect, a wide range of human impacts on the environment must be more fully integrated into our views of global change if we are to understand the relation between population and environment. That integration is also necessary if we are to adjust our behavior sufficiently to ensure the continued survival of the human species, and perhaps of the entire global ecosystem as well.

As noted in the Introduction, we must recognize that population policies are some of the most conflictual we know of in the history of modern public policy. Population policies touch deeply held values of race and ethnic identity, human sexuality, gender roles, fundamental values of individual rights and of life itself. Wherever a state's population is ethnically divided, population policies will be faced with deep, primordial fears of population decline, which accompany changing relative numbers of different ethnic groups. By contrast, many other environmental policies may be subject to at least some economically rational argument. Further, many environmental controversies can be resolved with scientific research and cost-benefit analyses. Many of the population controversies, however, are moral and ethical and cannot be resolved with scientific evidence or economic calculations. Although the world community as a whole has made massive strides in treating population issues with greater understanding, it still generates deep fears and conflicts. This situation will undoubtedly persist. Our task cannot be to provide the scientific evidence to resolve the population debates. It is, however, our task to provide the evidence and information that can make those moral and philosophical debates more fully informed.

Chapter IV

Constraints on Sustainable Growth in Agricultural Production

Vernon Ruttan

Introduction

By the end of the twentieth century, one of the most remarkable transitions in the history of agriculture will have been completed. Prior to this century almost all of the increase in food production was obtained by bringing new land into production. There were only a few exceptions to this generalization in limited areas of East Asia, the Middle East, and Western Europe (Hayami and Ruttan 1985). By the first decade of the next century, almost all of the increases in world food production must come from higher yields from increased output per hectare. In most of the world the transition from a resource-based to a science-based system of agriculture is occurring within a single century. In a few countries this transition began in the nineteenth century. For most of the currently developed countries, it did not begin until the first half of this century. Most of the countries of the developing world have been caught up in the transition only since mid-century. Among developing countries those countries of South, East and Southeast Asia have proceeded further in this transition than most countries in Latin America or Africa (Ruttan 1986).

This paper explores the constraints on sustainable growth in agricultural production into the first decades of the twenty-first century. Although the population issue is not directly addressed, a number of agricultural, resource, environmental and health concerns are examined which will condition the capacity of the agricultural sector to respond to the demands of population and income growth particularly in the developing countries of Latin America, Asia and Africa. The effects of population growth on the rate and direction of technological change in agriculture has been addressed in earlier papers (Hayami and Ruttan 1987).

The historical trends in production and consumption of the major food grains could easily be taken as evidence that one should not be excessively concerned about the capacity of the world's farmers to meet future food demands. World wheat prices, corrected for inflation, have declined since the middle of the last century (fig. 4.1). Rice prices have declined since the middle of this century (fig. 4.2). These trends suggest that productivity growth has been able to more than compensate for the rapid growth in demand, particularly during the decades since World War II.

For future generations, however, the sources of productivity growth are not as apparent as they were a quarter century ago. The demands that the developing economies place on agricultural producers from population growth and growth in per capita consumption arising will be exceedingly high. Population growth rates are expected to decline substantially in most countries during the first quarter of the next century. Per capita incomes are expected to increase. The effect of growth in per capita income will be more an increased demand for animal proteins and for maize and other feed crops. During the next several decades growth in the demand for food arising from growth in population and income will run upwards of 4 percent per year in many countries. Many will experience more than a doubling of food demand before the end of the second decade of the next century.

Biological and Technological Constraints on Crop and Animal Projection

It seems apparent that the gains in agricultural production required over the next quarter century will be achieved with much greater difficulty than in the immediate past. Difficulty is currently being experienced in raising yield ceilings for cereal crops which have experienced rapid yield gains in the past. The incremental response to increases in fertilizer use has declined. Expansion of irrigated area has become more costly. Maintenance research, the research required to prevent yields from declining, is rising as a share of research effort (Plucknett and Smith 1986). The institutional capacity to respond to these concerns is limited, even in the countries with the most effective national research and extension systems. Indeed, there has been considerable difficulty in many countries during the 1980s in maintaining the agricultural research capacity that had been established during the 1960s and 1970s (Cummings 1989). It is possible that within another decade, advances in basic knowledge will create new opportunities for advancing agricultural technology that will reverse the urgency of some of the above concerns. Institutionalization of private sector agricultural research in some developing countries is beginning to complement public sector capacity (Pray 1983). Advances in molecular biology and genetic engineering are occurring rapidly. But the date when these promising advances will be translated into productive technology appears distant.

Almost all increases in agricultural production over the next several decades must continue to come from further intensification of agricultural production

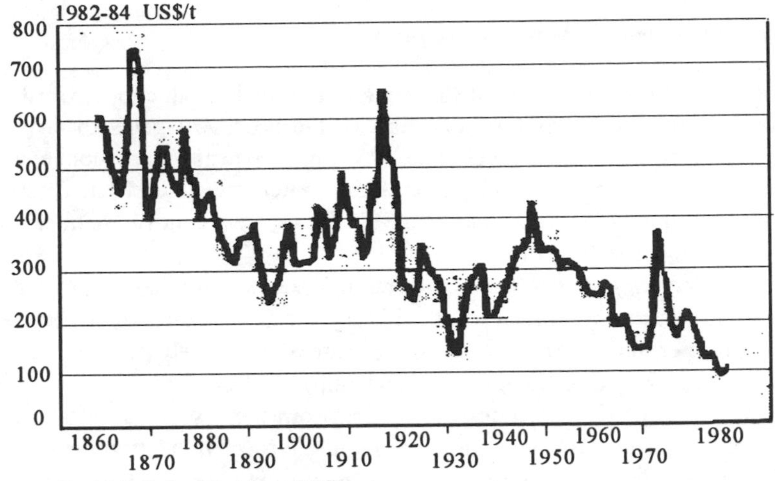

Fig. 4.1. Real wheat prices since mid-1800s.
Source: US Bureau of Census. 1975. Historical Statistics of the United States: Colonial Times to 1970. Washington, D.C.: US Dept. of Commerce (for data 1860-1960). US Dept. of Agriculture.n.d. Agricultural Outlook. Washington, D.C.: USDA (for data 1961-1988).

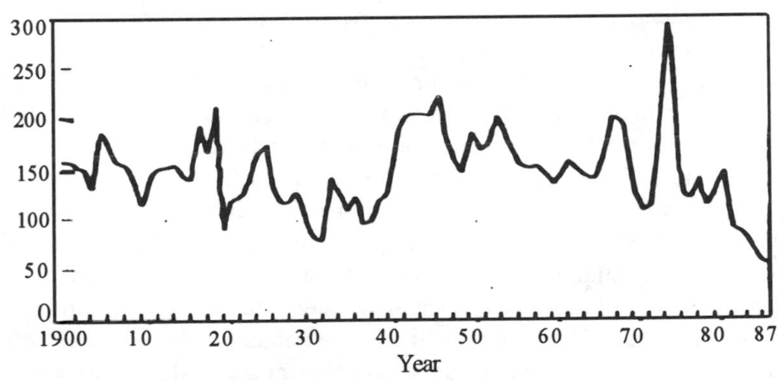

Fig. 4.2. Real Rice Prices, 1900-1987.
Source: Prabhu Pingali, 1988. Intensification and Diversification of Asian Rice Farming Systems. Los Banos, Philippines: IRRI Agricultural Economics Paper 88-41.

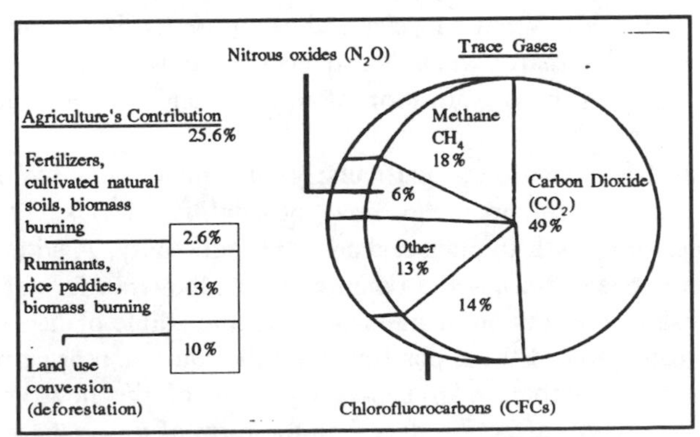

Fig. 4.3. Contributions to increaseses in radiative forcing in the 1990s.
Source: J. Reilly and R. Bucklin. 1989. Climate Change & Agriculture. *World Agriculture Situation and Outlook Report.* Washington, D.C. USDA/ERS. WAS-55.

on land that is presently devoted to crop and livestock production. Until well into the second decade of the next century the necessary gains in crop and animal productivity will be generated by improvements from conventional plant and animal breeding and from more intensive and efficient use of technological inputs including chemical fertilizers, pest control chemicals and more effective animal nutrition.

The productivity gains from conventional sources are likely to come in smaller increments than in the past. If they are to be realized, higher plant populations per unit area, new tillage practices, improved pest and disease control, more precise application of plant nutrients, and advances in soil and water management will be required. Gains from these sources will be crop, animal, and location specific. They will require closer articulation between the suppliers and users of new knowledge and new technology. These sources of yield gains are extremely knowledge and information intensive. If they are to be realized, research and technology transfer efforts in the areas of information and management technology will be increasingly important sources of growth in crop and animal productivity.

Advances in crop yields have come about primarily by increasing the ratio of grain to straw rather than by increasing total dry matter production. Advances in animal feed efficiency have come by decreasing the proportion of feed consumed that is devoted to animal maintenance and increasing the proportion used to produce usable animal products. There are severe physiological constraints to continued improvement along these conventional paths. These constraints are most severe in areas that have already achieved the highest levels of productivity--as in Western Europe, North America, and parts of East Asia. The impact of these constraints can be measured in terms of declining incremental response to energy inputs--both in the form of a reduction in the incremental yield increases from higher levels of fertilizer application, and a reduction in the incremental savings in labor inputs from the use of larger and more powerful mechanical equipment. If the incremental returns to agricultural research should also decline, it will impose a higher priority on efficiency in the organization of research and on the allocation of research resources.

Advances in basic science, particularly in molecular biology and biochemistry, continue to open up new possibilities for supplementing traditional sources of growth in plant and animal productivity. A wide range of possibilities have been discussed, ranging from the transfer of growth hormones into fish to conversion of lignocellulose into edible plant and animal products. The realization of these possibilities will require a reorganization of agricultural research systems. An increasing share of the new knowledge generated by research will affect producers in the form of proprietary products or services. This means that incentives must be created to draw substantially more private sector resources into agricultural research.

Within the public sector, research will have to increasingly move from a "little science" to a "big science" mode of organization. Examples include the collaborative research program on the biotechnology of rice sponsored by the Rockefeller Foundation and the University of Minnesota program on the biotechnology of maize. In the absence of more focused research efforts, it seems likely that the promised gains in agricultural productivity from biotechnology will continue to recede.

Crop and animal productivity levels in most developing countries remain well below the levels that are potentially feasible. Access to the conventional sources of productivity growth--from advances in plant breeding, agronomy, and soil and water management--will require the institutionalization of substantial agricultural research capacity for each crop or animal species of economic significance in each agro-climatic region. In a large number of developing countries this capacity is just beginning to be put in place. A number of countries that experienced substantial growth in capacity during the 1960s and 1970s have experienced an erosion of capacity in the 1980s. Even a relatively small country, producing a limited range of commodities under a limited range of agro-climatic conditions, will require a cadre of agricultural scientists in the 250-300 range. Countries that do not acquire adequate agricultural research capacity will not be able to meet the demands that they will place on their farmers as a result of growth in population and income.

Research underway in the tropical rain forest areas of Latin America and in the semi-arid tropics of Africa and Asia suggest the possibility of developing sustainable agricultural systems with substantially enhanced productivity even in unfavorable environments. It is unlikely, and perhaps undesirable, that all of these areas become important components of the global food supply system. But enhanced productivity is important to those who reside in these regions--now and in the future. It is important that research investment in soil and water management and in farming systems be intensified in these areas.

There are a series of basic biological research agendas that are important for applied research and technology development for agriculture in the tropics that continue to receive inadequate attention in the temperate region developed countries. There is also a need for closer articulation between training in applied science and technology and training in basic biology. When such institutes are established they will need to be more closely linked with existing academic centers of research and training such as the series of agricultural research institutes established by the Rockefeller and Ford Foundations and the Consultative Group on International Agricultural Research (CGIAR).

Resource and Environmental Constraints on Sustainable Growth

An examination of the prospects for the next century reveals the increasing impact of resource and environmental constraints that may seriously impinge on the capacity to sustain the required growth in agricultural production (from 2 to 4 percent per year) for most developing countries. One of the most important issues concerns the impact of agricultural production practices that will be employed in those areas which make the most progress in moving toward highly intensive systems of agriculture production. These include loss of soil resources due to erosion, water-logging and salinization, groundwater contamination from plant nutrients and pesticides, and growing resistance of insects, weeds and pathogens to present methods of control. If agriculture is forced to continue to expand into more fragile environments, such problems as soil erosion and desertification can be expected to become more severe. Additional deforestation will intensify problems of soil loss as well as degradation of water quality and contribute to the forcing of climate change.

A second set of concerns stem from the impact of industrialization on global climate and other environmental changes (Reilly and Bucklin 1989). The accumulation of carbon dioxide (CO_2) and other greenhouse gases principally methane (CH_4), nitrous oxide (N_{20}) and chlorofluorocarbons (CFC's) has set in motion a process that will result in a rise in global average surface temperatures over the next thirty to sixty years. And there continues to be great uncertainty about the climate changes that can be expected to occur at any particular date or location in the future. It is almost certain, however, that the climate changes will be accompanied by rises in the sea level and that these rises will impinge particularly heavily on islands in Southeast Asia and the greater river deltas of the region. Dryer and more erratic climate regimes can be expected in interior South Asia and North America. To partially balance this, some analysts have suggested that higher CO_2 levels may have a positive effect on yield (Rosenberg 1986).

The bulk of the carbon dioxide emissions come from fossil fuel consumption. Carbon dioxide accounts for roughly half of radiative forcing. Biomass burning, cultivated soils, natural soils, and fertilizers account for close to half of nitrous oxide emissions. Most of the known sources of methane are a product of agricultural activities--principally enteric fermentation in ruminant animals, release of methane from rice production and other cultivated wetlands, and biomass burning. Estimates of nitrous oxide and methane sources have a very fragile empirical base. Nevertheless, it appears that agriculture and related land use could account for somewhere in the neighborhood of 25 percent of radiative forcing (fig. 4.3).

The alternative policy approaches to the threat of global warming can be characterized as *preventionist* or *adaptationist*. A preventionist approach could involve five policy options: (a) reduction in fossil fuel use or capture of CO_2 emissions at the point of fossil fuel combustion; (b) reduction in the intensity of agricultural production; (c) reduction of biomass burning; (d) expansion of biomass production; and (e) energy conservation. Of these only energy efficiency and conservation are likely to make any significant contribution over the next generation. And the speed with which either will occur will be limited by the pace of capital replacement. Any hope of significant reversal of agricultural intensification, reduction in biomass burning, or increase in biomass absorption is unlikely to be realized within the next several decades. The institutional infrastructure or institutional resources that would be required do not exist and will not be in place rapidly enough. There will be no technological fix to the global warming problem. The fixes, whether driven by preventionist or adaptionist strategies, must be both technological and institutional.

Given such limitations, the *adaptationist* appears to be most conducive in assessing the implications of global climate change for future agricultural research agendas. Thus, in this context, an adaptionist strategy implies moving as rapidly as possible to design and put in place the institutions needed to remove the constraints that intensification of agricultural production are currently imposing on sustainable increases in agricultural productivity. For example, the implementation of this strategy would include: a) commodity policies--such as those of the United States, the EEC countries, and Japan--that encourage excessive use of chemical inputs as substitutes for land; and b) resource policies, such as those that inhibit the rational conservation, allocation, and use of surface and groundwater. If the development of policies and institutions needed to deal with existing resource constraints is successful, future policymakers will be in a better position to respond to changes that will emerge as a result of future global climate change.

The first research priority is to initiate a large-scale program of research on the design of institutions capable of implementing "incentive compatible" resource management policies and programs. Incentive compatible institutions can be defined as institutions capable of achieving compatibility between individual, organizational, and social objectives in resource management. A major source of the global warming and environmental pollution problem is the direct result of the operation of institutions which induce behavior by individuals, and public agencies that are not compatible with societal development--some might say survival--goals. In the absence of more efficient incentive compatible institutional design, the transaction costs involved in ad hoc approaches are likely to be enormous.

A clearer picture of the demands that are likely to be placed on agriculture over the next century and of the ways in which agricultural systems might be able to meet such demands has yet to be produced. World population could rise from the present five billion level to the ten to twenty billion range. The demands that will be placed on agriculture will also depend on the rate of growth of income--particularly in the poor countries where consumers spend a relatively large share of income growth on subsistence--food, clothing, and housing. The resources and technology that will be used to increase agricultural production by a multiple of three to six will depend on both the constraints on resource availability that are likely to emerge and the rate of advancement in knowledge. Advances in knowledge can permit the substitution of more abundance for increasingly scarce resources and reduce the resource constraints on commodity production. Past studies of potential climate change effects on agriculture have given insufficient attention to adaptive change in non-climate parameters. But application of advances in biological and chemical technology, which substitute knowledge for land, and advances in mechanical and engineering technology, which substitute knowledge for labor, have, in the past, been driven by increasingly favorable access to energy resources, particularly, the declining prices of energy. There will be strong incentive, by the early decades of the next century, to improve energy efficiency in agricultural production and utilization. Particular attention should be given to alternative and competing uses of land. Land use transformation, from forest to agriculture, is presently contributing to radiative forcing through release of CO_2 and methane into the atmosphere. Conversion of low intensity agricultural systems to forest has been proposed as a method of absorbing CO_2. There will also be increasing demands on land use for watershed protection, and biomass energy production.

It is a matter of serious concern that only in the last decade and a half has it been possible to estimate the magnitude and productivity effects of soil loss, even in the United States. Even rudimentary data on soil loss is almost completely unavailable in most developing countries. The same point holds, with greater force, for groundwater pollution, salinization, species loss and others. It is time to design the elements of a comprehensive, agriculturally related, resource monitoring system and to establish priorities for implementation. Data on the effects of environmental change on the health of individuals and communities is even less adequate. The monitoring effect should include a major focus on the effects of environmental change on human populations.

It is essential to have closer collaboration between production-oriented agricultural scientists, biological scientists with training in ecology, and the physical scientists that have been traditionally concerned with global climate change. This effort should be explicitly linked with the monitoring effects currently being pursued under the auspices of the International Geosphere-Biosphere Programs (IGBP).

During the next century water resources will become an increasingly serious constraint on agricultural production. Agricultural production is a major source of decline in the quality of both ground and surface water. Limited access to clean and uncontaminated water supplies is a major source of disease and poor health in many parts of the developing world and in the centrally planned economies. Global climate change can be expected to have a major differential impact on water availability, water demand, erosion, salinization, and flooding. The development and introduction of technologies and management systems that enhance water use efficiency represents a high priority both because of short and intermediate run constraints on water availability and the longer run possibility of seasonal and geographical shifts in water availability. The identification, breeding, and introduction of water efficient crops for dreamland and saline environments could be an important aspect of achieving greater water use efficiency.

Research on environmentally compatible farming systems should be intensified. In agriculture, as in the energy field, there are a number of technical and institutional innovations that could have both economic and environmental benefits. Among the technological possibilities is the design of a new "third" or "fourth" generation chemical, biorational, and biological pest management technology. Another is the design of land use technologies and institutions that will contribute to reduction of erosion, salinization, and groundwater pollution.

Immediate efforts should be made to reform agricultural commodity and income support policies. In both developed and developing countries producers decisions on land management, farming systems, and use of technical inputs (such as fertilizers and pesticides) are influenced by government interventions such as price supports and subsidies, programs to promote or limit production, as well as tax incentive and penalties. It is increasingly important that such interventions be designed to take into account the environmental consequences of decisions by land owners and producers induced by the interventions.

A food-system perspective should become an organizing principle for improvements in the performance of existing systems and for the design of new systems. The agricultural science community should be prepared, by the second quarter of the next century, to contribute to the design of alternative food systems. Many of these alternatives will include the use of plants other than the grain crops that now account for a major share of world food production. Some of these alternatives will involve radical changes in food sources. Rogoff and Rawlins (1987) have described one such system based on lignocellulose--both for animal feed and human consumption.

Health Constraints on Agricultural Development

Since the mid-1960s a number of commonly used health indicators such as life expectancy and infant mortality experienced substantial improvement for almost all developing countries. Concerns about nutritional deficiency as a source of poor health has receded in a large number of developing countries in the last several decades (BOSTID/IOM 1987; Commission on Health Research for Development 1990).

Yet there are a number of other indicators that suggest that health constraints could become increasingly important by the early decades of the next century. Daily calorie intake per capita has been declining for as much as two decades in a number of African countries. While dramatic progress has been made in the control and reduction of losses due to infectious disease and in the control of diarrheal disease, little progress has been made in the control of several important parasitic diseases. The sustainability of advances in malaria and tuberculosis control are causing serious concern. The emergence of AIDS, combined with other health threats, could become a major threat to economic viability in both developed and developing countries.

There is also a second set of health concerns arising out of the environmental consequences of the intensification of agricultural and industrial production. As the environmental impacts of agricultural and industrial intensification become clearer, it appears that they are already imposing significant health burdens on some countries, particularly in parts of the former former USSR and Eastern Europe, which may become more burdensome in the future.

If one visualizes a number of these health threats emerging simultaneously in a number of countries, it is not too difficult to construct a scenario in which there are large numbers of sick people in many villages around the world. The numbers could become large enough to become a serious constraint on food production capacity. The evidence suggesting that health represents a serious constraint on agricultural development is at best ambiguous. Scattered data from countries such as India, Indonesia and Ivory Coast indicate loss of days worked due to sickness in the 5-15 percent range. In the former USSR and Poland substantial numbers of days of work are lost due to respiratory disease associated with atmospheric pollution.

There have been major "plagues" in the past that resulted in mortality levels sufficient to seriously impinge on food supply. In the fifteenth century following the Spanish conquest, the American-Indian population in the basin of Mexico declined approximately 90 percent. Most of the decline was due to a series of epidemics--smallpox, measles, typhus, and plague. Famine, associated with the high dependency to working adult ratio, probably accounted for 10-15 percent of the population loss. The population loss from most historical plagues in Europe and Asia were concentrated in the younger and oldest age groups rather than among the adult population of working age. Many adults had survived earlier attacks and had acquired some degree of immunity. The incidence of death from the European and Asian diseases introduced into the Americas was spread more evenly across the age distribution because everyone was equally susceptible. The AIDS plague is unique in that it is killing people who would be at their most productive age. The result will be a rise in the dependency ratio--the ratio of the old and young relative to workers in the more productive age groups. There are important questions that have not yet been sorted out in the relationships between AIDS and other diseases. One apparent consequence of AIDS in East Africa is a rise in tuberculosis. The World Health Organization has an active program of cooperation with African and other high incidence AIDS countries in estimating HIV infection and AIDS incidence. A further step should be to model the direct and interaction effects of the simultaneous incidence of HIV infection and tropical parasitic and viral diseases on morbidity and on mortality.

The systems that are in place in most countries can be more accurately described as sickness recovery systems rather than health systems. A major deficiency is the lack of a system for providing families and individuals with the knowledge needed to achieve better health with less reliance on the health care system. Almost all countries have been able to design reasonably effective agricultural extension or technology transfer systems to provide farm people with the knowledge about resources and technology needed to achieve higher levels of productivity. However, an effective system has not yet been developed to provide families and individuals with the knowledge in the area of human biology, nutrition, and health practice that will enable them to lead more healthy lives.

The most serious impact is occurring in the centrally planned economies of Eastern Europe, the former USSR and China. Levels of atmospheric, water and soil pollution have resulted in higher mortality rates and reductions in life expectancy. The effects are evident in the form of congenital malformation, pulmonary malfunction and excessive heavy metals in soils and in crops grown on contaminated soils. Many of the health effects of agricultural and industrial intensification are due to inadequate investment in the technology needed to control or manage contaminants. Rapid industrial growth in poor countries, in which investment resources are severely limited, will continue to be accompanied by under investment in the technology needed to limit the release of contaminants. The situation that exists in Eastern Europe presents the vision of the future for many newly industrializing countries unless better technology can be made available and more effective management of environmental spillover effects can be implemented.

It is no longer possible to maintain the position that health related research results can simply be transferred from developed country research laboratories or pharmaceutical companies to practice in developing countries. Local capacity is needed for the identification and analyses of the sources of health problems. It is also needed for the analysis, design and testing of health delivery systems. The international donor community has been much slower in supporting the development of health research systems than agricultural research systems in the tropics. For example, there is now in place a network of more than a dozen international agricultural research centers (IARC's), sponsored by the Consultative Group on International Agricultural Research which plays an important role in backstopping national agricultural research efforts. The only comparable internationally supported center in the field of health is the Diarrheal Research Center in Bangladesh. Furthermore, the capacity to conduct research on tropical infectious and parasite diseases that was supported by the former colonial countries--the United Kingdom, France, Netherlands, and Belgium--has been allowed to atrophy.

The demographic transition--from high to low birth rates--has in the past, usually followed a rise in child survival rates. This suggests that improvements in health, particularly of mothers and children, is a prerequisite for decline in population growth. But high population growth rates, particularly in areas of high population density, are often associated with dietary deficiencies that contribute to poor health and high infant mortality rates.

The issue of how to achieve high levels of health and low birth rates at low cost in poor societies remains an unresolved issue. Several very low income countries have achieved relatively high levels of health--as measured by low infant mortality rates and high life expectancy rates--but often a high cost relative to per capita income. Other societies that have achieved relatively high incomes continue to exhibit relatively high infant mortality rates and only moderately high life expectancy levels.

More effective bridges must be built, both in research and in practice, between agricultural and health communities. At present these two "tribes", along with veterinary medicine and public health, occupy separate and often mutually hostile "island empires". But solutions to the problem of sustainable growth in agricultural production and improvement in the health of rural people and the consumers of agricultural commodities requires that each of these communities establish bridgeheads with each other. Multi-purpose water resource development projects have contributed to the spread of onchocerciasis. Successful efforts to control the black fly have reopened productive lands to cultivation. The introduction of improved cultivars and fertilization practices have helped make the productivity growth sustainable. But it is difficult to find examples of effective collaboration either in research or in project development.

Many of these problems are transactional in scope. This means that many of the institutions that will be needed to enable societies to respond to the constraints on sustainable increases in agricultural production will also have to be transactional. Slogans such as "think globally and act locally" are no longer sufficient. The capacity to respond to scientific, technical, resource, environmental and health constraints must be institutionalized. In the area of health, for example, it seems clear that almost every source of illness or poor health that exists somewhere--whether the source is an infectious organism or environmental change--will exist everywhere else. This statement may be an exaggeration, but it is only a slight exaggeration.

Another important issue is the limited capacity to design the institutional infrastructure required to sustain rates of growth in agricultural production into the first decades of the next century. The development of institutional infrastructures that facilitate more effective collaboration among engineers, agronomists and health scientists is imperative to deal with issues of production, environmental change, and the health of food producers and consumers. The social science disciplines and related professions (law, management, social service) have not demonstrated great capacity in the area of institutional design. Plant breeders have been much more effective. They do not just analyze the sources of yield differences--they utilize the agronomic and genetic knowledge that is obtained from their analyses to design improved cultivars--plants and animals that are responsive to management and that are resistant to the assaults of nature. In the social sciences once the analysis is complete the job is considered finished. Acquired knowledge is rarely brought to bear on institutional design.

More attention needs to be given to the design of both technologies and institutions that will broaden the options for choice or action. The highest incidence of AIDS is likely to occur, at least during the next several decades, in those parts of the world where the technologies and institutions needed to sustain food production are exceedingly weak. Wider technical options will be needed in both food production and utilization. The capacity to monitor the sources of productivity and environmental change is clearly inadequate. Very little is known about either the levels or the trajectories. There is much discussion about soil erosion but there is still no effective monitoring system to measure the extent to which it is weakening our capacity to produce. Scientists and policymakers are fighting a defensive battle against the health effects of the contamination of our food supply rather than anticipating the sources. One of the puzzling aspects of the available data is that the health effects of increased use of fertilizer is less than expected in spite of high levels of nitrate in surface and ground water. Neither the developed or developing countries have in place adequate surveillance systems for disease.

In the final analysis, the perspective for agricultural futures is cautiously optimistic. The challenges posed by the constraints on crop and animal productivity and by the resource, environmental and health constraints on sustainability should not be interpreted as a completely pessimistic assessment. The global agricultural research system, the technology supply industry, and farmers are much better equipped to confront the challenges of the future than they were when confronted with the food crises of the past.

It cannot be emphasized too strongly, however, that the challenges are both technical and institutional. The great institutional innovation of the nineteenth century was "the invention of the method of invention." The modern industrial research laboratory, the agricultural experiment station, and the research university were a product of this institutional innovation. But it was not until well after mid-century that national and international agricultural research institutions became firmly established in most developing countries.

The challenge to institutional innovation in the next century will be to design the institutions that can ameliorate the negative spillover into the soil, the water, and the atmosphere of the residuals from agricultural and industrial intensification.

The capacity to achieve sustainable growth in agricultural production and income will also depend on the changes that occur in the economic environment in which developing country farmers find themselves. The most favorable economic environment for releasing the constraints on crop and animal productivity and for achieving sustainable adaptation to the resource and environmental constraints that will impinge on Asian agriculture is one characterized by slow growth of population and by rapid growth of income and employment in the nonagricultural sector. Failure to achieve sustainable growth in the non-farm sector could place developing country farmers in a situation in which they can make adequate food and fiber available to this sector only at higher and higher prices thereby reversing the long-term trend. At the same time, the resources available to generate investments in technology development for sustainable growth will become increasingly inadequate.

The importance of favorable growth in the non-farm economy is particularly important for the landless and near landless workers in the rain-fed upland areas which have been left behind by the advances associated with agricultural technology of the last quarter century. Rapid growth in demand arising out of higher incomes, rather than from rapid population growth, can generate patterns of demand that permit farmers in these areas to diversify out of staple cereal production and into higher value crop and animal products. It may also permit the release of some of the more fragile lands from crop production to less intensive forms of land use.

Chapter V

Population as Concept and Parameter in the Modeling of Deforestation

Alan Grainger

Introduction

Widespread changes in land use and vegetation cover are currently taking place in both the humid and dry tropics, with serious environmental implications. In the humid tropics tropical rain forest is being cleared rapidly for a wide range of agricultural and other land uses. In the dry tropics overcultivation, overgrazing, poor irrigation management and deforestation are leading to extensive land degradation. These two phenomena, generally referred to as tropical deforestation and desertification respectively, are usually treated separately but their causal mechanisms possess a number of similarities (Grainger 1990a). One thing they have in common is a tendency to be linked with causes related to the way in which land is used, e.g., shifting cultivation and pastoralism. Underlying causes such as population growth are sometimes acknowledged but mainly in a general, conceptual, way. So far, relatively few empirical or theoretical studies of tropical land use have treated population as a quantifiable parameter that could be included in models in conjunction with other socioeconomic factors, land uses and physical factors.

If population-environment dynamics is to be established as a distinct and rigorous field of study in which empirical research is supported by, and in turn extends, the scope of theoretical analysis then it is clearly imperative to move from a situation in which population is only a general concept to one where it is a quantifiable parameter whose involvement in environmental change can be fully tested. This paper assesses the possibilities for doing this with respect to modeling land use change and deforestation in the humid tropics. Its particular focus is an examination of which population parameter is most suitable for models of this kind. The relative merits of two population-derived parameters, population density and forest area per capita, are discussed.

Population as a Concept in Land Use Change

As population rises so does the general need for agricultural land, and there is ample historical evidence for such trends in temperate countries (Simmons 1989). With populations now rising rapidly in tropical countries it is understandable that some commentators, taking a neo-Malthusian approach, have identified this as a major cause of such critical environmental problems as tropical deforestation and desertification and (either implicitly or explicitly) called for improved family planning to help solve those problems (Ehrlich and Ehrlich 1972). From a wider neo-Malthusian perspective, the tropical rain forests might also be said to represent one of the last remaining areas in the world which are, in principle, available for settlement by the human race as a whole. Indeed, the rapid agricultural settlement of Brazilian Amazonia, which began in the 1960s, was prompted, in part, by fear that the region could be invaded by people from overcrowded areas of other countries, even from countries outside Latin America (Fearnside 1986). However, the widespread poor soil fertility presents a major constraint on agricultural development in Amazonia and in the humid tropics generally.

The neo-Malthusian perspective was criticized by Boserup (1965, 1981), who claimed that rising population density in an area would lead inevitably to agricultural innovations which would increase the human carrying capacity of the land. Despite the virtues of her case, it appears to be contradicted by the all too common experience of high population density leading to over-intensive land use followed by land degradation (Cassen 1976).

Population is not the only factor involved. The economic conditions in which people live are also important. A more technocentric alternative to neo-Malthusianism than Boserup's would proclaim that the use of high-yielding crop varieties, fertilizers and pesticides should enable agricultural production in tropical countries to easily keep pace with rising populations (Simon 1981). Unfortunately, many farmers in tropical countries are just too poor to afford to invest in such practices. Indeed, their poverty is so dire that it severely constrains the sustainability of much less sophisticated practices and is just as important a cause of land misuse and degradation as any overcrowding (Blaikie and Brookfield 1987).

Economic conditions also influence population growth rates. In the early stages of economic development mortality rates are preferentially reduced relative to fertility rates as a result of improved health measures. This forms the basis for the so-called "demographic transition," during which a rise in net population growth rate is inevitable until economic conditions improve sufficiently for fertility rates to decline (Notestein 1945). Even then the decline in population growth rates will be delayed by the younger age structure. Some countries pass through the transition more rapidly than others, but Teitelbaum (1975) has warned against assuming that completion of the transition is inevitable, arguing that the pace of economic development is too slow in many countries to ensure a sufficiently rapid reduction in fertility rates. Demographic transition theory has been modified to take account of such objections (Caldwell 1976) yet its relevance is still the subject of controversy (Ness and Ando 1984; Schultz 1988).

Published views on the causes of deforestation rarely take into account such debates between human demographers. They tend to range between two extremes: at one end of the spectrum deforestation is seen as simply a socially induced phenomenon; at the other it is primarily caused by poor land use. Studies pointing to population growth as the main cause are much less common than those blaming a particular land use, although this is perhaps more indicative of a general lack of sophistication in analysis which excludes socioeconomic factors. The land use which foresters often blame for causing deforestation is shifting cultivation (Lanly 1982). On the other hand, environmentalists tend to focus on logging, cattle ranching or other forms of natural resource exploitation promoted by governments in the cause of economic development (Plumwood and Routley 1982; Hurst 1987). Finally, a few studies mix socioeconomic factors and land uses, such as the expansion of shifting cultivation in response to population growth (Myers 1980).

Most explanations of deforestation are still largely conceptual (and often anecdotal) in nature since theoretical work undertaken on the subject during the 1980s has still to permeate the wider arena. However, for those wishing to understand, in more detail, and to model the processes leading to tropical deforestation, discussing population only as a concept is inherently unsatisfactory. So there is an inevitable urge to move beyond this to develop models which include population as a parameter. As with all new ventures, this raises as many problems as it solves. Some of these problems are addressed in this chapter, following an outline in the next section of a theoretical framework for modeling deforestation which includes population, other socioeconomic factors, land uses, physical factors, and government policies.

Modeling Deforestation: A Theoretical Framework

Definitions

Deforestation is the process whereby forest cover is removed from an area of land and is replaced by another land use. Deforestation has been formally defined as the temporary or permanent clearance of forest for agriculture or other purposes (Grainger 1986). This definition is compatible with one used by the United Nations Food and Agriculture Organization (FAO) in estimating rates of deforestation (Lanly 1981).

Deforestation and Logging

The key word in this definition is "clearance," for if forest is not cleared then according to this definition deforestation does not take place. Instead of the clearfelling common in temperate forests (which would constitute deforestation), selective logging is the almost universal practice in the tropical rain forests. Only a few of the thousands of tree species are commercially marketable and so from one hectare (ha) of forest just two to ten trees are removed out of a total of up to three hundred good sized trees. Selective logging disturbs the forest canopy, and also (depending upon the skill of loggers) causes varying degrees of damage to those trees which are not felled, but the forest nevertheless retains its integrity so as to regenerate for the next harvest. Apart from this small amount of clearance, which is needed for roads and landings, selective logging does not therefore lead to deforestation directly. But it does have an indirect role in causing deforestation, since farmers may use logging roads later to gain access to forests which they would not otherwise have been able to reach.

Logging occurs in response to economic demand for tropical timber in both developed and developing countries. Although this is partly related to population growth, logging and deforestation are best treated as two distinct processes when assessing the extent of human impact on the tropical rain forests and analyzing causal mechanisms. The focus here is therefore on deforestation rather than on logging, which is discussed in more detail elsewhere (Grainger 1986, 1993)

Closed forests in the humid tropics, which covered an estimated 1,081 million hectares in 1980, had a mean rate of deforestation between 1976 and 1980 of 6.1 million hectares per annum (table 5.1) (Grainger 1983). The corresponding figures for all tropical closed forests were 1,201 million hectares and 7.3 million hectares per annum respectively. The 734 million hectares of open woodlands in the dry tropics had an estimated deforestation rate of 3.8 million hectares per annum (Lanly 1981). The mean rate of logging in the humid tropics was about 4 million hectares per annum (Grainger 1986).

The Causes of Deforestation

There have been various attempts to trace the causes of deforestation (Myers 1980, 1984; Grainger 1980). Most tend to be partial in scope, being restricted to certain parts of the humid tropics; or superficial, in focusing on land uses alone. For example, much attention is given to cattle ranching as a cause of deforestation, but the practice is mainly confined to Latin America and hardly found at all in the African and Asian humid tropics.

TABLE 5.1 Forest Areas: Areas of Tropical Closed Forests, Tropical Moist Closed Forests, and Tropical Dry Open Woodlands (1980).

	FOREST AREAS			DEFORESTION RATES		
	Closed		Open[c]	Closed		Open[c]
	All[a]	Moist[b]		All[a]	Moist[b]	
Africa	217	205	486	1.3	1.2	2.3
Asia-Pacific	306	264	31	1.8	1.6	0.2
Latin America	679	613	217	4.1	3.3	1.3
Total	1201*	1081*	734	7.3*	6.1	3.8

Deforestation Rates: Annual rates of deforestation (1976-1980).
*Totals are not the sum of regional figures due to rounding.
Sources: a. Lanly 1981; b. Grainger 1983, based on Lanly 1981; c. Lanly 1982

To fully understand the causes of deforestation a comprehensive theoretical framework is required. The one outlined here is based upon work by this author, presented first in Grainger (1986), summarized in Grainger (1990b) and developed further in Grainger (1993). The theoretical framework has four main components: (a) socioeconomic factors, such as population and economic activity; (b) land uses which replace forest, such as shifting or permanent agriculture; (c) physical factors, such as ease of access and topography; and (d) policy factors, such as the policies of governments and international agencies.

National Land Use Morphology

Each country has a distinct national land use morphology, a spatial distribution of different land uses which include various types of forest, other natural vegetation cover, agriculture, urban and industrial settlements. For most countries in the humid tropics, the national land use morphology may be roughly divided into the forest and agriculture sectors. Figure 5.1 portrays the various factors which influence national land use morphology as a complex system including both land uses and socioeconomic factors.

The model in figure 5.1 is necessarily simplified to make its structure legible and focus on the key links between the socioeconomic driving forces and land uses. Government policies which influence both sets of causes of deforestation are omitted (see below); so too is external demand for food

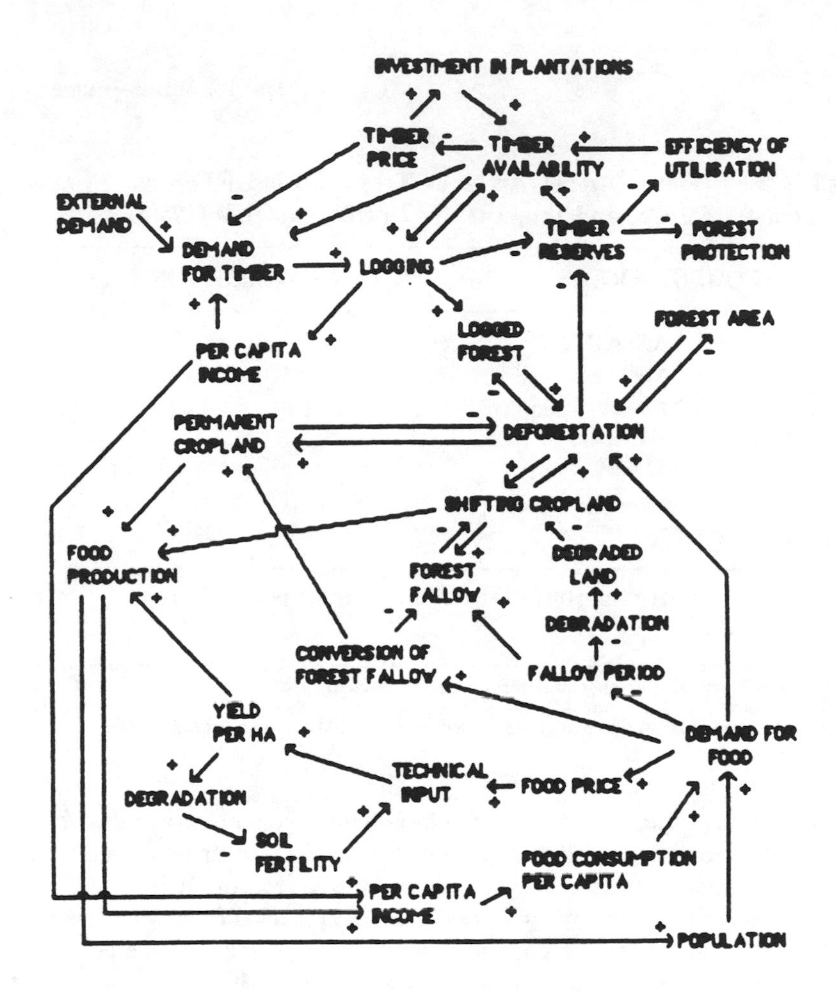

Fig. 5.1. A systems model of national land use.
Source: Grainger 1986.

and cash crops, although timber is included. In reality, the two main land use types mentioned, shifting agriculture and permanent agriculture, can be disaggregated into various subtypes (table 5.2).

Socio-Economic Factors

Socio-economic factors, particularly population growth and economic development, are the driving forces for change in national land use morphology. These lead to increased demand for food and timber, which are supplied by increases in agricultural area and/or intensity and in the logging rate respectively. Due to poor soil fertility and lack of capital to invest in fertilizers and other inputs, sustainable increases in agricultural yields in the tropics are limited. Thus, there is a continuing need for more agricultural land. In most countries in the humid tropics, large areas are still under forest and so the majority of new agricultural land requires forest clearance.

By increasing food consumption per capita, economic development tends to confirm the trend in demand for food due to population growth, increasing the need for agricultural land which therefore leads to deforestation.
But it also provides the capital to invest in more productive agriculture. Higher yields per hectare mean that less agricultural land is needed for a given level of food production, and this helps to control deforestation. The balance achieved between these two contrasting effects is clearly quite crucial in determining deforestation trends.

A National Land Use Transition

These socio-economic factors lead to selective changes in the wide range of land uses found in each country as listed in table 5.2 under the main categories of Shifting Agriculture, Permanent Agriculture, and Other (mainly non-agricultural) Land Uses such as mining and hydroelectric power schemes. Land uses may be thought of as the immediate causes of deforestation since they are the form in which it becomes apparent, but the underlying causes are the socio-economic factors, which translate into demand for various goods and services from the land.

Each country's land use morphology will respond to domestic and external demands for goods and services from the land in a unique way, preferentially expanding and/or contracting different land uses relative to others. This, along with the different scales of forest clearance required by each land use, the inherent instability of land uses practiced at too great an intensity on the generally poor soils of the humid tropics (which eventually leads to further deforestation) and each country's unique pattern of historical development, means that the national land use morphology's response to similar forcing factors will differ from country to country. Thus, even if two countries have similar population and economic growth rates, their deforestation rates may be quite different. The role of land uses in deforestation in this regard is not discussed in detail here, for convenience. Further analysis is given in Grainger (1986, 1990b, 1993).

TABLE 5.2 Land Uses Which Replace Forest

A.	**Shifting Agriculture**
	Traditional shifting cultivation
	Short-rotation shifting cultivation
	Encroaching cultivation
	Pastoralism
B.	**Permanent Agriculture**
	Permanent field crop cultivation
	Government sponsored resettlement schemes
	Commercial ranches
	Cash-crop plantations
C.	**Other Land Uses**
	Mining
	Hydroelectric power schemes
	Narcotic plant cultivation

In most humid tropical countries the many feedback loops which make up the national land use system are currently responding to rising demand for food by combining to drive the system towards a net transfer of land from the forest sector to the agriculture sector; the process which we call deforestation. If this continues unchecked, it eventually leads to a major change in national land use morphology from one dominated by forests to one in which the role of agriculture is more important. This shift can be termed a national land use transition. All forested countries, including those in the temperate world (e.g., the United States and the United Kingdom), will undergo this major transition at some point in their histories. Concern arises because it is currently taking place in so many countries in the humid tropics simultaneously, and at the expense of ecologically valuable forests.

Physical Factors: the Spatial Dimension

Physical factors are also an important cause. Deforestation is a spatial phenomenon which might be viewed simply in terms of the diffusion of people into forested areas from existing centres of settlement combined with an expansion of the latter. But the diffusion process is not random, nor is it indifferent to the environment. It is instead channeled by various physical factors, such as ease of access by rivers and roads, topography, soil type, etc. Some of these factors promote the diffusion of agriculture; others constrain it; many will have a secondary economic component. Thus, the distance by road and/or river to the nearest market determines the cost of transporting produce; this, in turn, affects the overall economic returns from farming and influences the choice of cash crops grown by farmers.

The Role of Government Policy

The final set of causal factors are government policies. These influence land uses, socio-economic factors and physical factors. In Brazilian Amazonia, for example, government agricultural and regional policies have actively promoted the expansion of cattle ranching; in Indonesia, forest policies have promoted the expansion of logging. However, land use is not just influenced by specific sectorial policies. Policies which promote population growth and economic growth can also have powerful indirect effects on land use change. If economic policies are inadequate to alleviate poverty then landless people may choose to migrate to forested areas to clear land to grow food to support themselves. Also, the ability provided by economic growth to invest in more productive agriculture is hampered if government policies keep food prices down. Farmers then have no incentive to invest in improvements and productivity will remain low.

The links between policies and changes in national land use morphology are therefore very complex and by no means deterministic. This is one reason why policies are not included as endogenous components of the model in figure 5.1. Another is that because one of the most important uses of models is as a planning tool for policy-makers they should be capable of simulating the alternative scenarios resulting from a range of different policies without any inherent bias.

Possible Future Trends in Deforestation

Assume that at a conceptual level the above provides a reasonable framework for explaining deforestation in terms of socio-economic factors, land uses, physical factors and government policies, and that the first two of these causative factors can be combined together in a national land use model.

What does the model tell us about likely future trends in deforestation? Bearing in mind the above qualifications about differences between countries, it would seem that such trends will very much depend upon how rising food production is shared between shifting and permanent agriculture and then between increased intensity and increased area in each type of agriculture.

In a world in which only shifting agriculture was practiced--and, in the humid tropics, this is mainly shifting cultivation--it would expand with population, clearing forest as allowed by accessibility, and becoming more intensive only under the constraints of growing population density and lack of additional forest to clear. The balance to be struck would seem to be that between current areas under cultivation and forest fallow. In the presence of logging activities, shifting cultivation would, in some areas, become spatially more confined, while in others it would expand due to the assistance of logging roads.

On the other hand, in a world in which only permanent agriculture was practiced, its expansion would be determined by such factors as food prices, proximity to markets and communications, and soil fertility. Forest land would be cleared, but land would also be taken from forest fallow, and hence reduce the extent of shifting cultivation. Rising production from permanent agriculture (due to both increased farmland and yield per hectare) would allow the reduction of deforestation rates and even cause net transfers of inferior cropland back to forest.

Assuming that as a country develops economically, some kind of trend from shifting to permanent agriculture takes place (although not the complete transfer which Boserup (1965) and others have proposed), then the area under permanent agriculture will rise at the expense of both forest and forest fallow and, for a time, account for a growing proportion of deforestation. As economic activity increases, there will be more investment in agriculture, average yield per hectare will rise, permanent cultivation will supply an increasing share of the food required for both domestic needs and export and deforestation rates will eventually decline. Cultivation will concentrate on the most highly productive lands and the remainder will be left for forestry or conservation purposes. Poorer quality lands cleared for permanent agriculture and some forest fallows will be abandoned and left to regrow to mature forest.

Actual trends could differ from those sketched here since:

1. The instability of both permanent and highly intensive shifting cultivation on poor fertility soil will cause further deforestation as soil is degraded and yields decline.

2. Displacement of shifting cultivation by permanent cultivation will also cause additional deforestation, either directly or due to the instability of shifting cultivation when it becomes too intensive.

3. Most soils are too poor to sustain permanent cultivation, so shifting cultivation will probably continue in some areas.

4. Rural to urban migration will favor intensification of permanent cultivation as farm labor becomes scarce. The reverse migration from urban to rural areas will give rise to encroaching cultivation and further deforestation.

5. Increasing external demand for cash crops will require the continued clearance of forest for plantations.

6. The needs of industry and urban areas for land, minerals and power generation (e.g., by hydroelectric schemes) will also grow as the economy develops.

The overall effect of the first four of these factors will depend upon the ability of government to provide suitable conditions for sustained economic growth and ensure not only the creation of jobs in urban areas (which will exacerbate the sixth factor) but also continued investment in the rural sector.

Moving Towards Equilibrium

The relative importance of such qualifications still needs to be determined by empirical studies. However, it would seem that the national land use system has implicit economic mechanisms which should eventually bring deforestation under control and national land use morphology into equilibrium.

This equilibrium hypothesis is central to discussions of deforestation trends in the remainder of the paper. It is also crucial with respect to population-environment dynamics. For if it were possible to increase agricultural productivity sufficiently quickly in a country, even if population continued to grow, then the deforestation rate would decline, ultimately to zero. Land use morphology would stabilize with all extra farm production gained by increased productivity rather than increased area. The limit on the number of people an area of land can support, a general facet of poorly developed agriculture, is thereby removed.

It is important to remember that humid tropical environments differ from those in the temperate world. Fertile soils are limited in area and soils with low fertility are widespread (Sanchez 1976). If agriculture is to be sustainable then the only feasible strategy is to concentrate intensive permanent agriculture on the most fertile soils, leaving low-intensity land uses such as shifting cultivation and agroforestry systems to predominate elsewhere. Raising yields on the most fertile soils could help to overcome the instability of permanent and shifting cultivation on poor soils which leads to more deforestation.

The trend towards equilibrium, as with all real-life complex systems, will probably not be a smooth one. The possibility of 'overshoot' is very real, if agricultural productivity does not increase at a sufficient pace and it takes time to abandon low productivity lands and practices. More deforestation may well occur than would be strictly necessary from a theoretical point of view. If land uses in a country do not come into balance soon enough by these means, and forest area continues to decline, then the government might, as a last resort, have to take action to protect remaining forests for timber production and conservation purposes. This has happened in the United States and the United Kingdom in the past (Williams 1989; Grainger 1981).

What happens after "equilibrium" has been reached? The equilibrium referred to here is that between the agriculture and forest sectors, since this is most apposite to the problem of deforestation. But this does not mean that once forest cover has been stabilized the various land uses within the agriculture sector and within the forest sector are fixed forever. Far from it. When the rapid decline in forest cover has been halted other changes are still possible. The increase in agricultural productivity may well lead to a continuing, albeit slow, net increase in forest cover on the less fertile lands. In addition, changes in productivity will continue to take place within both the agriculture and forest sectors as management is intensified and becomes more sustainable, new techniques and crops are introduced and adopted, and institutional structures are modified (Cain and McNicoll 1988). Both agricultural and forest sectors will, of course, lose land to the growth of settlements, roads and other infrastructure. National land use morphology is never static, but at the sectorial level changes are likely to be relatively minor once the national land use transition has been accomplished.

Two conclusions emerge from this discussion. First, because of the many interacting factors driving the expansion of the agriculture sector and leading to deforestation, it seems that improving agricultural productivity and sustainability should be the key ingredient of any integrated deforestation control strategy. Relying solely on attempts at increased conservation of forests within national parks and improving the protection of forests intended for long-term commercial timber production is unlikely to be successful. Second, the move towards equilibrium in national land use morphology will not depend upon land uses alone. Socio-economic factors will also be important in determining land use patterns and their role is the focus of discussion in the remainder of this chapter.

Population as a Parameter In Land Use Change

Quantifying the Model

Conceptual discussions are all very well, but they beg a number of questions. When exactly should the national land use morphology come into balance by itself? Is the control mechanism solely dependent on socio-economic checks or is government action an essential component? To answer such questions in specific terms requires that the model be quantified, and here the limitations of population growth as a mere concept become apparent.

The national land use model in figure 5.1 is presented in a system dynamics modeling format (Richardson and Pugh 1981; Gordon 1969) and so it could be quantified and used for trend simulation. If run for a given country then the point at which it comes into equilibrium will vary according to the values set for key parameters. But to make the model operational and produce meaningful results will require detailed empirical studies so that parameters can be initialized and functions specified for each country. Possibly each national model will also need to be divided into a number of sub-national (provincial models). This will take time and effort. Can anything be done to short-circuit this process and obtain some initial indications while basic research proceeds?

One solution would be to construct a much simpler model using just a few of the main parameters, ideally those that can be easily quantified. The results of such an exercise are reported below. Another approach would be to use available information to determine expected trends in some key parameters. These would, in any case, be needed to serve as "reference modes": general trends used to validate the model so that it could be made operational and used for simulations. Various parameters could be used for this purpose, a number of which include population, e.g., population increment, population density or forest area per capita. The luxury of choice creates difficulties. In practice, the optimal parameter will depend upon theoretical analysis, modeling requirements, data availability and other factors. This section discusses the problems involved in choosing a population parameter, and the insights which may be obtained thereby.

Current Rates of Deforestation

The simplest relationship which we might identify between changes in land use and population involves the contemporaneous rates of deforestation and population growth. Since, other things being equal, the greater the number of people in a given country the greater the area of agricultural land needed, it seems possible that the annual rate of deforestation (in hectares per annum) should be related, in some way, to the annual population increment in numbers of persons.

In fact, cross-sectional correlation analysis involving a number of countries does show such a link (Allen and Barnes 1986; Grainger 1986). Plotting the Logarithm of Average Annual Deforestation Rate (1976-80) in hectares per annum against the Logarithm of Average Annual Population Increment (1970-80) in number of persons for a set of forty-one countries in the humid tropics, produces a reasonable linear relationship with a correlation coefficient (r) of 0.56 (fig. 5.2). The relationship is by no means perfect, but the value of the correlation coefficient was the highest of any of the pairs of variables tested by Grainger (1986).

The low value of the correlation coefficient can be explained by the influence of a wide range of other factors, and by the inaccuracy of estimates of both deforestation rates and population growth rates. Because it takes time for rising populations to lead to increased demand for food, the relationship between deforestation and population growth is also probably a lagged one, i.e., deforestation rates, in a given year, are probably proportional to population growth rates for one or a number of preceding years rather than that year alone. Unfortunately time-series data on deforestation trends are generally insufficient for this hypothesis to be tested with any rigor.

Long-Term Trends in Forest Area and Deforestation Rate

Attempting to identify broad general trends in forest area or deforestation rate over time is a far more difficult task. The reason for the difficulty is the need to generalize for a large number of countries, each with unique histories, environments, and economic conditions. This problem can be partially overcome by using derived parameters that compensate for differences in absolute populations, land areas and forest areas (e.g., national percentage forest cover, population density and forest area per capita).

Forest Cover vs. Population Density

The impact of population on land use depends upon the size of the country involved. Thus, fifteen million people in a small island country such as Sri Lanka will have a greater impact than if they were living in a country such as Thailand which is eight times as large. A key derived social parameter should therefore be population density--the ratio between population and land area. The higher the population density, the greater the need to clear forest for agricultural use, and so the lower the forest area. A suitable derived vegation parameter to use in this regard might be national forest cover, the percentage of national land area covered by closed forest.

Population densities in the humid tropics vary enormously from country to country. The very fact that large areas are still covered by dense forest leads, as expected, to many countries having quite low population densities. A comparison of the population density in units of persons per thousand hectares for countries in the humid tropics shows that the value of 25 for Surinam is one of the lowest in the world (comparable with 21 for the largely desert country of Mauritania). Congo, Gabon and Guyana all have values between 40 and 60; Brazil, Peru, Madagascar and Zaire 100-200; Colombia, Ecuador, Venezuela, Cameroon and Ivory Coast 200-400; Costa Rica, Burma, Malaysia, Ghana and Sierra Leone 500-700; Cuba, Indonesia and Uganda, 800-1,000;

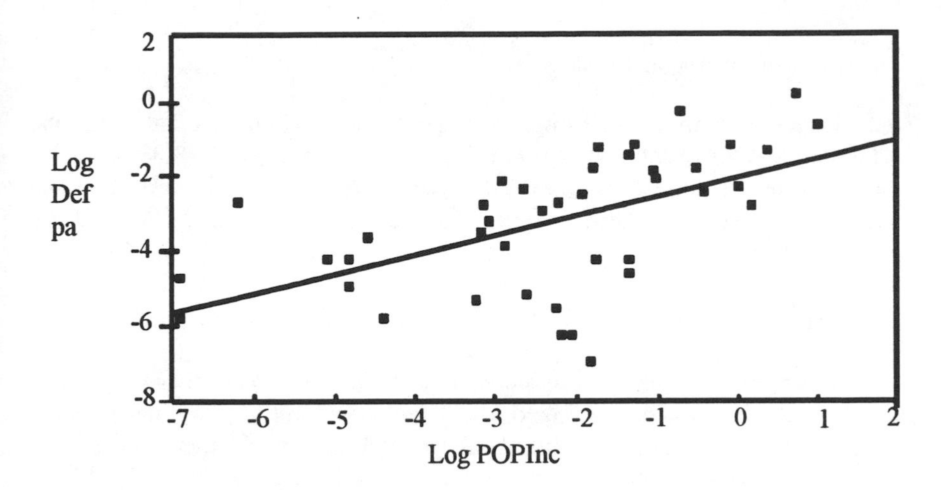

Fig. 5.2. Logarithm of average annual rate of deforestation 1976-80 versus logarithm of population increment of 1970-80.
Source: Grainger 1986

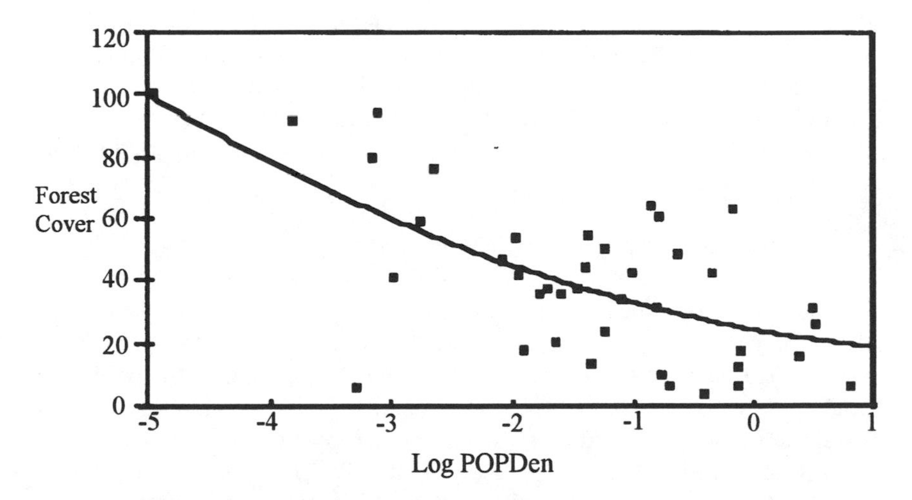

Fig. 5.3. National forest cover versus logarithm of population density for 43 countries in the humid tropics.
Source: Grainger 1986

Dominican Republic, Thailand and Nigeria, 1,000-1,500; El Salvador, the Philippines and Vietnam 2,000-3,000; and Bangladesh (8,389). By comparison, the population densities of some temperate countries in the same units are: Canada 29; United States, 269; United Kingdom, 2,326, and the Netherlands 4,335.

International Cross-Sectional Analysis

It might be expected that as population density rose a number of positive feedback loops in the national land use system would initially combine to cause forest area to decline. But this trend would be reversed later, when forest resources became depleted and most of the extra demand for food would have to be supplied by increased farming intensity rather than increased agricultural area. The negative feedback loops would then be dominant. Another control mechanism would be economic development which, together with government family planning programs, would tend to reduce the birth rate and hence the rate of net population growth.

A plot of National Forest Cover versus the Logarithm of Population Density for forty-three countries in the humid tropics (fig. 5.3) does indeed show a decline in forest cover as population density increases. The rate of decline in forest cover trails off, apparently moving towards a limiting value. This international cross-sectional function might be considered to represent the trend in forest cover in a single country as its population density increases, and is compatible with the above expectations about a move towards equilibrium in national land use morphology.

Using the Function for National Modeling

Since the curve appears to confirm expected general trends, might it not also be universally applicable in an absolute sense, i.e., capable of predicting the expected value of forest cover for a given value of population density? This would allow simulation of possible trends in national forest cover, and therefore in forest area and deforestation rate, as population density increases.

This has been attempted by Palo et al. (1987), who used a regression equation linking forest cover and population density to project future rates of deforestation.

Unfortunately, this approach leads to some problems. First, the relationship between forest cover and population density differs from region to region. Plotting the same variables on a regional rather than global basis shows sharp falls in Africa and Latin America, moving towards forest cover limits of below 10 percent, but for Asia-Pacific the downward trend is more gradual and forest cover appears to move to a higher limit of between 15 percent and 20 percent (Grainger 1986).

Second, most countries will have begun their history of deforestation with different national climax forest covers, so the same areas are not being compared when such curves are used in an absolute way. Thus, a country such as Brazil which encompasses a variety of climatic zones within the humid and dry tropics would have had a lower initial closed forest cover than a country like Malaysia situated entirely within the humid tropics. The difference between the apparent forest cover limit in Asia and those in Africa and Latin America is probably due to the generally higher climax closed forest cover in Southeast Asia.

The above reservations might be manageable if separate trends were plotted for each of the major regions (Palo et al. 1987) , and if only countries wholly or partially within the humid tropics were selected for trend analysis. But if all tropical countries, ranging from those wholly within the humid tropics to those wholly within the dry tropics, are included in the same plot then this could produce misleading trends, which is a criticism that may be made of the Palo study. The climax forest cover of countries which lie mainly within the dry tropics, such as Burkina Faso, would be very low even without the presence of human beings to cut it down. Using data points for such countries to infer a low limiting forest cover for countries in the humid tropics is therefore questionable.

Such problems are mainly of a technical nature, in that they refer to the way that modeling is practiced. Other problems are more fundamental. In the absence of effective forest management the approach of forest cover to the limiting value which represents an equilibrium point for national land use morphology will be largely determined by processes in the agricultural sector. The basic criterion for equilibrium is therefore probably the sustainability of agriculture rather than the sustainability of forest management, but since each country has a different population supporting capacity, according to land capability (based on soils, topography, etc.), agricultural techniques, fertilizers and other inputs, an absolute relationship between forest cover and population density would not be expected from an agricultural point of view. Similarities may exist between countries in the early stages of agricultural development, but in the later stages, the spread of forest cover values would be expected to increase as population density increases.

Intra-National Variation

Even within a country, land capability varies from one area to another, and so the relationship between forest cover and population density will vary too. This is evident from both static and dynamic analysis.

Static analysis. Static analysis considers population density merely from the point of view of land capability and its population supporting capacity.

Some countries in the humid tropics can, and do, support very high population densities on parts of their territories blessed with fertile alluvial or volcanic soils. The Indonesian island of Java is a good example of this, having a population density of 6,904 persons per thousand hectares compared with a density for the whole of Indonesia of only 985 persons per thousand hectares. Such differences within countries obviously bias the national figures for population density. Thailand's overall population density of 1,071 persons per thousand hectares is similar to Indonesia's but their forest covers in 1980 were 18 percent and 63 percent respectively. Deforestation and population density may therefore increase more rapidly in some parts of a country than in others. There is a great contrast, for example, between population density/forest cover values in southeast Brazil and in Amazonia.

Dynamic analysis. Static analysis can only indicate possible ultimate human numbers in an area for a given level of agricultural inputs, but reaching these prospective limits will take time and may never actually happen. This is a key obstacle when dealing with temporal trends. Dynamic analysis is therefore important. It suggests that in the initial stages of deforestation, population and settlements will tend to concentrate near coasts, along rivers and in other accessible places with fertile soils. The expansion of agriculture, and therefore of deforestation, will be influenced by various physical/infrastructure factors as described earlier in this paper. Thus, within Brazilian Amazonia the minimal deforestation at the centre of the region differs from the situation on the southern edges where deforestation is concentrated. This is partly due to the slightly better soils there, but a major reason is the ease of access via the highway system to leading population centers in southeast Brazil.

As populations expand there will be both migration to underpopulated marginal areas and intensification of agriculture on the better soils. Richer farmers will dominate the best lands, a significant proportion of which will be devoted to non-food cash crops. In marginal areas the use of artificial inputs (such as fertilizers) will be minimal and limits on yields fairly restrictive. But those forced to cultivate such lands will probably be poor and disadvantaged, and apparent physical limits to population supporting capacity may be breached as these people strive to scratch an existence, however meagre, from the land.

Population densities should, therefore, continue to increase in all parts of a country: both on highly productive agricultural lands and (at least initially) on marginal lands. Land use (and hence forest cover) should stabilize in areas with better soils, but stabilization will take longer in marginal areas where a considerable proportion of the remaining forests are to be found. However, every agricultural system has limits on the number of people it can support in a given area. Fearnside (1985, 1986) showed the importance of measuring the probability of crop failure when estimating the carrying capacity of marginal lands such as those along the TransAmazonian Highway in Brazilian Amazonia.

Implications for National Modeling

This discussion suggests that predictive modeling of land use change for individual countries using the relationship between forest cover and population density is possible but if it is to be valid the areas modelled should be highly disaggregated. At the very least, modeling should take place on a sub-national rather than a national scale, and even this may be too coarse. This raises other problems. For example, if rural population is used in modeling instead of total population there is a danger that rising demand for food by urban populations may be neglected.

The groundwork for a disaggregated modeling approach was laid in a major joint study by FAO and the International Institute for Applied Systems Analysis (IIASA). This study assessed the population supporting capacity of areas throughout the tropics, based on the spatial distribution of climate, soil, topography, and three different levels of agricultural inputs, such as the use of improved crop varieties, fertilizer, pesticides, herbicides, fallowing, soil conservation techniques, etc. A distinction was made, for example, between the ability of Kalimantan (Indonesian Borneo) to support five hundred to one thousand persons per thousand hectares at low levels of inputs and that of Java to support two thousand to five thousand persons per thousand hectares (Higgins et al. 1983). But, as Blaikie and Brookfield (1987) have pointed out, whether most farmers would be able to afford to use high levels of inputs is doubtful.

Deforestation Rates vs. Forest Area Per Capita

Modeling is a very pragmatic practice. It seeks to construct simple representations of complex phenomena so that our understanding of those phenomena is increased and the models can be used for various practical applications. But when modeling geographical and social phenomena, besides the stochastic nature of relationships there is also the inevitable problem of lack of sufficient data to match the degree of detail required by theory. Modellers therefore have to build simpler models which require less data, and/or use other parameters for which data are more readily available.

The previous section showed that there are limitations in using national forest cover and population density to predict deforestation trends and the possible limiting of national forest areas at which the agricultural and forest sectors in the national land use morphology should come into equilibrium. Such models essentially focus on the agricultural sector, assuming that the agricultural area of a country is dependent upon the food requirements of the population living in a fixed national territory. An alternative approach, discussed in this section, is to focus on the forest sector, and the ability of a country's forest resources to satisfy the timber requirements of its population.

At the same time that the agriculture sector is developing according to the combination of its own feedback loops, the forest sector will be developing too. Usually, this will consist initially of an extensive and largely uncontrolled exploitation of forests for timber, accompanied by deforestation for agriculture which reduces overall timber reserves and especially those in logged forests. But when depletion of reserves becomes apparent, then timber prices will rise and eventually the combination of forest sector feedback loops will switch to give remaining logged and unlogged forests greater protection from deforestation (figure 5.1). Imports of forest products will increase, as will the use of non-wood substitutes, but a premium will still be placed on the need for substantial national forest resources.

In practice this means that when forests are plentiful in a country, whether this be the United States or Malaysia, the natural human reaction is to consider timber supplies to be limitless. Deforestation and logging therefore proceed with abandon, until finally, there comes a point at which forest resources are so scarce, relative to current needs, that action has to be taken to conserve those that are left and bring forests under sustainable management. Such action would occur as a result of either normal economic mechanisms of supply and demand, or because governments make a decision (perhaps prompted by economic signals) to prevent further deforestation, by protecting within reserves a certain minimum area designated for timber production and conservation. Most countries will also have a certain residual area of forest which remains uncleared because it is either commercially unproductive or inaccessible.

Critical Forest Limits

At what point in the history of deforestation would further loss of forest be prevented in this way? Is there some critical limit of forest area beyond which the introduction of sustainable forest management and rigorous protection is inevitable? If so, how can this be determined? Clearly it must be related in some way to the ability of a country to supply its population's needs for timber and environmental services.

One possible parameter to use as an independent variable in place of population density might be Forest Area Per Capita. This would avoid some of the problems noted above in relation to forest cover and population density. For example, it allows for variations in climax forest cover and agricultural productivity from one country to another, since the key relationship is between absolute (rather than percentage) forest area and the number of people in a country. However, differences between countries in respect of the historical development of their economies, forest industries etc. could still remain.

One could postulate a critical minimum value of forest area per capita below which land uses should generally stabilize and deforestation fall to a low level as forest is protected in order to sustain domestic wood supplies. The area of forest might even be increased subsequently by reforestation or natural regeneration. This author would argue that it is domestic demand, rather than all demand (which would include that for imports by other countries), which is likely to prove of paramount importance to the functioning of local timber markets and the actions of governments. Of course, there are some problems with this approach, e.g., overall forest growth rates and timber volume differ from country to country, and those for commercial timber differ even more. However, growth rates are not fixed, and could be modified by the use of plantations or silvicultural techniques.

To make some estimate of the order of magnitude of the critical value of forest area per capita, let us arbitrarily assume an average wood consumption per capita of 1 cubic meter per annum (in roundwood equivalent volume) and a tropical rain forest growth rate of 2 cubic meters per hectares per annum. (Average wood consumption per capita in 1986 varied quite widely between countries, from 0.1 cubic metres per annum for the United Kingdom to 0.6 for Japan, France, and the Philippines, 0.9 for Indonesia, 1.0 for Ivory Coast, 1.9 for the United States, and 7.0 for Canada (author's calculations, based on FAO, 1988)). Given these two values, a possible critical limit of forest area per capita to supply all domestic needs might therefore be of the order of 0.5 hectares. On the other hand, forest plantations in the humid tropics capable of supplying sawlog-grade timber, rather than fuelwood or pulpwood, typically have average growth rates of about 10 cubic metres per hectare per annum, in which case, the critical value of forest area per capita would be reduced to about 0.1 hectare. Forest area per capita could therefore decline to between 0.1 and 0.5 hectare before sustainable management was introduced.

International Cross-Sectional Analysis

To test this hypothesis, a cross-sectional plot of Log Average Annual Deforestation Rate against Log Forest Area Per Capita was made for forty-three countries in the humid tropics (fig. 5.4). This suggests that the deforestation rate initially rises as forest area per capita decreases and possibly peaks at some point, but because of the high scatter the evidence is not convincing.

A number of countries have a value of forest area per capita close to or even below 0.1 hectare. Those below it in 1980 include El Salvador (0.02), Dominican Republic (0.08) and Nigeria (0.07). Nigeria, once a major timber exporter, became a net importer some time ago and now has a vigorous forest plantation program. Close to the limit are Cuba (0.13), Vietnam (0.14), Ghana (0.15), Thailand (0.18), and the Philippines (0.19).

The examples of Thailand and the Philippines show the benefits of using forest area per capita as a model parameter. Both were formerly major timber exporters, but are now suffering from forest depletion, with all the commercial and environmental consequences which this entails, although this is not immediately apparent from their forest covers of 18 percent and 32 percent respectively. Thailand is now a net timber importer too. A study of trends in forest area in the Philippines by David Kummer (Boston University, Personal Communication) showed that, by 1980, forest area per capita had fallen to 0.17 hectare and, by 1987, to about 0.13 hectare. The limiting value of 0.1 hectare had still not been reached by that time, but the deforestation rate was decelerating.

Critical limiting values of forest area per capita are not perfect indicators of land use stabilization because conditions differ from country to country, but they may serve a useful purpose in the early stages of modeling and provide a basis for improvement as research progresses. To place this arbitrary limit in a wider perspective, the average per capita area of forest and woodland for all developing countries is 0.7 hectare and for all developed countries 1.6 hectare but it drops to 0.3 hectare for Western and Eastern Europe and to 0.2 hectare for Japan (Peck 1984).

A Simple Deforestation Model

Model Structure

To simulate possible future trends in deforestation this author designed a simple model, based on the principles of the national land use model and the assumption of a critical limiting value of forest area per capita (Grainger 1986). In the model, farmland area A varies as a function of population P, per capita food consumption and yield per hectare. "Farmland" includes both cropland and permanent pasture, as defined by FAO. Population grows logistically, initially at average 1970-80 rates but only up to a limit of P_{max}, the hypothetical estimated stable population (World Bank 1982). If the annual growth rates of population, per capita food consumption and yield per hectare are π, α, and β respectively, and k_1, k_2 are constants, farmland area A in year t is given by:

1. $$A_t = \frac{k_2 . P_{max} \cdot e^{(\alpha-\beta)t}}{(1 + k_1 . e^{-\pi t})}$$

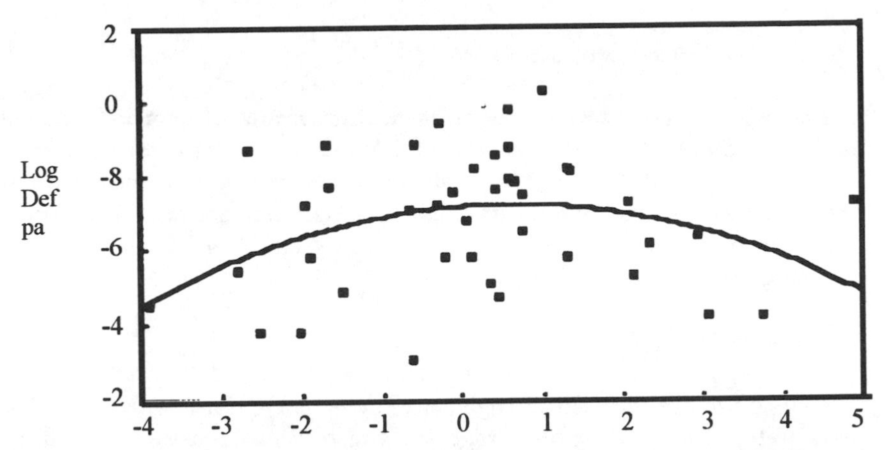

Fig. 5.4. Logarithm of average annual rate of deforestation 1976-80 versus logarithm of forest area per capita 1980 for 43 countries in the humid tropics.
Source: Grainger 1986

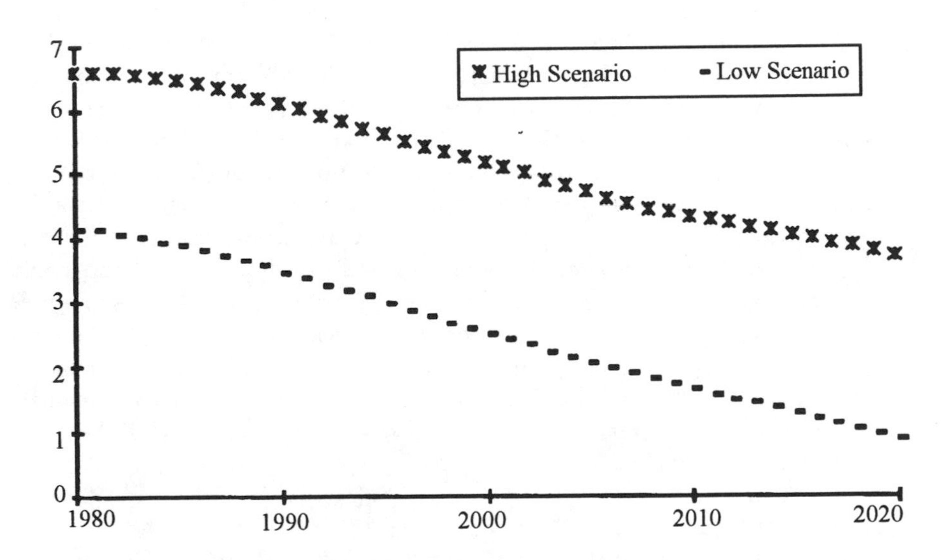

Fig. 5.5. Two scenarios for deforestation rates in the humid tropics 1980-2020 (million ha per annum).
Source: Grainger 1986

Deforestation rates normally equal the additional area of farmland required each year. Since all countries are expected to retain a certain minimum area of forest it is assumed that the deforestation rate becomes zero when forest area per capita reaches the arbitrary limit of 0.1 hectare (see above). Afterwards, increased food production can only be gained by raising yield per hectare or obtaining extra farmland from non-forest land.

Scenario Design

Two alternative scenarios, "High" and "Low", were tested with the model, using initial population growth rates (π) which were the same as the mean values for 1970-80, and growth rates in food consumption per capita (α) and yield per hectare (β) estimated on the basis of mean regional values for 1970-80 (table 5.3). For Africa, in the Low Scenario the growth in food consumption kept pace with population growth and exceeded it by 0.5 percent per annum in the High Scenario, but in both scenarios average yield per hectare only increased by 1 percent per annum. It was thought that continued economic growth in Asia would allow food consumption per capita to rise at up to 1.5 percent per annum, but that yield per hectare would only rise at 2.0 percent per annum in the Low Scenario and 1.5 percent in the High Scenario. These rates are still high, because capital should still be available for agricultural investment and new agricultural land will become increasingly scarce. In Latin America, the Low Scenario assumed food consumption per capita rising at 0.5 percent per annum and yield per hectare at 1.5 percent per annum. In the High Scenario a more optimistic view was taken of economic growth, with these parameters set at 1.5 percent and 2.0 percent respectively.

The model was simulated for forty-three individual countries in the humid tropics from 1980 to 2020. These countries are a subset of the sixty-three nations containing tropical moist forests (tropical rain forests plus tropical moist deciduous forests) listed in Sommer (1976). The same parameter values were used for all countries in a region except in the case of five countries (Brazil, Colombia, French Guiana, Gabon and Madagascar) which had slightly different values (table 5.2).

Results of Simulations

Simulations with the model gave deforestation rates in 1980 for the tropical moist forests as a whole of 4.1 million hectares per annum in the Low Scenario and 6.6 million hectares per annum in the High Scenario (table 5.4). This compares well with estimates based on Lanly (1981) for the same forty-three countries of 5.6 million hectares per annum between 1976-80. In the High Scenario, the deforestation rate fell from 6.6 to 3.7 million hectares per annum between 1980 and 2020 (fig. 5.5). By the end of the period

TABLE 5.3 Assumed Increases in Per Capita Food Consumption (α) and Yield Per Ha (ß) (% per annum)

	High Scenario			Low Scenario		
	α	ß	α-ß	α	ß	α-ß
Africa	0.5	1.0	-0.5	0.0	1.0	-1.0
Asia-Pacific	1.5	1.5	0.0	1.5	2.0	-0.5
Latin America	1.5	2.0	-0.5	0.5	1.5	-1.0

Note: Modified values of α-ß for the High and Low Scenarios respectively: Gabon -0.2, -0.5; Madagascar -1.0, -1.5; Brazil -0.5, -0.8; Colombia -0.5 -0.8; French Guinea +0.5, -1.0.
Source: Adapted from Grainger (1986)

deforestation was still occurring at a substantial rate in all three regions, with Latin America accounting for half of the total, as in 1980. In the Low Scenario the overall deforestation rate fell from 4.1 to just 0.9 million hectares per annum over the same period. It ended close to zero in Latin America, where in a number of countries increased agricultural productivity in the Low Scenario could make possible a net return of land to forest, mostly towards the end of the simulation period. Projections were made on an annual basis (figure 5.5), but tables 5.4 and 5.5 contain only the simulated values for 1980, 2000, 2010 and 2020. The scenarios summarized in tables 5.4 and 5.5 are regional aggregates of individual simulations made for forty-three countries, and although regional deforestation rates declined throughout the simulation period, national deforestation rates did rise initially in about half of the countries. In Africa, deforestation rates fell to zero in Ghana, Guinea, Ivory Coast and Sierra Leone as forest area per capita fell to 0.1 hectare and would start at zero for Nigeria and Uganda which were already below this limit. For the same reason deforestation rates fell to zero in the Philippines, Thailand, Vietnam, Cuba, Dominican Republic, and El Salvador.

In both scenarios a substantial area of tropical moist forest remained by the year 2020. The High Scenario predicted a reduction of about 20 percent in total forest area to 831 million hectares compared with a fall of less than 10 percent to 936 million hectares in the Low Scenario. If deforestation rates were to continue unchanged at current estimated levels the overall reduction would be 23 percent (table 5.5). Thus, on the assumptions stated and beginning with deforestation rates of the same order as those estimated to be currently occurring, the effect of deforestation on tropical moist forest would not be as devastating as some commentators have feared.

Palo et al. (1987) simulated future trends in forest cover with a linear regression equation in which the sole independent variable was population density. When applied to sixty tropical countries (in groups) it led to a reduction of 16 percent in the area of all tropical forests (moist and dry, closed and open) by 2025. Alternatively, using two different forms of non-linear equations with the same variables gave reductions of 30 percent and 45 percent over the same period.

TABLE 5.4 Regional Trends in Deforestation Rates in the Humid Tropics 1980-2020 (million ha per annum)

	High Scenario					Low Scenario				
	1980	1990	2000	2010	2020	1980	1990	2000	2010	2020
Africa	1.552	1.486	1.207	0.885	0.866	1.036	0.935	0.670	0.550	0.427
Asia-Pacific	1.731	1.505	1.219	1.192	1.149	1.128	0.940	0.707	0.526	0.417
L. America	3.275	3.065	2.705	2.221	1.670	1.971	1.599	1.111	0.575	0.036
Total	6.558	6.056	5.131	4.298	3.685	4.135	3.474	2.488	1.660	0.880

Source: Grainger 1986.

TABLE 5.5 Regional Trends in Forest Area in the Humid Tropics 1980-2020 (million ha)

	High Scenario					Low Scenario				
	1980	1990	2000	2010	2020	1980	1990	2000	2019	2020
Africa	198.9	183.5	170.3	160.4	151.6	198.9	188.9	181.1	175.0	170.1
Asia-Pacific	239.4	222.8	209.5	197.5	185.8	239.4	228.9	220.8	214.8	210.2
L. America	598.0	566.2	537.5	513.0	493.8	598.0	580.1	566.7	558.6	555.8
Total	1036.3	972.6	917.2	870.8	831.1	1036.3	997.9	968.7	948.5	936.1

Source: Grainger 1986.

Discussion and Conclusions

This paper has demonstrated that the transition from concept to parameter for the role of population in environmental change in the tropics can be achieved. The key step was to create a theoretical modeling framework which provided a believable representation of events on the ground and was capable of quantification. It formed the basis for a systems model of national land use, and in turn for a simpler quantifiable model which was used to simulate possible future trends in deforestation.

Tropical land use change was explored from two perspectives, one based on agriculture, the other on forestry. These led to alternative population-derived parameters, population density and forest area per capita respectively, which could be used in modeling future land use trends. The pros and cons of each parameter were discussed. In each case modeling was constrained by a lack of detailed data, which led to a highly aggregated national treatment with relatively few parameters. Instead of a complex systems model, the approach to land use equilibrium was interpreted on the basis of trends in the two derived parameters.

Ultimately, this author decided on forest area per capita as the best parameter to use in the simple simulation model, given the particular modeling task and the data available. Another author chose population density instead. A diversity of approaches is valuable and not to be dismissed. However, it does indicate the embryonic nature of population-environment dynamics theory at the level at which it is discussed here. This is not to say that there is a theoretical or a technical void--the modeling approach described in this paper has built upon techniques employed in human geographical modeling (e.g., Hagerstrand 1952; Cliff et al. 1975)--merely that a lot more remains to be done.

One topic which has not been treated here explicitly is land degradation. Both for reasons of space and also to impose some boundaries on the theoretical analysis, the main focus has been on the human side of the people-land equation and on the discussion of possible equilibrium positions. However, it has been pointed out above that the instability of land uses in the humid tropics is an important contributor to deforestation and one which could lead to much higher deforestation rates than simulations with the simple model would indicate. Instability results when land is unable to support human use at the intensity at which it is practiced. At the moment, in most countries in the humid tropics there is still the option to move elsewhere when yields become too low. Ultimately, however, if agricultural productivity on the better soils does not rise at a sufficient rate such options will be foreclosed. Continued farming will be inevitable on lands with depleted fertility, with serious consequences for land degradation.

Currently, environmental degradation is particularly serious in the dry tropics, where a large but undetermined area of land is lost to productive use every year due to desertification (Dregne 1983; Grainger 1990a, 1992). There the two main components of environmental degradation--soil degradation and vegetation degradation--are much more evenly balanced in intensity. Desertification is reversible up to a point (Dregne 1983) but beyond that the land is lost for good. Human poverty is both a cause and effect of such processes, and exacerbated by the frequent droughts whose current persistence in some areas like the Sahel could be due to the effects of environmental degradation on rainfall formation.

Will environmental breakdown occur in the humid tropics too? The possibility certainly exists, but some of the more dramatic prognoses, such as widespread soil laterization following deforestation, were greatly exaggerated. Land degradation should be incorporated in theoretical treatments, but while conceptual integration is easy, realistically assessing quantitative links is difficult because of a lack of empirical data (Grainger 1986). This is as true for modeling the feedback between logging and deforestation as for the drain in soil fertility due to overfarming. Deterioration may be either gradual or catastrophic, but a sound basis of knowledge is needed before choosing which one to include in a model. The one thing which can be predicted with certainty is that analyses of the phenomena of tropical deforestation and desertification will gradually converge as the overall similarities between them become more widely recognized.

Suggestions for Future Research

Whether greater unanimity will be attained in the theoretical analysis of population-environment dynamics in the tropics as more research is carried out and the field matures is difficult to say, especially bearing in mind the interdisciplinary nature of these investigations. But a common frame of reference is needed within which different approaches would be compatible.

Two specific priority objectives for research can be identified. First, land use modeling of this kind is vital now for understanding the processes involved in tropical deforestation and improving our ability to control it. Second, such modeling will also become increasingly important in assessing the causes and possible impacts of the greenhouse effect. The terrestrial component of global climate modeling has long been the poor relation of the atmospheric component, but this will change in the next ten years because of the need to study the net impact of tropical land use change on atmospheric carbon dioxide levels and the impact of climatic change on both natural ecosystems and agriculture.

It is difficult to give an exact prescription for the future research needed to achieve these objectives. In addition to the variety of theoretical research which may be expected, five priorities for empirically-based research and data collection are listed here.

First, extensive empirical studies of the processes of tropical land use change, both on the ground and using remote sensing techniques, are needed now, while these processes are still available for study over a range of stages of the development of national land use morphology.

Second, obtaining more spatial data on tropical forest cover and land use at a global level is an urgent requirement. This can be achieved by the use of available remote sensing techniques, particularly satellite imaging, but there is also a great need for a continuous monitoring programme for tropical forests which can give comprehensive and regular reports on forest areas and rates of deforestation (Grainger 1984).

Third, improvements are also needed in the quality and quantity of data on global population distributions. This will be a far more difficult task than improving deforestation monitoring.

Fourth, improving the resolution of modeling will depend upon the continued expansion of land capability classification studies and the complementary development of more sophisticated models of farming systems (e.g., Trenbath 1989).

Fifth, using these data for modeling will require the development of advanced techniques which can interactively link land use models with large-array spatial data stored using geographic information system (GIS) techniques.

Section II

The State as Actor: Population-Environment Dynamics in Large Collectivities

Introduction

The chapters of this section explore issues of population-environment dynamics found in large collectivities, such as nation-states and larger politically-defined regions. As the section title suggests, the distinctive character of these large-scale collectivities is that their political institutions, especially the modern state, play a major role in shaping the transformations and transitions of the modern world. This indicates, among other things, that population-environment interactions are bounded less exclusively by natural conditions and more by political boundaries, or by conditions that are essentially *socially constructed.* As socially constructed entities, nation-states and their related policies and actions emerge as new, distinctively human, and highly decisive elements in the population-environment dynamic.

All of the states covered in these chapters have colonial histories. Those in Africa, South and Southeast Asia, and Latin America all emerged as independent states only in the past few decades. George Martine's Brazil went through the process of independence about one and a half centuries ago. Later we may see that both the colonial past, and the difference in the timing of independence will have an impact on the population-environment relationship.

The population-environment dynamic itself is treated quite differently in all the chapters, based more on the specific interests of the authors than on any overriding theoretical formulation. As in the previous section, the chapters considered here use different *tracers* or *indicators* of the character of the dynamic. Both the observations of the impact of different state characteristics, and the use of different tracers will provide materials for generating hypotheses and a research agenda, to which we turn in the final chapter.

Ness begins with an historical overview of the social and economic transformation of Southeast Asia with particular reference to rice and population growth from the sixteenth century to the present. State policies and new technologies have been responsible for the recent transitions to high rice output and a successful completion of the demographic transition. The variance among states in Southeast Asia, however, also permits him to show that both state and technology are limited by local cultures. In Zimbabwe, McIntosh attributes the severe erosion of communal lands of the Black majority to high population density, which is itself the result of a complex set of social, economic, and political factors. She concludes that traditional population policies which focus on fertility reduction to slow population growth will do little to relieve population pressures on eroding communal lands in the near future. Any effective solution must include a policy of population redistribution.

Kartomo explores Indonesia's remarkable recent economic development, showing how it has produced great stresses to which the government and the society must react. Finally, Martine analyzes agricultural development in Brazil from 1965 to the present. Three phases were evident. He shows that each phase, a product of internal policies and world events, has had a unique set of demographic, social, and environmental consequences.

Population

The authors of this section go beyond sheer numbers to deal with many dimensions of variance in population. In addition to size and rates of growth; fertility levels, migration, density, and distribution, as well as race, ethnicity and class are considered. Quality of life issues are raised in each chapter of this section.

In Ness's review of Southeast Asia, the standard population references of size and rates are used and compared over time. Population size in this region grew slowly but steadily for centuries, followed by rapid growth, a pattern that is commonly referred to as the demographic transition. Concerns over fertility levels are identified as a focus of modern government programs, which are related to goals of economic development. Issues of population quality, especially health, are also related to development goals. He also finds, however, that race or ethnicity and religion are important determinants of the population-environment dynamic, largely by affecting state policies.

Population density and race are important variables in McIntosh's review of land use in Zimbabwe. High density is related to the seriously uneven distribution of Zimbabwe's Black majority population located on limited native lands. Migration is another variable as Blacks leave communal lands for the cities, or other less populated rural areas in the Western part of the country. Finally, McIntosh notes that reducing the level of fertility through government-sponsored family planning is considered a modern population control strategy.

Discussion of population in Kartomo's chapters is more general. He mentions issues of numbers, and fertility rates within Southeast Asia and his native Indonesia. Martine focuses more than Kartomo on population quality and the welfare of small farmers, and on the important condition of class.

Both authors raise issues of rural out-migration; Martine in reference to Brazil's small farmers who are essentially squeezed off the land as a consequence of government policies. Kartomo discusses Indonesia's transmigration policies with the government's attempt to reduce population pressures on Java by moving people to the less densely settled outer islands.

Environment

Each of the authors considers slightly different aspects of the environment, but all view it in general terms as a set of basic resources for use by human population within and across national boundaries. In this section the issue of units of analysis and their boundaries becomes central. Now the human community becomes a major player in drawing boundaries around a territory. This raises questions of the distinction between natural and political boundaries, and their interrelations. The natural resource base, in reference to political boundaries of the nation-state, is a principal determinant of national wealth. Ness frames the environment as a natural resources base for domestic use and international trade. Countries with valuable natural resources were prizes during the colonial period. Rice production itself is seen as a product of land and water as well as human organization and technology. State policies or political conditions are argued to be much more influential in determining rice output than variations in the natural resource base.

McIntosh focuses on Zimbabwe's rural farm and grazing lands and specifically on variation in land fertility, which has become stratified by race. Environmental degradation can be clearly seen in soil erosion, which is a product of government policies as well as of natural population dynamics. Martine describes a similar situation in Brazil. However, the social and environmental consequence of Brazil's agricultural modernization seems to be rooted in differences of social class and political power rather than race. Here again the human community becomes a major focus of analysis, involving variations in the power to control resources and thus to determine the land use of specific politically bounded landscape ecosystems.

Two basic kinds of environments emerge in this section, defined by two different types of boundaries--ecological and political. Ecological, or natural, boundaries change with soils, climate, and location. Watersheds, islands, and mountain chains are distinctive environments defined by natural physical conditions. Political environments are defined by socially created boundaries. The most important of these for the current population-environment interactions are clearly the boundaries of nation states.[1] Political boundaries

---

1. There is a long debate over the role of states and state boundaries, especially in relation to international economic activity. The Marxian tradition tends to see the state as a relatively dependent tool of international capital. Non-Marxian perspectives see the state as a more

are often not coterminous with major natural boundaries. Consequently actions and activities of one area or country can affect the environmental and social welfare of another area or country. States are hurt by flooding or toxic effluents that originate in other states. Natural environmental conditions also have often been the subject of political conflict. States fight over oil and water, as well as over land and access to ocean fishing access. It is precisely this division of environments that leads today to urgent efforts of an emerging international community to create a new type of boundary, which will provide some form of political control over the entire world ecosystem.

The Relationship

One of the most striking observations from these chapters is the critical role the state and its policies have played in defining population-environment relationships, especially over the last fifty to one hundred years. Policies of developing nations geared towards stimulating economic development have rapidly reshaped how people interact with their environments. Industrialization, urbanization, and modern commercial agriculture have radically altered the means of production, supporting ever larger numbers of people. Technology, too, has helped to redefine the relationship dramatically. Modern medicines, chemical inputs with miracle seeds, and innovative machines have on the whole reduced mortality, and improved our health, our ability to produce food, and to care for ourselves and families. At the same time we have created new problems. Polluted air and water and endless pavement detract from these improvements. New illnesses emerge and biological diversity diminishes.

In chapter 6 Ness demonstrates the importance of political institutions in mediating population-environment relationships, including the use of available technology. The rise of population in Southeast Asia since the 1800's has been the product of sustained economic development that is the result, first of trade, and later colonialism. Since attaining independence, these countries have continued to build upon the foundation established earlier, with the centralized political structures and commercial agriculture playing very prominent roles. The recent successes in raising rice yields and output, and in reducing population growth rates are largely a product of state policies and new technologies. The variance among the states of Southeast Asia in both areas attests to the impact of state policies and their interaction with local cultures. This variance can be traced to a combination of factors, including cultural, ethnic and religious differences, the availability of technology, but most

---

powerful and independent player in the international scene. We cannot enter into this debate here, but will return to the issue in the final chapter.

importantly, to the specific political choices of governments and their ability and willingness to implement the policies they adopt.

In Zimbabwe, the former segregationist policies of White rule required the country's vast Black majority to reside within "native lands", which account for only about 40 percent of the country's arable land. Even with Independence it has been difficult to dismantle the effects of this earlier policy. Many Blacks continue to live and farm their relatively less productive land and their numbers are growing very rapidly. The resulting high densities and increasing numbers of people on marginal lands have contributed to high rates of soil erosion. The consequent loss of soil productivity has contributed to poverty of the rural Blacks and to an array of other social, economic, and environmental problems that is commonly found among the rural poor in developing countries. The loss of soil also hinders broader economic development efforts by silting in hydroelectric dam reservoirs and clogging vital irrigation canals. The erosion, poverty, and the increase in population can be mitigated, to some extent, by an array of appropriate government-sponsored interventions.

In Indonesia, natural resources interact with human institutions to produce a somewhat typical pattern of development, stresses, and adjustments. Health measures have been very successful in reducing mortality. This introduced a stress from rapid population growth, which led in turn to a reaction in the form of public programs to reduce fertility. Rich volcanic soils, tin mines, massive stands of tropical hardwood, rubber plantations, and, more recently, oil provided the basis of the wealth that locked Indonesia into the world economy and promoted its economic development. This also produced great stresses on its environment. Slowly, however, a new set of institutions is emerging to deal with the domestic problems and also to connect the country more effectively to the rest of the world. There may be no complete solution, but Indonesia's progress in fertility control, rice improvement and in at least addressing the environmental issues has placed it ahead of many developing countries in what can be called ecosystem health. Kartomo also shows how internal variations in human institutions can have an important impact on this health.

In Martine's study of agricultural modernization in Brazil, population-environment relationships were largely determined by a combination of government policies, especially in the form of available credit and other subsidies and incentives. World conditions, especially a global recession and problems related to international debt, along with modern technology, were also critical factors in the equation. Before the financial crisis of the early 1980's, government policies stimulated agricultural development programs that encouraged only the wealthy to invest heavily in modern commercial agricultural activities, producing much suffering among the large number of impoverished small farmers. Many were forced to migrate to the cities to look for work. The modernization program also encouraged the deforestation of large tracts of tropical forests, with only a fraction proving to be productive investments. Also, modern agriculture requires application of man-made chemicals which have harmed resources, wildlife, and urban as well as rural residents. The global financial crisis of the first five years of the 1980's forced the scaling back of these development efforts, which reduced their negative consequences, while not significantly affecting agricultural output. Many of the small farmers forced to migrate to the city returned to farming and improved livelihoods. Here state policies affected the population-environment relationship, with especially sharp class divisions in what can be called the environmental health outcomes.

In this section on large collectivities, boundaries emerged as a critical determinant of population-environment relationships. Political boundaries were the most immediately obvious. Within these boundaries we see the impact of the character and the policies of the state. There are, however, other boundaries that are more subtle and less visible, but perhaps of equal importance. These boundaries have mostly socioeconomic roots. At least four surfaced from the chapters of this section: *culture, religion, race, and social class*. Ness points out the importance of culture and religion both in shaping state policies and setting the limits on those policies. Local groups, often with vague and shifting boundaries, have different relationships to the land, and different views about technology and the morality of population control. These boundaries, even as vague as they are, define environments that can greatly enhance or limit the efforts of government to promote change.

McIntosh shows the consequences of racial boundaries. This boundary and its political-economic manifestations, perhaps more than any other, most clearly defines the different population-environment relationships in Zimbabwe. Finally, Martine draws attention to class boundaries. Government agricultural policies greatly favored the country's wealthiest class, increasing the already large gap between the rich and the poor, and subsidizing wholesale destruction of large sections of tropical forests. Here we can infer the proposition that the *health of the ecosystem* requires a greater equality in the distribution of resources than we currently see in Brazil or Zimbabwe, or in many other ecosystems as well.

These observations from the individual papers will provide the basis for some questions for future research, to which we turn our attention in the final chapter.

Chapter VI

The Powers and Limits of State and Technology: Rice and Population in Southeast Asia

Gayl D. Ness

Introduction

The recent period of rapid population growth reviewed in chapter three, represents a new and radical disequilibrium in the population-environment relationship. There are two major changes in energy technology which were especially important in producing this new disequilibrium. Of course, more than just a change to sails and fossil fuels have been involved. Technological change has been broad and deep, touching almost every arena in which populations and environments interact. Human institutions, in particular the character of what can be called the state, were critical determinants of the technology which was developed, used and diffused. The role of the state and technology can be clearly perceived by examining the long history of rice and population growth in Southeast Asia.

Three different periods in Southeast Asian history saw three distinct forms of states emerge: the indigenous states prior to 1800, the colonial state imposed by the West from 1800 to 1950, and the modern independent state which has existed since 1950. Each form had different capacities to organize and apply the technologies of rice production and mortality-fertility control. Thus each had the capacity to set the parameters of population growth in its specific environment. Within each period the state varied considerably in its capacity to import and apply the new rice and population technologies, which have great powers to alter the population-environment relationship. But the states and these technologies have also been limited in their power by specific conditions of local groups. Thus, the story of the population-environment dynamic in this distinctive region is a story of the powers and limits of state and technology.

There are two parallel processes of state induced technological use in managing the population-environment relationship. This chapter will examine the general process of state power, and its limits, in technology use. The specific relationship between rice output and population growth, though it is certainly a legitimate area of enquiry, will not be dealt with extensively, although similar state conditions are found to affect technology in both rice production and mortality-fertility control. It can be argued that there are general political conditions at least partly independent of the specific technology in question. This strategy of enquiry is facilitated by the observation that both sets of technology have similar organizational requirements. Both require the development of a new technology, and the diffusion of that technology to many different *location-specific* users. Thus, there must be an organizational capacity for both development and diffusion.

The Setting: Southeast Asia

Southeast Asia shares with the rest of the world a familiar pattern of population-environment dynamics. This pattern is characterized by many centuries of slow, at times negative, population growth, followed by two centuries of rapid growth in both population and environmental resource mobilization. At the same time, the region represents a unique ecological system with distinctive internal and internal-external dynamics. Examining both the similarities and the differences of this ecosystem can provide a better understanding of the dynamic relationship between population and environment.

Jutting from the underbelly of Asia, Southeast Asia's mainland is cut by rivers and broken by the seas into an extensive archipelago. It is a riverine and sea-based ecosystem characterized for two millennia by wet rice agriculture and water-based trade. Its social systems are marked by considerable linguistic diversity and some diversity of religions--Theravada Buddhism in the mainland and Islam and Christianity in the islands. More important, the entire region is marked by an equality in gender roles, or rather by a degree of female autonomy, that is very different from the larger portions of South and East Asia which it abuts.

This overall ecosystem experienced centuries of slow population growth and relatively weak exploitation of resources. The region's population may have reached twenty-two million by 1600, reflecting centuries of very slow growth, and often decline. In the sixteenth century the situation began to change as the region was drawn into the emerging world system. By 1800 the region probably reached thirty-two million, representing two centuries of slow, but perhaps steady growth at about 0.2 percent per year. The change was decisive by the beginning of the nineteenth century.

In the nineteenth century, the average annual growth rate rose to about 1 percent, bringing the total population to 83.5 million in 1900 (table 6.1). Despite the devastation of World War II, by 1950 the population had reached

TABLE 6.1 Population of Southeast Asia: 1500-2000 (millions)

Country	500	1600	1700	1800	1900	2000
Burma	4.0	4.5	5.0	6.0	12.5	51.1
Indochina	4.0	4.5	5.3	6.5	15.5	98.3
Indonesia	7.8	8.5	9.5	12.5	38.0	209.2
Malaysia	0.4	0.5	0.6	0.8	2.5	23.9
Philippines	0.5	0.8	1.3	2.5	8.0	77.4
Thailand	2.0	2.2	2.5	3.0	7.0	63.7
Total	18.7	21.0	24.1	31.3	83.5	523.8

Source: Data for the period 1500-1900 are from McEvedy and Jones (1978); the figure for the year 2000 are medium variant estimates provided by the United Nations (1988).

about 182 million, representing a half century of growth at about 1.6 percent per year. In the next two decades, the rate of growth continued to increase, rising to near 2.5 percent per year, and the population reached 286 million before its growth rate began to decline after 1970. In 1990, the population stood at 441 million. It is expected to grow by about eight million per year throughout the rest of this century, even though the rate of growth declined to about 1.74 percent by 1990. In the year 2000, the region is expected to have about 524 million people, and may reach over 700 million by 2025, even with what is expected to be a continued decline in the rate of growth to 1.17 percent in between 2000 and 2005 and to 0.93 percent in between 2020 and 2025.[1]

Although the total population figures show exponential growth since about 1800, the growth rates display two or three discontinuities. These three discontinuities are marked by significant changes in both death and birth rates. The first occurred in the nineteenth century as past slow growth rates jumped by five to ten times to about 1 percent. The next break occurred after 1950, when the rates more than doubled again, climbing to 2.5 percent. The beginning of the decline in the growth rate can be seen as a third major change.

As happened in much of the rest of the world, the rapid population growth since 1800 has been accompanied by substantial economic development. Spices have been a major production and export crop for centuries, forming

1. Population figures for 1600 and 1800 are taken from Reid (1988). For 1900, I have used McEvedy and Jones (1978). Reid's figure of 32.4 million for 1800 is close to the McEvedy and Jones estimate of 31.3 million. For the period 1950-2025, I have used the United Nations 1988 appraisal, which gives a figure of 182 million for 1950. This is slightly higher than the McEvedy and Jones estimate of 177 million, but the proximity of the two estimates gives one more confidence in the longer time series provided by McEvedy and Jones.

much of the early Western attraction for the region. In the nineteenth century sugar, coffee, tobacco, hemp and copra were important export crops. They were joined by rubber and palm oil in the twentieth century.

Rice has been the staple food grain throughout the region, and its growth since 1800 has been remarkable. By the late nineteenth century, Burma, Thailand and Indochina had become major rice exporters. In 1880, they were exporting almost 1.5 million metric tons, accounting for almost half of all the world's rice exports. Half a century later, in the 1930s, those three countries alone were exporting more than eight million metric tons, accounting for three-quarters of the entire world's rice export.[2] Total rice production figures before 1900 are difficult to obtain, but it is reasonable to assume that output kept up with population growth in the nineteenth century, though some parts of the region were rice importers.

Here again are discontinuities. In the nineteenth century growth rates rose from what must have been very low levels to about 1 percent. After 1950 they jumped to about 3 percent per year. These discontinuities are clearly reflected in the pattern of changing yields and outputs. Whether or not the current growth rate will decline in the future remains to be seen. At least since the nineteenth century, the staple cereal has matched and often exceeded population growth.

Overall this two centuries of combined population and economic growth established the modern period as distinctive and highly discontinuous from the past.[3] But there are in fact two or three separate discontinuities here, and all of the changes described had different trajectories in the various societies within the region. Tracing both discontinuities and some of the internal differences within the region can bring about a fuller understanding of the dynamic interactions of population and environment. The conditions and dynamics of the three periods will be summarized and set apart by the two major discontinuities: the coming of the West and the "modernization" that was ushered in with political independence after 1950. Although it is the longest, we shall have least to say about the early period, the centuries up to about 1800. There is more to examine about the colonial period, roughly from 1800 to 1950. In the modern independence period the greatest discontinuity has occurred and the most detailed information is available for study.

---

2. Randolph Barker, Robert W. Herdt and Beth Rose, The Rice Economy of Asia, (Washington DC, Resources for the Future, 1985), p 187, citing Owen (1971).

3. Anthony Giddens (1990) is only the most recent to note the distinctive break with the past that is marked by what we call modernity. William McNeill (1963) and Emmanuel Wallerstein (1976) also provide important current analyses of this distinctiveness. It should also be noted, however, that the rise of modern sociology and anthropology also derive from a perception of the great discontinuity of modernization.

The Old World: To 1800[4]

For the past two thousand years, Southeast Asia has been trading with China and India. The earliest routes came from India to the Kra Isthmus, crossing at its narrowest point, hugging the coast through Central Thailand and around the south coast of Vietnam, and then moving up to China. As trade and maritime technology advanced, routes crossed more open water, running down through the Straits of Malacca, touching on Sumatra and Java, then running with the summer monsoons up to Vietnam and China.

Wet rice cultivation came to the region from its birthplace in Southern China, though its extensive cultivation spread only slowly in the first half of this period. Still, an early maturing variety was transported from Champa in central Vietnam to China as early as the twelfth century. This permitted double cropping, which spread through China during the twelfth and thirteenth centuries, giving rise to high population densities in the Yangtze basin (Barker et al 1985, 16-17).

Throughout these early centuries, centers of power were often located on estuaries, which linked sea trade with up-river agricultural settlements. Rice was grown in river valleys either in swamplands requiring little construction or in bunded fields which may have involved extensive terracing of slopes and construction of water courses. While some of the rice growing areas were relatively isolated from external trade, as were Bali, Pagan in Burma, Angkor in central Kampuchea, and Ifugao land in Northern Luzon, most of the larger political centers emerged as major sea trading centers. These changed over time as political centers rose and declined. Funan, at the southern tip of Vietnam, was one of the earliest. It was followed by Dvaravati at the mouth of the Chgo Phrya River, then Tambralinga on the Kra Isthmus, Sri Vijaya in Southern Sumatra, and finally Mataram and Majapahit in Java.

What is striking and as yet not fully explained is the rise and decline of these political centers, and the possible connection between this phenomenon and the region's slow population growth. Trade, agricultural growth and population growth were at times mutually stimulating.[5] Political power came to be concentrated in local chiefs, who developed substantial kingdoms through managing warfare, taxation, and public construction especially of canals and port facilities. This attracted trade, which increased the demand for

4. This section draws on Hall and Whitmore, 1976, Hutterer, 1977, Marr and Miller, 1986, Hall, 1985, and Reid, 1988. I am also especially indebted to Victor Lieberman for his recent work on the pattern of Southeast Asian history (1990 and 1991).

5. There is a long debate in Southeast Asian history over the source of the stimulus for state formation and social structure. Earlier authors, such as Wheatley (1975) and Coeddes (1964), were impressed with the weight of the external influence of Indian traders and religious scholars. A younger group of scholars, taking the lead from Harry Benda, emphasize the independent forms of social and political life that emerged in the region. This debate need not detain us, though it is difficult to avoid the conclusion of the independent generation, especially given the striking character of gender roles in the region, and the equally striking difference the region shows with both South and East Asia.

rice, both to support sea voyages and as a trade item itself. This in turn increased immigration and probably increased local human fertility as well.[6] The increased population produced trade goods, provided labor for trade, and maintained the construction that trade required. This stimulated more trade, leading to more population and agricultural growth. This mutually reinforcing process apparently had limits, however.

Janice Stargardt (1986) provides a suggestion of some of those limits in her analysis of the Satingpra Peninsula on the Kra Isthmus over a period that runs from the sixth through the thirteenth centuries. Here a major center arose around the trade that came across the Isthmus. Canals were dug from the large inland lake, the Thale Sap, eastward through the town to the South China Sea, and parallel to the sea. The canals provided drainage and irrigation for the highly productive agriculture. Canals were also dug westward from the Thale Sap through two estuaries (at Old Paliang and Old Trang) to the Bay of Bengal. This permitted a rich activity in trade across the isthmus. An early phase of Satingpra's development, from the sixth through the eighth centuries, came to an end with a major attack on the city in A.D. 835. But within half a century the city was rebuilt, the extensive trade canals were dug across the isthmus, and the city enjoyed four more centuries of prosperity. There is some suggestion of Indonesian cultural and political dominance in this period. From the twelfth century, however, strains began to appear.

Labor for digging and maintaining the canals of commerce drew too heavily on that needed to maintain agricultural canals and rice production, with a consequent decline in agriculture. This weakened the center, making it less capable of defending itself from the inevitable external attack, and also less capable of rebuilding after such an attack. One came in the late thirteenth century, from which there was only a weak recovery. The final devastation came with an attack in about A.D. 1340, which destroyed the city and brought to an end this major population center. A critical limiting ingredient in this process was apparently the internal competition between commerce and agriculture. The supply of labor was apparently insufficient to meet both needs. The competition also involved, however, elite orientations and preferences. Commerce had replaced agriculture as the preferred source of wealth.

Note, however, the importance of external military ventures. Throughout the region, as trade centers grew, they became rich targets for attack. Indeed,

---

6. Reid (1988) suggests that one reason for the slow population growth in Southeast Asia lay in low fertility, which resulted from exceptionally heavy work of women in swidden agriculture, in the practice of "roasting" women after childbirth, and from women's deliberate avoidance of many births. The latter is certainly congruent with the unique autonomy of Southeast Asian women, though this relationship between fertility and female autonomy has not been examined in the early period. It is, however, one of the apparent causes of the rapid decline of fertility after 1975 in Thailand (Knodel 1988). Reid also notes that the move to wet rice agriculture, with its lower work loads for women, and the move to Islam or Christianity, which restricted the autonomy of women, may also have led to increased fertility.

one of the means for increasing political power in one locale was the successful attack on another. This provided the conqueror with the wealth and charisma to attract more supporters, which gave him the power to protect against internal violence and external attack. The rise of one center often meant the decline of another. In addressing the basic question of the reasons for the slow population growth in the region up to 1800, Anthony Reid (1988) gives greatest weight to warfare.

After the fourteenth century, there appears a divergence between political developments on the mainland and on the islands. Lieberman (1990, 1991) shows how state development moved steadily ahead in three major river systems: Burma's Irrawaddy, Thailand's Chao Phrya, and Tonkin's (North Vietnam's) Red River. Central power grew partially due to control of river trade and penetration of local autonomous regions. These were drawn increasingly under central control through military conquest and administrative expansion. The cultural forms of the center were also adopted by local elites, providing a symbolic as well as structural penetration and an increasing cultural homogenization. Populations were also deliberately moved to frontier areas both to cultivate and to protect outlying areas, providing an ethnic as well as cultural homogenization. As central power grew and expanded, both population and rice output grew as well, though it is not possible to be precise about the figures or their distribution. Lieberman compares this state development with that in Tokugawa, Japan and early modern Europe. By the nineteenth century it had advanced considerably, and was essentially aborted and reshaped by Western colonial conquest.

The outside world occasionally entered into this internal struggle of forces in the region. In the fifteenth century the Chinese admiral, Zhang He, for example, protected Malacca against a Thai attack. For the most part, however, Southeast Asian power struggles worked themselves out internally. Local political conditions determined the growth of both agriculture and population in specific centers in the region, but the anarchy of the larger political system of the region as a whole, placed severe limits on that growth. This was especially true in the archipelago. Even the greater advances in state development on the mainland, however, produced only very modest expansions of the human population throughout this environment. That would change greatly in the nineteenth century.

The Colonial Period: 1800-1950

The intrusion of the West at first only added to the internal anarchy of the region. Portuguese, Spanish, Dutch and English engaged in what is euphemistically called "armed trade" from 1511 when the Portuguese seized Malacca through the Napoleonic Wars, when Holland temporarily gave up its Indies holdings to Great Britain. The Dutch had been especially vigorous in using native rulers against one another in their struggle to control trade. Anarchy was also fueled by European competition, first between the Spanish and Portuguese, and finally between Dutch, British and French powers.

The Dutch and British Accords, following the Napoleonic Wars, resolved one major source of conflict, and left each relatively free to advance its own interests. In the early nineteenth century their interests expanded from trade alone to land control. Through three wars in Burma and a number of treaties and police actions in Malaya, the British became paramount colonial powers in the area. Through a series of wars that lasted into the early twentieth century, the Dutch took control of Indonesia. In the late nineteenth century, the French also seeking colonial expansion, gained control of Kampuchea, Laos, and Vietnam. Through skillful manipulation of external powers and the luck of geography, Thailand remained independent, forming a buffer between French and British interests which was useful for all parties. The Spanish colonial system laid down in the Philippines in the sixteenth century, began to lose control in the nineteenth century as a result of both liberal movements in Spain and the loss of their American colonies. The region's first major nationalist liberation movement began in the Philippines at the end of the century, offering the Americans their opportunity to become a world colonial power by annexing the Philippines.

Thus the nineteenth century signalled a major change in political organization. The region came to be ruled by outside powers. The *Pax Imperica* was established. Its impact on rice and population growth were remarkable, at least by past if not by future standards. In the half century from the last quarter of the nineteenth through the first quarter of the twentieth, the area under rice cultivation increased dramatically. The lower portion of Burma, the great central plain of Thailand, the Mekong Delta in Vietnam, the central plains of Luzon, river valleys in Java and Madura, and the smaller Krian and Kedah plains in Malaya all experienced extensive development. Land was cleared, drained, bundled, and sometimes terraced to increase rice production. By 1930 there were twenty million hectares of rice land in the region, representing at least a doubling of the area planted half a century earlier.[7] Production increased accordingly to just over twenty million metric tons by 1930. Burma, Thailand, and Vietnam became major rice exporters, sending some five to eight million tons a year into the world market.

Much the same growth was found in populations. The thirty-one or thirty-two million in 1800 grew to more than eighty-three million in 1900,

7. Barker et al (1985) estimates 6.3 million hectares in Burma, Thailand and the Philippines for 1901-1910. Estimates are not given for Indonesia, Indochina and Malaya because of the weakness of the records. The Burmese figure of 3.7 million hectares, however, already represents almost a four fold increase from the one million hectares estimated for 1855 at the close of the second Anglo-Burmese War (Steinberg 1987, pp. 230-31).

representing an average annual increase of 1 percent. This, of course, masked a much more turbulent process of growth, as massive influxes of people in one region were matched by population declines from natural disasters and new epidemics of diseases. Cholera was introduced into the region from Bengal in 1817, and the clearing of land for rice and other plantation crops brought an expansion of malaria. Increasing population densities and mixing of people also advanced the march of smallpox and other infectious diseases.

Western capitalist colonialism represented a new form of political organization that made possible a new relationship between population and environment. As it did elsewhere, this new organization broke down barriers to the movement of the factors of production. The region was flooded with labor from southeast China and India, with capital from the West as well as from India and China, and with entrepreneurship from the West, China, India, and the Middle East.

In some areas, the new political organization implied a new technology of production. Plantation agriculture emerged as a new and vastly more productive technology for rubber, palm oil, tobacco, sugar, and copra. The mining of tin, iron, and gold greatly expanded, in part under the new capital-intensive technology of the West. In rice, however, production expanded largely through the mechanical extension of the older technology and organization. Small peasants dominated the scene throughout the period, only expanding in area and volume what they had been doing for centuries. Clearest evidence for this lies in yields. Traditional levels of one to two metric tons per hectare have been characteristic of the region for as long as any record or estimate exists. One ton per hectare represents peasant-level production, with a traditional technology. This is also the level obtained in Japan at the beginning of the Meiji era, and in Taiwan before the Japanese investment in rice improvement early in this century.

New forms of social organization also began to emerge. Major urban centers eventually grew, with tentacles that reached through smaller towns to the most remote rural areas. This emergence represented a far more dramatic change in the population-environment relationship. Education and communication spread, albeit slowly, but also with profound implications. Subsistence agriculture was commercialized, drawing peasants into the erratic movement of the world economy. This connection produced great stimuli for growth in some periods, and equally devastating stagnation in others, such as the Great Depression of the 1930s.

The population growth of the period came in large part from increased in-migration. Death rates remained high throughout the nineteenth century, and at the time of few epidemic diseases, could produce a negative natural increase, countered only by greater migration. It is possible that in some areas the birth rate may have increased, as the move to wet rice agriculture and the more settled political situation removed pressures for women to restrict births.

Political stability also meant a change in the sex ratio of the immigrants; the arrival of the Chinese and Indian women certainly could have contributed to an increase in the birth rate. This change was still gaining momentum in the early twentieth century.[8]

Thus in both rice and population, the change in the population-environment relationship can be termed more mechanical than organic. The *Pax Imperica* permitted both rice output and population to grow without radical changes in technology. Old forms of social and economic organization could be expanded into new lands. The technology for producing rice and dealing with fertility and mortality remained much as they had been for centuries. This would change with the upheavals of the mid-twentieth century, which would usher in new forms of political organization and new technologies. These would have the most dramatic implications ever seen for the relationship between population and environment.

The Modern Period: 1950 To The Present

Although World War II ushered in a period of radical political, social and technological change, the changes were in some respects merely extensions of forces set in motion by the colonial system that grew up in the preceding one and a half centuries. Increasing urbanization and commercialization gave the political centers new administrative capacities as well as new burdens and responsibilities. As unresponsive as they were to human social problems, the colonial powers began to assume more responsibility for health and welfare. Thus, the new smallpox vaccination discovered in England just before 1800, was being gradually applied to Southeast Asian populations by the end of the nineteenth century. The colonial powers brought the dreaded cholera epidemics, but this also ultimately led them to undertake public health projects to secure drinking water and deal with sewage. If clearing land for plantations brought an increase in malaria, the more advanced colonial powers sought to control malaria, if only to reduce their labor costs. In effect, the urban colonial centers came to be points from which the new public health organization would develop. This organization would soon prove capable of importing the new death-controlling technologies which were greatly advanced by mid-century.

Radical changes were also brought about through the spread of education. Motives varied greatly, from religious and altruistic values of promoting human enlightenment to the more restrained desire to reduce administrative

8. Immigrants were at first mainly males. Late in the nineteenth and early in the twentieth centuries, women came as well. In Malaya, where the Chinese and Indians were proportionately the largest minorities of any country in the region, the Chinese sex ratio was 178 women per 1000 men in 1850 (Ness 1967). Even as late as 1930, the sex ratio was only 500 females to 1000 males. To be sure, early in the migration Chinese immigrants took local wives, but the gradual increase in Chinese female immigrants would be expected to increase the birth rate.

costs by using native clerks for the burgeoning colonial bureaucracies. Regardless of the motives, education did bring a kind of enlightenment, spreading new ideas of equality and human welfare, creating important new responsibilities for governments. Perhaps most important was a new theory of the legitimacy of government, which rested on the consent of the governed. Accepted by colonial masters and subjects alike, this would ultimately make the old colonial system untenable.

Differences within the region, as well as proximate external influences, also served to undermine the colonial system. In taking the Philippines by force, the Americans were placed in an intense moral dilemma. How could a revolutionary government justify the subjugation of another people? The dilemma led the Americans in 1934 to promise the Philippines independence within ten years. This scarcely provided support and comfort to other colonial powers in the region, who seemed to be bedding down for a lengthy rule. Events in India produced the same tensions. The growth of the Indian nationalist movement and the Congress Party brought the British colonial contradictions into full light. A metropolitan government espousing the divine right of the masses[9] to choose their own government at home could not easily hold a foreign territory against the demonstrated wishes of its people.

These contradictions in the theories of political legitimacy go a long way to explain how it was that the colonial system that had taken three centuries to construct could be dismantled in a mere twenty-five years. Most of South and Southeast Asia were conquered relatively easily by Western powers, roughly beginning in 1750. Often Western armies prevailed against far larger indigenous forces in the field. In 1945, however, the metropolitan governments were vastly more powerful in the region militarily than at any time during their conquest, yet they had neither the ability nor the desire to hold on to their territories (Ness and Stahl 1977). Metropolitan voters were in no mood to shed the blood of their children for this cause. When colonial powers used indigenous troops (e.g., the British use of Indian troops in Burma and Indonesia) the surge of nationalist sentiment among those colonial troops made them unwilling to carry on the fight for the metropole.

These contradictions also largely explain the orientation of the new political organizations that replaced the colonial rulers. New political orientations represented relatively strong conceptions of the state. If these fit well with traditional Southeast Asian views of kingship, they were, more importantly, supported by a new sense of political responsibility. From the historical interplay of capitalism and revolution, and especially from the rise of Soviet central planning and the collapse of the world market economy in the 1930s, the early laissez-faire orientations came to be replaced by a strong sense that

---

9. The term is from Eric T. Stokes, The English Utilitarians in India (Oxford: Clarendon Press, 1957).

the state had a responsibility to promote economic development and popular welfare.[10] Along with this value change came a new technology for economic planning, and for promoting human welfare. It is this organizational and technological change that makes the current period so discontinuous. This new organization and technology have radically changed the population-environment relationship and are clearly evident in the movements of population growth and rice production.

Mortality Decline

The first change to occur was a fall in death rates. As table 6.2 shows, the average crude death rate for the region was 24.4 (per one thousand population) in 1950, only slightly lower than typical rates reported throughout good times in the past. They were highest in the less developed states (Burma, Indonesia, Kampuchea, Laos, and Vietnam), and considerably lower in the other states. Singapore's more advanced urban society and extensive administrative system had already achieved what amounted to the modern low death rates, toward which all the rest would move. By 1970-75, all but Kampuchea and Laos had rates well below twenty. Currently, most of the states have reached the low levels typical of advanced industrialized countries.

This rapid decline of mortality was the direct result of the application of the world's new mortality-controlling technology. This technology was distributed through the medical and public health systems that had been laid down in the colonial period, and were greatly expanded under the new independent states. Where internal violence has obstructed the expansion of this welfare administration, the new technology has not been as widely distributed, and mortality has fallen more slowly. In effect, the capacity of the state to produce internal peace, and its commitment to welfare as well as economic growth, were major determinants of the extent to which the new mortality-controlling technology would be diffused through the population.

There are, however, many local variations even in states where substantial advances took place. Here one can see how local cultures can place limits on what the state and technology can accomplish. West Sumatra provides a good illustration. It has been a rich, well watered rice producing province for the past four decades. Self sufficient in rice, it has been an exporter to the rest of Indonesia. Nonetheless, West Sumatra has consistently had one of the highest levels of infant mortality in Indonesia. This is the more unusual, since West Sumatra is the home of the matrilineal Menangkabua, and has shown some of the lowest male-female status difference in the country.[11] Caldwell (1986) has

---

10. See Nigel Harris (1986) for a good review of the varying history of role of the state in economic development.

11. For example male-female school attendance ratios for 1980 are nearly equal only in North Sulawesi (99.4), West Java (100.6) and West Sumatra (106.4). In the remaining 23

shown that the high status or relative autonomy of women is closely associated with rapid mortality decline in less developed countries. Thus we should expect West Sumatra to have low infant mortality both because of female status and agricultural wealth. It appears that both the province's dietary custom, which eschews green and thus reduces Vitamin A content, and its birthing practices, may be responsible for its deviant status.[12] The Indonesian government is currently reorganizing its rural health services to place greater emphasis on health education and service delivery at the provincial level. This has probably helped the province to achieve striking declines in infant mortality in the past few years, but the experience still shows how local culture can limit the impact of the new state and health technology.

TABLE 6.2 The Decline of the Death Rate in Southeast Asia 1950-1985

COUNTRY	1950	1955	1960	1965	1970	1975	1980	1985
Burma	23.7	21.8	19.5	16.2	14.2	12.4	11.0	9.7
Kampuchea	23.8	22.1	20.4	19.4	22.5	40.0	19.7	16.6
Indonesia	26.1	24.3	21.5	19.3	17.3	15.1	12.6	11.1
Laos	25.3	22.8	22.7	22.6	22.7	20.7	18.7	16.4
Malaysia	19.9	16.5	13.3	10.4	8.8	7.2	6.0	5.6
Philippines	19.5	16.1	13.1	10.7	10.5	9.1	8.5	7.7
Singapore	10.6	8.6	7.1	5.6	5.1	5.1	5.4	5.6
Thailand	19.2	16.9	13.4	11.4	9.3	8.3	8.0	7.0
Vietnam	28.5	25.6	21.2	16.6	14.3	11.4	11.2	9.5
All SE Asia	24.4	21.8	18.8	16.1	14.4	12.8	11.1	9.7

Source: United Nations (1988)

Increased Rice Production

The new organization and technology for rice production also came from the outside, building on local agricultural administrative systems that had been laid down by the colonial powers. In 1959-60 The Ford and Rockefeller Foundations came together to establish the *International Rice Research Institute* (IRRI), the first of what has come to be a substantial network of international agricultural research stations (Baum 1986). The Institute was

provinces the mean ratio is 133.6, with a standard deviation of only ten. (Hugo, Hull, Hull, and Jones, 1987, p. 67).

12. Personal communication, Dr. Peter Fajans, Department of Population Planning and International Health, University of Michigan.

built on the grounds of the University of the Philippines College of Agriculture at Los Banos.[13] The goal of the Institute was to use modern agricultural science, especially the techniques of plant breeding, to develop new high yielding varieties of rice. The resultant increased output would help to solve pressing problems stemming from rapid population growth. The Institute was eagerly welcomed by national and international development officers, who saw this as an important step to promoting economic development.

By all measures IRRI has been immensely successful. By 1965 it had developed new high yielding varieties of rice that could as much as quadruple yields. These varieties rapidly spread about the region, tripling rice output from 1950 to 1984 (IRRI 1986). As table 6.3 shows, rice output increased from 32.5 million metric tons in 1950 to 99.3 million metric tons in 1984. Table 6.4 shows that overall yields have also risen from 1.4 to 2.7 metric tons per hectare, indicating a major shift to a new rice growing technology.[14] Even more, IRRI had established an organization and a process for continued research to improve rice yields. This work spanned all of the relevant agricultural sciences, promoting basic and applied research, and training a large cadre of scientists, educators, and extension workers, who would staff national agricultural stations throughout Southeast Asia and the world.

TABLE 6.3 Rice Output in Southeast Asia 1950-1985 (Million Metric Tons)

COUNTRY	1950	1955	1960	1965	1970	1975	1980	1985
Burma	5.39	5.70	6.85	8.26	8.16	9.21	13.10	14.50
Kampuchea	1.58	1.49	2.34	2.50	3.81	1.50	1.47	1.30
Laos	0.58	0.51	0.50	0.74	0.90	0.91	1.05	1.32
Indonesia	8.85	11.04	14.95	15.06	19.32	22.33	29.65	37.50
Malaysia	0.87	0.69	0.94	1.26	1.68	1.98	2.01	1.78
Philippines	2.62	3.27	3.71	4.04	5.34	6.16	7.84	8.28
Thailand	8.23	8.90	9.51	11.16	13.57	15.30	17.37	19.20
Vietnam	5.41	6.29	9.17	9.82	10.72	12.00	11.68	15.42
All SE Asia	32.95	37.38	47.47	52.10	62.60	68.48	83.12	97.98

Source: IRRI (1986)

13. The University was established by the new American colonial government in 1903, with a college of agriculture for research and extension. These were built on the lines of the American agricultural research and extension system.

14. Note that these are both overall regional averages. In many places, rice yields are consistently six to seven tons per hectare. The overall average for all of Indonesia is near four tons.

TABLE 6.4 Rice Yields in Southeast Asia 1950-1985 (Metric tons per hectare)

COUNTRY	1950	1955	1960	1965	1970	1975	1980	1985
Burma	1.46	1.41	1.69	1.70	1.70	1.83	2.73	3.10
Kampuchea	0.95	0.85	1.09	1.07	1.59	1.43	1.08	0.94
Laos	1.16	1.06	0.80	0.81	1.36	1.34	1.44	2.17
Indonesia	1.55	1.68	2.05	2.06	2.38	2.63	3.29	3.87
Malaysia	1.67	1.68	2.06	2.22	2.41	2.64	1.44	2.66
Philippines	1.16	1.19	1.16	1.31	1.72	1.72	2.15	2.49
Thailand	1.55	1.66	1.69	1.87	1.98	1.79	1.91	1.98
Vietnam	1.44	1.44	1.99	1.99	2.14	2.26	2.11	2.74
All SE Asia	1.44	1.49	1.70	1.74	2.01	2.08	2.41	2.78

Source: IRRI (1986)

There is, however, a limit to what the new state and technology can do to increase rice output. The range of conditions that determine the extent to which the new technology will be adopted and will raise output is broad and exceedingly complex. Natural ecological conditions, such as water, soil, climate, insect and disease conditions, vary geographically and temporally. Local cultural preferences for specific tastes also play a role in acceptance of the new rices. Political-economic conditions, including government prices and subsidies for rice, fertilizer and chemicals, water policies and market control are also important determinants. Highly specific combinations of these conditions produce considerable variance in the extent to which specific local groups adopt and prosper from the new technology. Illustrations of this immense variance and basic figures for this analysis are seen in the change of rice yields, shown in table 6.4.

Overall, Indonesia has shown the most rapid and dramatic increase in rice yields and output in the region. For deeply embedded cultural and political-economic reasons, the government gave rice improvement high priority. Extensive resources were allocated to importing new rice seeds and to producing them locally. Resources were also lavished on imports of chemical fertilizers and pesticides. Agricultural extension services were greatly increased, and investments were made in local irrigation systems. The government cooperated fully with international aid agencies, and received a great deal of external support for rice improvement. Agricultural officials and local political leaders cooperated and competed with one another to increase their province's rice output. This implied heavy pressure on farmers to adopt the new technology and cooperate with government policies. The results were quite dramatic. Average yields rose from about 1.5 metric tons per hectare in 1950 to 3.87 in 1984; total output rose from just under nine to over thirty-seven million metric tons during that period. As one of the major rice importers in the region throughout history, with values reaching one million tons a year, Indonesia has recently gained near self-sufficiency in rice.[15]

Burma represents the other extreme. It was the last to adopt the new technology and its own internal agricultural policies have done much to discourage rice production.[16] Farmers were forced to sell to the state marketing board at less than world prices, with no allowances for transportation costs. Further, the closure of the country after 1962, as well as the previous ineffective economic planning, meant that there was little in the way of consumer goods to trade for agricultural surpluses, further dampening farmer incentives. These conditions explain in part why it took the country until 1970 to reach its prewar levels of production. Even as late as 1975, production was only at nine million tons and the average yield was 1.8 metric tons per hectare (compared with Indonesia's 2.6). Then in 1975, the government launched a new agricultural development campaign (Win 1990), increasing allocations of both fertilizer and fuel for those districts that would follow the instructions for the new technology (high yielding seeds, intensive cropping, precise water control, and fertilizer applications). For this, Burma could draw on its own agricultural scientists and the only mildly decayed agricultural research and extension services the British had developed during the colonial period. It also drew on the scores of scientists and technicians who had been trained by IRRI over the past decade. As in Indonesia, the results were striking. Yields rose to 3.1 metric tons per hectare in 1985 and total output rose from eight to over fourteen million tons. Unfortunately, this did not last. The government curtailed investments in agricultural supports. Oil and fertilizer costs rose rapidly to their unprotected levels. Farmers were taxed erratically, and there were no consumer goods to trade for rice. Production has since stagnated or fallen, though not to the much lower pre-1980 levels.

Thailand remains the region's largest exporter, and has tripled its output along with the rest of the region. In 1950, it was producing just over six million metric tons and exporting 1.4 million tons. By 1984 it was producing over nineteen million tons and exporting 4.6, making it the world's largest rice exporter.[17] Unlike Indonesia, however, Thai rice yields have remained some

15. Indonesia has not become self-sufficient in all cereals, however. It has become a major importer of wheat.

16. See Jonathan Levin (1960) for an early account of the negative impact of government policies on agriculture. Steinberg (1981 and 1982) provides an account of the overall development of government policies and their impact on all aspects of life.

17. Note, however, that only a very small portion of the world's rice output is internationally traded. Of the 470 million metric tons produced in 1984, only 12.5 tons, or about 2.5 percent,

of the lowest in the region. In effect, much of Thailand's great increase in output has come from an extension of the land under cultivation, rather than from the adoption of new technology. The high quality of traditional Thai rice varieties has long brought premium prices on the market. Thus, Thai farmers have been reluctant to adopt the new rices, whose taste and quality fail to match the traditional rices. This, plus the availability of new land for production, help to explain the increase in output with only limited adoption of the new technology.

There is another instructive difference between Thailand and Burma which sheds more light on the population-environment dynamics. The political-economic organization of expanding rice production during the colonial period was quite different in the two countries. Under British capitalist colonialism, Burma relied on imported, Chettyar, capital to open the new rice lands in Southern Burma. This produced a situation similar to absentee-landlordism, with a high turnover in Burmese rice producers, working as highly indebted peasants under urban-based Chettyar money-lenders. This produced a highly explosive political situation, giving rise to substantial revolutionary pressures.[18] Burmese rice producers working for ethnically differentiated urban capitalists fueled the national liberation movement, and led both to Burma's rejection of Commonwealth status after independence, and the imposition of the government- controlled production program which has since hindered agricultural development. Thailand's development during the same time period brought local political control, and the expansion of Thai rice production which remained in the hands of Thai peasants as owner-operators. This far more egalitarian system led to a more conservative approach both to the new winds of independence and to the economic development strategies adopted after the war. As in population, state orientations and capacities greatly affected the adoption and diffusion of the new technology. But there are also local limiting conditions, such as tastes for particular rice flavors, that affect the final outcome. These local cultural conditions emerge even more strongly in the case of fertility decline.

The Decline of Fertility

Perhaps the most dramatic change in the population-environment nexus in Southeast Asia came in the form of fertility decline. It is difficult to overestimate the profound character of this change. Throughout human history the control of death and the production of food have been valued activities.

---

were exported. The United States is the second largest exporter with just over two tons, followed by both China and Pakistan, each exporting just over one million tons.

18. See Stinchcomb (1961) for an analysis of rural class relations in agriculture. It provides insight into both the Burmese revolutionary pressures, and the more conservative pressures in Thailand.

Reducing mortality and increasing food production have been supported by deep human values as well as by the government. But human history has also valued sustaining high or relatively high fertility.[19] Thus the fertility decline that has been pervasive in industrial societies and is now occurring in the less developed countries, represents a radical change in human values.

In the developing countries, it also represents a radical change in government policies. In the past, when fertility fell in the industrialized countries it was often against government aims and desires. Since India made its momentous decision in 1952 to reduce population growth by reducing fertility, most governments in the less developed countries have adopted roughly similar policies. The new anti-natalist policy is thus a truly revolutionary change.

This is a change for which new technologies (modern contraceptives and safe abortions) and a new organization (medical delivery systems that range from the private market to government family planning programs) are now readily available. Whether that organization and technology will be used, however, depends on the character and policies of the state, and the willingness of the population at large to curtail reproduction. Thus modern fertility decline shows most clearly both the powers and the limits of state and technology. The extensive variance in Southeast Asia provides a fruitful setting for examining this dimension of the population-environment dynamic.

Table 6.5 provides basic data and shows a general decline in fertility rates throughout the region. There is, however, extremely wide variation from rapid fertility decline and the completion of the demographic transition to a very sluggish and somewhat erratic decline in fertility with sustained high rates of population growth. The nine countries in table 6.5 fall into three somewhat distinct categories. In 1985, Singapore and Thailand were the most advanced in fertility decline with total fertility rates of 1.65 and 2.60 respectively. Five other countries (Burma, Kampuchea, Laos, Philippines and Vietnam) all remain above 4.0. Indonesia (3.30) and Malaysia (3.50), show moderate and declining rates of fertility.

These variations can be explained to a large extent by political, cultural, and economic conditions. Political conditions affect the application of the new organization and technology. Cultural and economic factors affect their acceptance by the general population.[20] Where governments have been effective in providing health and education, and have been committed to fertility reduction, the new delivery systems have made the contraceptive

19. The upper limit of human fertility appears to be around twenty births per woman during a reproductive life. This is a level experienced by only a few societies. The Hutterites are the most common current example. But for most societies, a total fertility rate of seven or eight has been near the empirical upper limit. This indicates widespread control of births in most societies throughout history. See Virginia Abernathy (1979) for a discussion.

20. See Ness and Ando (1984) for a discussion of many of these political and organizational conditions.

technology readily available, even in rural areas and to poor and less educated families. Greater use is made of the new fertility limiting technology in areas where women are relatively free to move about in public life and more access to education, and where economic changes have made children more of a liability than an asset.

TABLE 6.5 The Decline of Fertility in Southeast Asia 1950-1985 (Total Fertility Rate)

COUNTRY	1950	1955	1960	1965	1970	1975	1980	1985
Burma	5.64	6.05	5.94	5.74	5.43	5.02	4.61	4.02
Kampuchea	6.29	6.29	6.29	6.22	5.53	4.10	5.12	4.71
Indonesia	5.49	5.67	5.42	5.57	5.10	4.68	4.10	3.30
Laos	6.15	6.15	6.15	6.15	6.15	6.15	6.15	5.74
Malaysia	6.83	6.94	6.72	5.94	5.15	4.16	3.91	3.50
Philippines	7.29	7.09	6.61	6.04	5.29	4.96	4.74	4.33
Singapore	6.41	6.00	4.93	3.46	2.63	1.87	1.69	1.65
Thailand	6.62	6.42	6.42	6.14	5.01	4.27	3.52	2.60
Vietnam	6.05	6.05	6.05	5.94	5.85	5.59	4.82	4.10
All SE Asia	5.99	6.08	5.89	5.79	5.26	4.79	4.28	3.58

Source: United Nations (1988)

The Singapore government provided extensive health and education services, and a powerful public housing program after the Peoples' Action Party government came to power in 1959. By 1965-70 the government had espoused a strong fertility limitation policy and had an effective family planning program in place. Further, the government promoted a program of rapid industrialization and economic development that has been one of the great successes of the developing world. Women responded by postponing marriage to gain greater education and advanced occupational positions. They also eagerly took up the new contraceptive technology, readily available through government clinics and, even more so, through the private sector. The power of the state and technology were clearly evident during this period.

More recently, however, policies have changed and the limits of the state have begun to show as the government becomes concerned about the potential decline of the population, and its implication for labor shortages and the state's ability to meet the costs of an aging population. In response to these concerns, it has attempted to promote higher fertility. It is doubtful, however, that this policy is having any appreciable effect, since Singapore is now a "modern" society in terms of contraceptive use.[21]

Thailand has been, in some respects, the most successful case in terms of promoting rapid fertility decline (Knodel 1988). Although it has experienced rapid economic growth over the past four decades, it remains highly rural and agrarian. At the same time, the government has been very effective in promoting the rapid expansion of health services and near universal primary education for both boys and girls. It has also developed one of the most successful national family planning programs to be found in the developing world. Finally, it is a society with a high degree of female autonomy. Political forces have thus supported the rapid extension of the delivery system and have made contraceptive technology highly available. Cultural and economic forces have also exerted strong pressures for fertility decline. Given these conditions, it is not surprising that Thailand has experienced perhaps the most rapid fertility decline of any modern society, without recourse to coercive government action.

At the other extreme, the five countries with total fertility rates above 4.0 show a variety of political, cultural and economic conditions that sustain high fertility. The slow pace of economic development and extreme levels of internal turmoil in Laos and Kampuchea provide little support for new organization and the availability of technology. Nor do these conditions produce pressures on individuals to reduce fertility. Not much change is expected in the near future. Current projections are that these societies will not reduce fertility to the replacement level (2.1 TFR) until well past the first quarter of the next century.

Although the government's commitment to fertility control now appears quite strong and the economic pressures to reduce fertility are considerable, Vietnam's internal turmoil has worked against the rapid expansion of the new technology. It is expected that fertility will eventually decline, with projections for the arrival of replacement level now set at just before the year 2020. It is also possible that the current expansion of contraceptive services combined with great economic pressures on individuals will produce a far more rapid fertility decline than expected (Nahn and Hannenberg 1990).

Burma and the Philippines represent cases of government unwillingness or inability to promote fertility decline. Burma has remained staunchly pronatalist, and formally forbids the distribution of contraceptives. Contraceptives are now generally available on the country's extensive black market, but this also implies both irregularity in supply and the lack of popular knowledge of their availability. The Philippines government has vacillated

21. In 1989 Singapore's total fertility rate rose from 1.7 to 1.8, suggesting to some that the new policies were having an impact. It is also possible, however, that the year of the dragon produced an increase that will only be temporary. In any event, the rise from 1.7 to 1.8 scarcely protends a rise to pre-controlled levels.

greatly over the past two decades. In 1970, it announced a national family planning program, actively promoted through 1978, then reversed to highly limited or active discouragement of family planning until 1989, when it again announced a policy similar to that it adopted in 1970. In both Burma and the Philippines, however, deteriorating economic conditions have exerted pressures on individuals to reduce fertility. Without an effective delivery system for the new contraceptive technology, however, these pressures have been reflected in the delay of marriage (Myint 1990) and in increased abortion. The medical profession in both countries reports serious problems with maternal deaths from illegal abortions.[22] Both countries are projected to reach replacement fertility five or ten years after Vietnam.

Indonesia and Malaysia are in the middle ground of fertility decline, showing a variety of political, economic and cultural conditions affecting reproductive change. Fertility is declining in both countries, and is expected to reach replacement level before the end of the first decade in the next century. The political contribution to this change will be very different in the two countries.

The Indonesian government has provided strong support for a national family planning program for the past two decades. It has also developed one of the most energetic and imaginative programs to be found in the less developed countries. It is especially effective in extending the delivery system deep into even the remote rural areas (Ness 1988). Given the problems of poverty and population distribution, the political system has also been relatively successful in providing health and educational services for most of the population (Crone 1988; Liddle 1990). Thus the political conditions have been supportive of the development of the delivery system, making the technology available to most of the population. Limits to its acceptance, however, lie in specific local cultures. Where more orthodox or conservative religious conditions prevail, persons tend to resist modern fertility control (West Java and North Sumatra); where the local political culture supports individual autonomy (West Sumatra), contraceptive use spreads more slowly. Contraceptive use spreads far more rapidly where high population density is combined with strong local organization (Bali) or a cultural acceptance of strong governmental authority (East Java).

In Malaysia, government policy has vacillated, as it has in the Philippines and Singapore, though for very different reasons. The government launched a strong and relatively effective family planning program in the mid-1960s. In 1982 this began to change as the government adopted what is, in effect, a two-part population policy, based on ethnic distinctions. For the Malays, there is

---

22. Sources: In the Philippines, personal communication from Manila doctors, with public television programs providing details of the problem. In Burma, the medical profession held a national seminar in 1988, indicating that abortion was a leading cause of maternal deaths, and that young women had little knowledge of available contraceptives.

what amounts to a pro-natalist policy; for non-Malays there is a passive anti-natalist policy. The national family planning program is quite weak and actually provides contraceptive services to less than 10 percent of contraceptive users (Ness 1988). At the same time, government policies and world conditions have together promoted rapid economic development and the expansion of health and education services. These have increased the demand for fertility limitation, which the private market sector is able to serve.

Political Systems and Technology

A great variety of locally and historically specific conditions affect the relationship between population and rice, a sector of the population-environment dynamic. This chapter has scarcely touched on all of the factors. For example, soil, topography, and water cycles, which are among the more important natural conditions, have been neglected. To a certain extent, these can be considered roughly constant even across the two hundred years of our review. The human conditions show far more changes over time and place, and thus can be used to explain the varying population-environment relationship. Although these human conditions fluctuate far beyond variables described here, this chapter has identified some of the more important human conditions that affect this relationship. These include political institutions, technology, and that great class of rich and complex conditions called culture.

Clearly the political systems of Southeast Asia have been major mediators between population and the environment. Two millennia ago small scale political systems were developing significant amounts of central power. They emerged, in part, under the stimulus of external trade. There was a mutually reinforcing relationship between trade and agriculture and population, both facets, stimulating and being stimulated by, the other. External stimuli and internal stimulation, which improved trade and agriculture, indirectly stimulated population growth. Thus, it is evident that population growth before the eighteenth century in Southeast Asia was closely tied to political centralization and technological improvement.

Political centralization appears especially important for mobilizing human resources to create and use the more productive technologies (e.g., bringing people together to dig canals, manage water storage and delivery systems). Such mobilization required some measure of coercion: to protect against external attack, to provide internal peace, and to extract resources from the population. This process is often referred to as state building. But the coercive power the state mobilized could also limit population growth. In the sparsely settled, benign environment of the region, a too heavy hand forced peasants to flee to other river valleys where they were less taxed. The state's power could also be used against other centers of power, thus producing a larger anarchic setting that severely limited population growth in any of the local emerging states. After the thirteenth century, there appears to have been more sustained success in this state building on the mainland than in the archipelago. It is possible that the river systems provided an environment in which state building was easier than along the trade routes of the islands. Thus the relationship between the environment and the institutions that promoted the growth of both population and rice was complex and multidirectional.

The growth of rice production and population in the nineteenth and early twentieth century was also clearly related to the emergence of new, larger forms of political organization, with greater capacities to coerce or induce increased labor and technological development. The *Pax Imperica* of the West brought increased exploitation of the region's natural resources, both permitting and inducing rapid population growth, largely through in-migration. The *Pax Imperica* brought new technologies as well. These were less dramatic in rice than in the control of mortality, but they can be found in both sectors. A new technology for collecting and distributing rice, i.e., the world market driven by fossil fuel power, greatly stimulated the increase of rice production, even as the basic technology of production remained much the same as it has for two millennia. The basic framework for future technological change was laid down in the central agricultural administrations that would soon be involved in the application of new science to rice production, which would lead to the Green Revolution in the second half of the twentieth century. State and technology would produce a new organization of rice growing, changing forever the long-standing traditional pattern of peasant rice production.

The new technology for reducing mortality, distributed through the recently developed organization of public health developed by the colonial powers, would ultimately bring a rapid decline in the death rate. If the organization and part of the technology were laid down in the late nineteenth and early twentieth centuries, they required the radical progression of that technology that came in the mid-twentieth century to produce the dramatic mortality declines which occurred in the past few decades. Again, the state administration moving a new technology had a profound impact on the population-environment relationship.

In all of these cases some form of greater political centralization and a new form of technology have been important for producing a change in the population-environment relationship. Both state and technology have, however, consistently been limited by local patterns of acceptance. Local cultures vary in their development or adoption of new technologies, regardless of whether or not these are supported by the state. There is continuing debate over the *rationality* of this acceptance. The economic perspective sees much rational calculation of marginal costs and benefits in this adoption. The other social sciences tend to see other non-rational *commitments* as more important.[23] Quite apart from the acceptance of a specific technology, however, is the more general acceptance of central authority and its leadership in promoting new technologies and new patterns of behavior. This is obviously determined by deep and, as yet, little understood cultural characteristics. East and West Java clearly exemplify deep value orientations that readily accept and just as readily reject central authority. Thailand and Burma also show considerable differences in this political culture, if the degree of state coercion and popular resistance are any measures. Acceptance of central authority is also partly determined, however, by how successful the central authority is in permitting local initiative and in rewarding performance rather than privilege.[24]

Where the state is committed to and successful in promoting modern economic growth, and where it can assure relatively broad sharing in the advantages of that growth through rising individual income and welfare, it can generally count on a fair degree of acceptance of its authority. When the state chooses to use that authority to promote a better balance between population and environment, as Singapore, Thailand, Indonesia, and Malaysia have generally done, there is a rapid growth of population and output, and a relatively rapid adjustment through reduced levels of population growth. Individual families adopt the new contraceptive technology more rapidly and thoroughly, and undergo a radical change to fertility control.

Other states have been more ambivalent about the direction, character, and means to achieve modern economic development. They have also been less willing or able to provide basic protection and social services to the population and apparently less concerned about the broad sharing of the advantages derived from development. When this is the case, the population-environment relationship is less benign. Human productivity rises more slowly, general mortality falls more slowly, producing rapid population growth. Finally, the rapid population growth is adjusted to less by fertility control and more by higher levels of infant and maternal mortality. If these movements of mortality and fertility reflect what some have called ecosystem health (Rapport 1989), then it is clear that the character and orientation of the state play a large role in determining the level of that health.

---

23. Cleland and Wilson (1987) review economic and non-economic theories of fertility decline, and come to the conclusion that the non-economic theories have more empirical support.

24. Edward Shils (1962) provides an excellent early analysis of the importance of permitting and promoting local initiative, which many central governments in the new states failed to do, and in their failure, weakened the central power itself. Earlier Rensis Likert (1953) provided the same analysis for modern organizations when he identified the supreme paradox of power: "if you wish to increase your power, you must give it away." Arthur Stinchcomb (1974) provided an equally penetrating analysis of the importance for modern organizational performance of rewarding performance rather than privilege.

Chapter VII

Population Policy and Environmental Impacts in Rural Zimbabwe

Alison McIntosh

Introduction

The relationship between soil degradation and density of population in the communal farming areas of Zimbabwe is as clear an example of a deleterious population-environment interaction as one is likely to find anywhere. Although awareness of the physical causes of erosion has existed in Zimbabwe, formerly Southern Rhodesia, since the turn of the century (Cripps 1909; Jennings 1921; Whitlow 1988a) recognition of the role played by land tenure and population density in communal areas is more recent. As Kay pointed out in his study of the human geography of Rhodesia, for much of the colonial era deforestation and degradation of the soil in the tribal areas was commonly attributed to the poor methods of land use employed by African farmers (Kay 1970, 86). In 1975, Kay proposed a model for changing patterns of land use in subsistence systems undergoing population pressures.[1] In 1980, Whitlow (1980a, 1980b, 1980c) discussed the implications of overpopulation in relation to some of the physical causes of erosion. However, it was not until the publication of the report of the National Erosion Study (Whitlow, 1988a; Whitlow and Campbell, 1988) that the overwhelming importance of population density and land tenure was unambiguously documented.

Population pressure in the communal areas has contributed to a variety of serious social, economic and environmental problems both for the people and the government. Large numbers of people are marginalized and live in poverty. Erosion has reached advanced levels in many parts of the country.

1. The model suggests that under increasing population pressure, the area devoted to cropping expands into regions previously used for hunting, fishing, gathering and grazing. As population increases still further, cropping extends into more marginal areas producing lower yields, while gathering and grazing are pushed into previously unused lands. Early signs of soil deterioration may appear. In the final stage, improved land use may increase yields in some areas but this is accompanied by the abandonment of other cropping areas. Grazing is confined to a narrow range, and gathering may no longer be possible.

One estimate suggests that the rate of soil loss in communal areas averages seventy-five tons per hectare per year on grazing land, and fifty tons on arable land. These rates far exceed the rate of soil formation and the rates on non-African farms (Stocking 1986). Recently, there have been reports that land in some areas has been abandoned (Whitlow, July 1990). Much of the soil eroded from hill slopes enters the main river systems which have become heavily silted. One survey of 132 dams in the lower reaches of the Save River showed that all were more than 50 percent silted and that 16 were completely silted (Ellwell 1985).

The picture for crop fields is less clear as the use of conservation farming methods and high yielding varieties of maize by some communal farmers has increased their yields. However, there are signs that yields of some traditional small-grain crops have decreased (Whitlow 1988a, 4).

This chapter outlines the responses of the post-Independence government to the joint problems of population density and environmental degradation and examines the potential role of population policy as a partial solution to the problem. The paper will review the complex of historical, cultural, political, and economic factors that have created the problem and shaped the perceptions of both people and government as they cope with it. The paper draws on the literature on population, agriculture, and land reform as well as on information obtained in interviews with officials in the relevant ministries, and with academics at the University of Zimbabwe and other research institutions during several visits to Zimbabwe between 1984 and 1989. First, however, it is necessary to say a few words about the origins of land apportionment and the National Erosion Survey to provide a context for the discussion.

Land Tenure and Erosion

The division of land between the races in Zimbabwe goes back to the earliest days of the colony. Unlike most other colonies in Africa, Southern Rhodesia was occupied by settlers who were governed, not by the British government, but by administrators of the British South Africa Company which operated under Royal Charter.[2] Later, from 1923 to Independence in 1980, the colony was governed internally by the settlers themselves under the supervision of the British government in London. From the start in 1894, large areas of apparently unoccupied land were alienated by the settlers and by the Company itself although, for many years, much of it remained unused. As early as 1896 and 1897, both the major tribes, the Ndebele and the Shona, rose against the appropriation of their land in what has been described as "the most violent, sustained, and highly organized resistance to colonial rule anywhere in

2. This section draws upon some of the excellent histories of Rhodesia and the division of land along racial lines. See especially, Ranger (1967); Kay (1970); and Palmer (1977).

Africa" (Palmer 1977, 55). The brutality with which the uprisings were put down prompted the Imperial government to establish a small presence in the colony, although at no time was it very large or interventionist.

One of the first problems facing the Imperial Resident Commissioner on arrival was securing the provision of land for African use. Since so much land had already been alienated, the obvious solution was to follow South African precedent and to set aside land as "native reserves" which would afford some protection against European greed (Palmer 1977, 57). The work of creating the reserves, later called Tribal Trust Lands (TTL's) and today communal lands, was entrusted to the Company's Native Commissioners who, generally speaking, set aside land that was already occupied by the majority of the African population. Much of it was in areas with the least fertile soils as they were the most easy to cultivate with the primitive tools of the African population. Although the creation of reserves was initially intended as a temporary stop-gap measure until more suitable arrangements were worked out, the reserves in fact became more and more firmly entrenched through increasingly segregationist legislation. This policy culminated in the Land Apportionment Act of 1931, which set aside little over half of the country for European settlement and denied Africans the right to purchase farms in these areas, and the Land Tenure Act of 1970 which brought about an equal division of the available farming land between Africans and Europeans (Zinyama and Whitlow 1986, 369).

Concurrently with the legislation on land apportionment, the settler government mounted a major research and extension effort to develop locally appropriate methods of conservation farming and to educate European farmers in their use. As a result, erosion on European land, once just as severe as on African land, was brought under control. While it was intended to mount a parallel educational effort in the TTLs, the African program was chronically starved of funds and conservation measures. Thus, consolidation of arable lands and their separation from grazing and residential areas, contour ploughing, the construction of contour ridges and storm water drains, and "destocking" (reduction in the number of cattle) had to be introduced by coercive rather than educational measures. This created enormous resentment among the African population who, not surprisingly, defined the problem in terms of land shortage. Nevertheless, some improvement in land use and farming methods took place during this time. Under the influence of growing civil unrest during the late 1960s and 1970s, the Ian Smith regime started to permit greater African participation in decisions on farming methods. However, progress was halted by the escalation of the guerilla war during the late 1970s, which dislocated parts of the population and disrupted the program (Whitlow 1988b).

In 1980, the newly created state of Zimbabwe therefore inherited a dual rural economy based upon differing patterns of land tenure. Although the segregationist legislation was repealed after Independence, the system is so deeply entrenched that it will undoubtedly take many years to change. There are still four main categories of land. First, areas now known as General Lands, or Large Scale Farming Areas, comprise the regions formerly set aside for Europeans. Secondly, there are the former TTLs, now known as Communal Areas, on which the system of tenure adheres closely to customary practice. In the communal areas, the land is "owned" by the community, which allocates plots to individuals for use as cropland and maintains grazing areas for the use of all. Despite the term "communal," farming in these areas is not a cooperative venture. Each farmer manages his own farm, and the right to the use of the land may be inherited by a wife or son while they remain on the land. In addition to the General and Communal Lands, approximately 3.5 percent of the available farming land known as Small Scale Commercial Farming Areas (formerly Native Purchase Areas) has been set aside for outright purchase by Africans.[3] Finally, some 13 percent of the land is designated as non-agricultural land and comprises of game parks, forest reserves and urban areas.

At Independence, general and communal lands each occupied approximately 40 percent of the area available for farming. However, there were marked differences in the size and density of the populations supported by each. The census of 1982 indicated that, at that time, the African population comprised over 97.5 percent of the total population of 7.54 million. Data derived from the census suggest that a population of some 4.2 million Africans were living on the communal lands, with an average population density of 25.5 persons per square kilometer. The general lands, by contrast, supported 1.2 million Africans, mainly as farm laborers on large scale commercial farms. Population density in those areas, at 7.6 people per square kilometer, was less than one-third of that on the communal lands. The small-scale commercial areas contained a population of only 166,600, with a population density of 12.4 people per square kilometer, intermediate between the two (Whitlow 1988a, 38-9; Weiner et al. 1985, 256). While all forms of tenure are to be found in all parts of the country, nearly three quarters of the communal areas are located on the poorest land, designated as suitable only for semi-extensive or extensive farming; by contrast, some 49 percent of commercial farms are located on the best agricultural land, considered suitable for intensive, semi-intensive, and specialized farming (Weiner et al. 1985, 259).

The National Erosion Survey was based on the analysis of 1:25,000 scale pan-chromatic serial photographs obtained from the 1980 to 1984 round of the systematic, country-wide blanket photography that has been flown every five years since 1963 (Whitlow 1988a, 16). The analysis was based on a stratified

---

3. Since Independence, Zimbabweans may now purchase land in areas formerly reserved for Whites.

random sample of grid cells of approximately forty-five square kilometers, a methodology that had been successfully employed in earlier work (ibid., 17). The author observes that because it is difficult to pick up areas of slight erosion using photography of this scale, his findings probably constitute a conservative estimate of the erosion present in the early 1980s (ibid.).

In general terms, the association of erosion with land tenure is quite clear. First, while communal lands now comprise just 46 percent of the national territory, over 80 percent of total erosion is found on them; by contrast only 15 percent of total erosion is found on the general lands, which comprise 39 percent of the national land total. Stated differently, with 4.7 percent of the total land area affected by erosion, the percentage of eroded land in communal areas averages 8.43 percent, while the average for the general areas is only 1.77 percent. Secondly, within tenure types, some 27 percent of erosion on the communal lands was rated severe or very severe while only 1.6 percent of the general lands were so rated. At the other end of the scale, over 85 percent of the general lands had negligible or very limited erosion, whereas just 40 percent of communal lands fell into these categories. (Whitlow 1988a, 21-3) By all available measures, the non-agricultural areas had significantly less erosion than either of the other land categories (ibid.).

The effect of population density on erosion is more complicated to assess as good farm management in erosion-prone areas can lessen the damage while poor management of good land will increase it. Moreover, it is not only the density of population that causes damage, but also the animals that people bring with them. This can be a serious factor in cultures like Zimbabwe's in which cattle have a ritual significance over and above their economic value and farmers use cattle to supply draught power for their ploughs. The concentration of communal land in the poorer agro-ecological regions of the country is another complicating factor.

Nevertheless, the National Erosion Survey confirms the higher concentration of population within the communal lands where 27 percent of the land has population densities of over thirty per square kilometer compared with less that 9 percent of the general lands. Conversely, only 34 percent of communal lands have densities of less than eleven per square kilometer compared with 42 percent of the general lands (Whitlow 1988a, 40). In both types of land use area, there is a linear relationship between amount of erosion and population density; however, the levels of erosion are much higher in the communal areas as compared with the general lands (ibid.).

The Impact of Population Growth

The maldistribution of population is possibly the most potent threat to the environment in Zimbabwe. Behind the distribution problem, however, is another--the rapid growth of the population--that both exacerbates the maldistribution problem and makes a solution more difficult to implement. The rate of growth of the population, which is a product of relatively low mortality and continued high fertility, was estimated at 2.9 percent per annum in 1989. This rate of growth, if continued, implies a doubling of the population in approximately twenty-three years. What this means in actual numbers can be seen in the difference between the 7.5 million total population counted in the 1982 census and the 1989 estimate of 8.9 million (ibid.). In the absence of significant change in the distribution of population, most of the anticipated increase will take place in the communal areas.

The impact of natural increase on the distribution and density of population between the 1969 and 1982 censuses is difficult to determine because the intercensal period witnessed a significant out-migration of population from densely populated areas in the north and east of the country to the cities and to less densely populated areas in the west. Part of this internal migration was brought about by planned resettlement of farmers in the pre-independence period, and part was spontaneous movement of population away from the north and east, the areas most affected by the war of independence, to places of safety in the cities. Another source of potential error is the 1982 census itself, which, though generally accurate, was taken at a time when population dislocation brought about by wartime conditions had not yet stabilized.

Based on an analysis of population change in the intercensal periods 1962 to 1969 and 1969 to 1982, Zimyama and Whitlow (1986, 374-5) discern two different levels of growth in different parts of the country, depending on whether they are areas of in- or out-migration. Areas that had received the resettled farmers mentioned above and which also received some of the sparser population that had to be moved prior to the construction of the Kariba Dam on the Zambezi River, had increased by around 60 percent between 1969 and 1982. Other areas of high growth are found around Harare where extensive irrigation schemes have been developed and where cotton has been introduced (ibid., 374). The authors note, however, that the source areas from which the migrants came also showed high levels of growth largely because the out-migration, which is significant in the areas of reception, represents only a small fraction of the natural increase experienced by the areas of origin (ibid.). Another cause of the continued high density in the areas of origin is the influx of refugees from Mozambique crossing the border to take refuge with their kinspeople in Zimbabwe.

Other parts of the country had grown by only around 40 percent in the intercensal period. These included areas located close to Lake Kariba in the northwest, in Matabeleland northwest of Bulawayo, northeast of Masvingo in the south central area, and in the south of the country. The authors comment that these are all areas that were badly affected by the war and that have already reached their capacity to absorb population. Evidence for this is seen in the influx of illegal squatters from the communal areas onto adjacent commercial farms (ibid.).

Population Policy in Zimbabwe: A Two-Pronged Approach

The discussion so far suggests that a multi-sectoral strategy will be required to solve the problem of soil degradation in Zimbabwe. Considerable success has already been achieved in increasing the productivity of smallholders through a combination of agricultural policy measures: extension services, attractive farm prices, continued development of drought-resistant crops and methods of conservation farming, access to credit, and numerous others (Financial Gazette 1984; Weiner et al. 1985, 278-81; Skalnes 1989; Bratton 1987). Nevertheless, it is clear that this approach needs to be backed by a population policy that will slow the rate of population growth and improve the distribution of the existing population. By the nature of demographic transformations, however, measures to reduce the rate of population growth will bring about significant change only in the long term.[4] Redistribution is, therefore, the more critical strategy for the short to medium term.

Zimbabwe does not yet have a formal population policy. Since Independence, however, there have been signs that the government is moving towards adopting one. There have been a number of seminars advocating the need for population policy.[5] In 1985, a significant step towards legitimizing the idea was taken when Zimbabwe agreed to host the All-Africa Parliamentary Conference on Population and Development sponsored by the United Nations Population Fund. A year earlier, Zimbabwe's statement at the United Nations International Conference on Population held in Mexico in August 1984, was supportive of the need to reduce population growth (UNFPA 1985, 179). The former prime minister, now president, Robert Mugabe, has never concealed his concern over Zimbabwe's population problems and in his address to the UNFPA-sponsored International Forum on Population in the Twenty-first Century, held at Amsterdam in November 1989, he announced that he would shortly appoint a population policy secretariat to draft a population policy and plan of action for the nation (Mugabe 1989). The meetings, the statements, and the occasions on which the statements were made, however, all imply that what is under discussion is a fertility reduction policy intended to reduce the rate of population growth.

---

4. If, improbably, fertility declined by the end of the century to "replacement level," the level at which the present generation is exactly replaced by the next one, and remained at that level, the population would still continue to grow throughout most of the next century.

5. These include the Population and Development Seminar, jointly sponsored by the Central Statistical Office, a constituent unit of the Ministry of Finance, Economic Planning and Development, and UNFPA in Harare, June 7-9, 1984; and the Seminar on Population in Development Planning, organized by the University of Zimbabwe, Department of Sociology, in Harare, October 28-30, 1987.

Fertility Reduction and Family Planning

Population policy is commonly defined as a set of government objectives to influence population size, growth, distribution, or composition, together with the instruments with which it may be used to achieve them (United Nations 1973, 631-2). The plan of action to implement a population policy is designed to act on one or more of the three demographic variables: fertility, mortality and migration. Although Zimbabwe does not yet have a population policy, it does have an active and, by the standards of the continent, effective family planning program. The origins of the program go back to the 1950s when a number of private family welfare agencies started to offer family planning counselling (Caldwell n.di, 8-16). More organized efforts stem from 1965 when these agencies merged to become the Family Planning Association of Rhodesia (Boohene and Dow 1987, 1). Increasing government support enabled the program to expand rapidly, augmenting hospital and clinic services in urban centers with mobile services in rural areas and, as early as 1967, with family planning outreach workers (ibid.). From the start, the program's objective was to improve the health and welfare of women and children. Only in the last three to four years has it started to move gradually to a more demographic orientation (Interviews with ZNFPC officials, July 1987).

In the context of the Unilateral Declaration of Independence from Britain (UDI) by Ian Smith's conservative white government in 1965, government efforts to recruit white immigrants, and escalating civil unrest, the program was increasingly perceived by the African population as having a genocidal intent (Boohene and Dow 1987, 1; Caldwell 1970, 17-18). At Independence, when the program was taken over by the Ministry of Health, it was questionable whether it would survive. By 1982, the tide had turned and the program once again started to receive strong government support. In 1983, under a new name, the Child Spacing and Family Planning Council (later Zimbabwe National Family Planning Council (ZNFPC)), it became a parastatal agency under the aegis of the Ministry of Health (Boohene and Dow 1987, 3).[6] Since that time the program has been significantly strengthened, especially in the area of community-based distribution which now reaches virtually every corner of the country. Significantly, by 1984, the government was providing over half of the Council's budget (Mugabe 1989), a circumstance that distinguishes the program from many, perhaps most, of the programs in Africa.

Data from the Demographic and Health Survey (DHS) (Government of Zimbabwe/Westinghouse 1989) indicate that over 95 percent of women in Zimbabwe had heard of at least one modern method of contraception. The proportion rises to nearly 98 percent among married women. Ninety-six

6. A parastatal agency is a semi-autonomous unit under the control of a government ministry.

percent of married women also know of a source from which modern methods can be obtained. Thirty-six percent of married women reported that they were currently using a modern method (27 percent of all women) and a further 7 percent said they were using a traditional method.[7] The survey showed that during the previous five years, the total fertility rate (TFR)[8] among women aged fifteen to forty-nine was 5.7.

There is some doubt about the validity of TFRs derived from various sources in post-Independence Zimbabwe. Compared with the DHS statistic of 5.7 in 1989, the TFR derived from the 1982 census was 5.6, while the statistic obtained by the Zimbabwe Reproductive Health Survey (ZRHS) in 1984 was 6.5. The reason for the discrepancies has not been fully established, although it is possible that the different methologies employed in the three surveys could explain at least some of the differentials. A more significant problem is that in both the ZRHS and DHS surveys, the TFRs obtained are considerably higher than would be expected given the contraceptive prevalence rates obtained in the same surveys. Working backward from the TFRs, a recent reanalysis (Adamchak and Mbizvo 1990) found that nearly half the difference between the observed and expected contraceptive prevalence rates can be explained by overlapping protection obtained by women who used contraception while still protected from pregnancy by post-partum amenorrhoea. This finding underscores a problem that may be faced by many family planning programs in Africa: where significant numbers of women still practice extended periods of breast feeding and post-partum abstinence, higher levels of contraceptive prevalence may be needed to bring about a decline of fertility than are needed to attain the same level of fertility in other cultural regions.

In addition to this problem, it is possible that the family planning program, as is often the case, may have reached a plateau. There is scope for introducing a more diverse set of approaches to contraceptive delivery: use of the private sector to deliver some services; more emphasis on services for special groups such as teenagers, men, factory and estate workers, and the introduction of social marketing of contraceptives. Indeed, a number of initiatives of this type are currently being introduced with funding from the United States Agency for International Development (USAID) and the United Nations Population Fund (UNFPA). It is possible that the adoption of a population policy might give a fillip to activities of this sort.

The outline of exactly such a policy has been circulating in Harare for the past two years (ZNFPC 1988). Submitted by the ZNFPC to the Ministry of

---

7. Primarily periodic abstinence and withdrawal.

8. The TFR is derived from the age-specific rates observed in the population under study. It is a synthetic measure of the number of children a woman would have during her life if current fertility rates prevailed during all of her reproductive years. In a population with declining fertility, the TFR tends to over-estimate the number of children that will be born to younger women in the sample. Subject to this caveat, the TFR gives an indication of the completed size of a woman's family on the average.

Health, the outline, not surprisingly, focuses on fertility reduction through family planning and supportive activities, and eludes the critical issue of maldistribution. In line with almost all the population policies elaborated in developing countries in the last two decades, the proposed policy bears the stamp of the international agencies that, through the provision of technical assistance, commodities and funding, have been influential in shaping the conventional wisdom on population policy.[9] To the extent that the adoption of such a policy would undoubtedly attract significant additional funds for international assistance, the policy proposal has merit; as an answer to Zimbabwe's population and environmental problems, it is inadequate.

Toward More Equitable Distribution

The outline of a policy published by George Kay (1980) immediately prior to the creation of the new state, provides a different perspective on population that places the distribution problem at the center of concern. The objective of Kay's policy was to improve, as rapidly as possible, the social and economic status of the African population while conserving the natural resources of the country. Fearing that the dual agricultural economy was so deeply entrenched that significant change could only occur in the relatively long term, Kay proposed the restructuring of the urban industrial economy so that it could absorb much of the under-employed population from the communal areas.

Kay suggested that the economy should be run on labor-intensive lines, with some admixture of intermediate technology into the existing capital intensive structure, and that an urban informal sector should be encouraged (ibid., 106-111). He argued that "the circulatory system of labor migration, for long obsolete, should be terminated as expeditiously as possible" (ibid., 108), together with the policy of exploiting large pools of cheap labor from neighboring countries. Kay advocated that the legislation preventing urban workers from bringing their families with them to reside in the cities be repealed, that employers be required to introduce pay scales that would make this possible, and, in fact, that urban workers be required to bring their families with them.[10] Finally, while Kay also saw the necessity for a strong family planning program, he believed that African high fertility norms and the

---

9. For exemplary statements about the role of major donors of population assistance see, Salas 1979, 229-76; USAID 1982, 1983, 1986; Wolfson 1983; Finkle 1973; Warwick 1982, 44-67; Symonds and Carder 1973; Piotrow 1973. It is thought that, if China and India are excluded, governments of all other developing countries and foreign donors each contribute approximately 50 percent of all spending on population policies and programs, with the government share tending to rise the longer a program is in existence. (World Bank 1984, 148; Ness, Johnson and Bernstein 1984). Through its bilateral and multilateral programs, the US is by far the largest donor, providing approximately 50 percent of all foreign population assistance. Japan, the second donor, provides about 10 percent. For the most recent data see UNFPA 1989.

10. All legal restrictions on movement and residence were repealed soon after Independence. However, many urban workers maintain a foothold in the communal areas either to augment their incomes or as insurance against unemployment.

small-holder's need for farm labor would make such a program more effective in the urban areas. In both parts of population policy, growth and distribution, Kay regarded the urban areas and the modern sector as the areas most susceptible to change.

Land Reform and Resettlement

Political imperatives at the time of Independence dictated a very different response to the land question than that envisioned by Kay. For most of the Zimbabwean peasantry, the return of the land to Africans was probably the most important objective of the independence struggle (Ranger 1985; Report of the Commission of Inquiry into the Agricultural Industry, cited in Munslow 1985, 43). Return of the lost lands and assurance that the people themselves would be able to decide how to manage their lands were, in Ranger's view, the promises that carried the Shona peasant areas overwhelmingly for ZANU-PF,[11] under the leadership of Robert Mugabe, in the first post-independence elections in 1980 (Ranger 1985, 287-88).

In the Transitional National Development Plan, published in 1982, the New ZANU-PF government announced an ambitious plan to resettle 162,000 families, approximately 800,000 individuals, on previously white-owned farm land during the three years of the Plan period. Three basic types of resettlement schemes were initially introduced.[12] On Model A schemes, settlers were allocated approximately five hectares of arable land for individual farming, with communally-managed grazing for five to fifteen stocking units according to the quality of the land.[13] Model B schemes were farmed collectively, using more capital-intensive methods. Groups of individuals formed themselves into collectives and were registered as cooperatives with the Ministry of Cooperative Development. Some of these cooperative farms were initially very large, occupying up to two thousand hectares, often in the better agricultural-ecological zones (Weiner et al. 1985, 258), and with as many as two hundred members. More recently the size of Model B farms has been significantly reduced. Model C schemes are state farms run by a parastatal organization, the Agricultural and Rural Development Authority (ARDA). They tend to be large, highly mechanized estates incorporated within a Model A settlement pattern (Weiner et al. 1985, 259-260; Munslow 1985, 47).

To assist farmers in some communal areas in Matabeleland, where no grass remained, an experimental Model D was later designed. In Model D schemes,

---

11. Zimbabwe African National Union-Patriotic Front is the party which ultimately forced the Smith regime to the negotiating table after mounting a successful guerilla insurgency.

12. Unless otherwise noted, information in this section was obtained in interviews with senior officials in the relevant ministries.

13. A stocking unit is a measure based on the average weight of the animal such that one unit equals fifty kilograms. Cattle, therefore, count as two units while smaller stock may be counted as one, or less than one, unit.

ranch land was purchased adjacent to the communal areas and was carefully managed to ensure a constant supply of feed and grass. Farmers were then permitted to rotate their stock through the Model D farms for specified periods in order to allow their own lands to recover. Game as well as cattle ranching are encouraged on Model D schemes both to give a more immediate return and to provide a more diversified range of economic activities as a hedge against recurrent drought (Munslow 1985, 47).

Establishing the resettlement program has proven to be more difficult than was foreseen. Numerous problems have arisen, some of them correctable and others more fundamental in nature. Among the less serious problems, it was found that the original blanket plan for Model A farms was inappropriate and required modification. Under the blanket scheme, cropping was intended to be the main source of income with grazing providing only supplemental income. Experience showed that the model needed to be more flexible to take account of climatic conditions and soil type. Thus the plan needed considerable refinement. Moreover, since under the Lancaster House agreement[14] land could only be acquired on a willing seller-willing buyer basis, white farmers in the more marginal areas were the most likely to offer their land for sale. Thus very little high quality land has been made available.

The level of interest in farming among the general population was also misjudged. The criteria for the selection of settlers--landless, poor, unemployed and, in the earliest stages, refugee or ex-combatant status--placed many on the land who showed little interest in or aptitude for farming. Wholesale felling of timber, overgrazing and general mismanagement soon reduced some farms to ruin and a shortage of funds has limited the amount of government assistance that has been available for equipment, training, and extension work. Gradually, some certified Master Farmers were resettled with more positive results but a recent proposal to give priority to Master Farmers has met with opposition from the more ideologically minded who see this as subverting the original intent of the program.

Undercapitalization and lack of management experience have been especially severe problems on Model B farms, where members were expected to provide their own capital with only limited government assistance, and to manage their own affairs. Many such farms have fallen into serious debt and declining membership (See G0Z/UK 1988, 1989). By 1989, less than fifty percent of the original target for Model B schemes had been achieved (Department of Rural Development 1989). Similarly, at least one official

---

14. The Lancaster House agreement was the outcome of the negotiated settlement between the Patriotic Front and other Rhodesian-Zimbabwean leaders, both Black and White, and the British government in December 1979. The agreement temporarily restored British colonial authority until elections could be held; formulated a constitution for the new country, introduced a "bill of rights" to preserve democratic procedures; and protected the political, economic, and property rights of Whites for a period of ten years. Under the agreement, twenty out of one hundred parliamentary seats were set aside for Whites until 1987.

interviewed by the author suggested that lack of management skills is partly to blame for the failure of Model C schemes to attain their target. In this case, it seems, overcapitalization contributed to the problems; the mode of farming was simply too sophisticated for the level of management skill available. While Model A schemes have now achieved the original target level, and the resettlement project as a whole can boast some success stories, the fact remains that progress has been slower and more beset by difficulties than was anticipated.

Many of these problems are clearly teething problems which can be, and are being, corrected. There are others, however, that are more deeply rooted and suggest that, in the absence of radical change in the orientation of the program, progress is likely to be as slow in the future as in the past. It is clear that the program is much more expensive than was anticipated and has been seriously underfunded. Part of the reason for this is that the government has had competing priorities--especially education, health, agriculture--and managed to appropriate relatively small sums to the program. In addition, foreign assistance has been minimal and most of it has been allocated to Model C--the state farm scheme. With the exception of a matching grant from Britain of 50 million pounds, very little international funding has been made available for Models A and B.

There is also a more fundamental question of whether resettlement is exactly what the peasantry had in mind when they fought for the return of their land. Ranger (1985, 284-338, esp. 337) strongly suggests that in Makoni District, admittedly an overcrowded and radicalized area in the east of the country, what the residents were expecting was the incorporation of some white-owned land into the communal areas. Additionally, Michael Bratton (1987, 187-193), an experienced political observer of Southern Africa, has argued that in underfunding the resettlement program while appropriating much larger sums to agricultural policy, the government was not only consulting its economic interests, but was also correctly reading the political signals coming from the countryside. Bratton argues that in a survey carried out in four districts in 1983, only one smallholder in five cited land as the most important production constraint.[15] Other inputs like draught power, implements, cash, and labor were cited as more important in several areas. Since these messages supported the government's perception of the need for caution in transforming the pattern of agricultural production, Bratton believes it felt confident in reducing the appropriations to resettlement.

Whatever may be the validity of these analyses, numerous commentators have remarked on the reluctance of some peasants to accept resettlement on the grounds that government restrictions on land use and farming methods in the resettlement areas curtail individual freedom and are reminiscent of colonial

15. But with a range of 0 to 31 percent.

practices. A number of senior officials have commented to this author that the resettlement program has not been successful in weaning settlers away from the communal areas, partly for cultural reasons--people do not want to move far from the graves of their ancestors--and partly because they do not get clear title to the land on which they are resettled. While the new farm may be inherited by a widow for her lifetime, children of the union and second wives in polygamous unions are forced to find other employment. For these reasons, many settlers make a point of keeping some stock on the communal lands or otherwise maintaining their claim to land in these areas.

This unwillingness to cut ties with the communal areas undoubtedly undermines the resettlement program from a population policy perspective. From the author's discussions with officials in seven or eight ministries and departments closely involved with resettlement, it seems likely that the program was never seriously considered from a population perspective. This would not be surprising because, while the Zimbabwean ministries concerned with agriculture, natural resources, water supplies, and related issues are strong, there is a dearth of highly trained social demographers capable of advising on the demographic implications of the resettlement plan.[16] Nevertheless, it seems evident that the resettlement program, if not accompanied by a shift of labor out of agriculture and into modern sector employment will take many years to reduce overcrowding in the communal areas.

Non-Agricultural Employment

At Independence, Zimbabwe inherited a stronger and more diversified economic structure than almost any other country in Sub-Saharan Africa, South Africa excepted. Under the settler government, Rhodesia had moved a considerable distance beyond the typical colonial economy heavily dependent on agriculture and the export of one or two basic commodities. (Stoneman and Cliffe 1988, 135-37) Diversification had been greatly stimulated by the import substitution strategy introduced by the Smith regime in response to sanctions imposed by the international community during UDI (ibid., 141-42). By 1980, manufacturing accounted for 24 percent of gross national product (GNP), three times the African average, while the value of the three major exports amounted to only 22 percent of the total compared with an average of 79 percent elsewhere on the continent (World Bank 1981, 156). While most of the benefits of economic activities accrued to the small White community, recognition of its potential to benefit the nation as a whole clearly underlay

16. When I asked my informants if their agencies had anything to contribute to a population policy, not one official replied in the affirmative. I interpret this to mean that population policy is generally perceived to mean fertility reduction.

Kay's proposal to base his population redistribution policy on the development of urban employment rather than on rural resettlement.

Although Zimbabwe attained Independence with a stronger and more diversified economy than most other countries in Sub-Saharan Africa, since that time economic growth has consistently fallen well below the levels anticipated in the various national economic plans.[17] An immediate cause of this shortfall has been the job creation program which has also failed to meet its quite modest targets--a projected 3 percent per annum growth in the Transitional Plan covering 1982 to 1983 and 1984 to 1985 (Govt. of Zimbabwe 1982, 2:31). As part of this plan, and presumably to control the numbers of young people flooding into the urban areas, the government proposed to establish small-scale industries at "growth points" in the rural areas. Like the employment plan itself, very few of the growth points have taken off; while many have been provided with schools and clinics, and have attracted a number of small traders, few industries have been started. It is reported that one reason for the reluctance of enterprises to establish themselves on these sites is the difficulty of getting clear title to the land.

Like other countries with a high birth rate, Zimbabwe also has the burden of a young age-structure. The United Nations estimates that approximately 48 percent of the population are under the age of fifteen (United Nations 1988, 181). Combined with slow economic growth, this means that Zimbabwe now finds itself with an unemployment crisis of some magnitude. *The Statistical Yearbook* for 1987, cited by Herbst (1989, 70), indicated that by 1990 Zimbabwe would experience an increase in the number of unemployed workers equal to 62 percent of the current labor force of 1.45 million workers. Moreover, much of the small increase in employment since 1980 has been in the public and service sectors; indeed, there has been a decline in the number engaged in industry and agriculture (Stoneman and Cliffe 1988, 126).

The situation is made more poignant by the success of the education program instituted soon after Independence. By the mid-1980s, primary enrollment was virtually complete and, in 1986, 55 percent of boys and 37 percent of girls in the appropriate age groups were enrolled in secondary schools (World Bank 1989, 275). Each year, therefore, growing numbers of increasingly well educated young people enter the job market with little prospect that they will find suitable employment. These numbers are expected to peak at about 250,000 a year in the early 1990s (Stoneman and Cliffe 1988, 127). *The Annual Economic Review of Zimbabwe 1986* estimated that only 10 percent of secondary school leavers would find employment in 1990, down

17. For example, in 1986 the economy expanded by only 0.2 percent. (Minister of Finance, Budget Speech 1987. cited by Herbst 1989, 70). Much of the blame for this poor performance lies in conditions outside of Zimbabwe's control; in the early 1980s, the global recession, severe drought, poor export prices, and a reduction in foreign investment meant borrowing, debt, the IMF and a currency devaluation. More recently, economic growth has picked up to around four percent per annum (Economist 7 April 1990, p. 48).

from 29 percent in 1985 (Herbst 1989, 70). It is probable that many of these young people would consider agricultural employment only as a last resort. In sum, it appears that a more active employment program would have to be an integral part of any population policy intended to reduce overcrowding in the communal areas.

Population, Land and Policy in the 1990s

In April 1990, the Lancaster House Agreement came to its scheduled end opening the way for a new constitution and terminating the constraints on the acquisition of land from white farmers. Two issues dominated the National People's Congress (party convention) in November 1989, and the campaign leading up to the elections in March 1990: the creation of a one-party state and land distribution (New York Times 1989; Arnold 1990; Financial Gazette 1990; Meldrum 1990; Economist 1990). ZANU-PF's election manifesto speaks of resettling "thousands" of families. (Economist 1990, 49). Indeed, redistribution of the land as part of a wider program of socialist redistribution has long been part of the populist ideology promulgated by ZANU, especially at election times (Libby 1984, 146-53; Sylvester 1986). This year, while ZANU received seventy-eight percent of the votes cast, just over 50 percent of those eligible actually voted, down from over ninety percent in 1980 and 1985 (Economist 1990, 49). The most contentious issue was undoubtedly the one-party state.

Does the removal of the Lancaster House constraints herald a marked change in the program of land acquisition and resettlement? Though it may be unwise to second-guess the Zimbabwe government, the answer is, probably not. There are a number of reason for this assessment. First Mugabe is well aware of the importance of agriculture to Zimbabwe's economy. Although it now constitutes only 13 percent of GNP, agriculture provides one-third of all formal wage employment, supplies 40 percent of the inputs into manufacturing, and earns 40 percent of foreign exchange. In addition, 70 percent of the population is directly dependent on agriculture (Weiner et al. 1985, 251-52). Secondly, during the past ten years, Mugabe has stuck scrupulously to the letter of the Lancaster House agreement, ignoring a clause that permitted the government compulsorily to acquire underutilized land for government use if compensation was paid fully in foreign exchange funds. (Stoneman and Cliffe 1988, 132).

Thirdly, whatever may be Mugabe's personal commitment to socialism, there is no doubt that his government has been guided by pragmatism rather than ideology. The most significant evidence for this is the appointment to key positions in the cabinet and government of Western trained professionals, white as well as black. Some of these men had spent the war years in exile in the West, and others had occupied senior positions in the former government or represented farming or business interests under that regime (Libby 1984, 145).[18] As Libby comments, "This faction takes the position that high output and high export earnings from modern sector industries are the overriding economic policy objectives for the country" (ibid., 146). Indeed, political observers see the split between the technocratic government and the ideological party as the most significant political cleavage in the country (Libby 1984; Herbst 1989).[19]

Finally, this economic pragmatism in government has mandated the conciliatory attitude of the regime towards the white community, both businessmen and farmers. It is widely rumored, also, that at Independence Mugabe was advised by his Mozambiqan colleagues not to repeat their mistake by precipitating a mass exodus of the whites, whose skills and experience were essential for the economic security of the country, during a transition period. There was also the fact that as the strongest member of the front line states, Zimbabwe could not afford to let its economy decline. Certainly, Mugabe has made strenuous efforts to retain the white population and to ensure their active participation in creating the new state. With the employment crisis in mind, this has most recently involved discussions with business leaders about possible measures to increase foreign investment.

If a radical redistribution program is an unlikely scenario, the most likely alternative is the compulsory purchase by government of unused white land. Whereas it has been commonly thought that most white-owned land was farmed to its maximum capacity, giving due regard to sound conservation farming methods, recent work suggests that only a small proportion, maybe as little as 3 percent, of such land has ever been farmed (Weiner et al. 1985). This proportion rises to 20 to 40 percent when based only on arable land. Weiner et al. suggest (269) "that roughly two-thirds to one-half of Zimbabwe's prime agricultural land is being neither cropped nor fallowed." It would be possible, therefore, for the government to acquire a large amount of land, of much better quality than most of what has been offered in the past ten years, without seriously affecting the productivity of white farmers.[20]

The enactment of such a policy would not, of course, eliminate the financial, managerial and technical difficulties that have been experienced by the resettlement scheme to date. To move the program forward more rapidly than in the past would likely call for much larger infusions of foreign

18. Notable among these are the highly-respected and influential Minister of Finance, Economic Planning and Development, Dr. Bernard Chidzero, who has retained his position since 1983, and also kept some of his staff.

19. Since this was written, President Mugabe has announced that he has abandoned his desire for Zimbabwe to become a one-party state (Africa Research Bulletin, Political Series 28(1), 1-31 Jan 1991, p. 9722. Further, the ruling party, ZANU-PF, decided in June 1991 to abandon its adherence to Marxism, Leninism, and Scientific Socialism (Africa Research Bulletin 1991).

20. On December 12, 1990, the House of Assembly approved an amendment to the Constitution to allow the state greater power to nationalize land ownership. The immediate policy is to purchase approximately 5 million hectares (approximately 50%) of white-owned land and to resettle 110,0000 peasant farmers (Keesings Record of World Events 36(12), 1990, p. 37,909).

assistance than has been forthcoming since Independence. The announcement of such a policy might, however, satisfy the populist demands of the Party and its rural constituency while not unduly frightening the white population. Nevertheless, resettlement is likely to remain a slow and difficult process and a parallel program of urban and rural non-agricultural job creation seems an indispensable addition if overcrowding on the communal lands is to be relieved.

Conclusion

In Zimbabwe, as in many other African countries, erosion is a serious economic problem brought about by the interaction of physical, climatic and behavioral factors. Zimbabwe differs from other countries, however, in that the aetiology of the problem is complicated by severe maldistribution of the population brought about by policies dating from the turn of the century. In addition to the economic question, therefore, there is also a problem of human justice and equity.

The solution of this problem calls for actions at many levels from land use planning and conservation farming to the reduction of the rate of population growth and dispersal of the concentrations of people and animals. As this paper has tried to show, orthodox population policy, focused on fertility and intended to reduce the rate of population growth, is unlikely to reduce population pressure in the communal areas except in the very long term. For the past quarter century, however, population policy has largely been defined as fertility policy intended to regulate the rate of growth of the population. The primary motivating force for the adoption of such policies has been the international population assistance establishment which has provided intellectual leadership and, especially in the early stages in any country, significant funding. In most places and for most of the time, this has been a realistic approach to the serious problem of rapid population growth in developing countries. It is not sufficient, however, to deal adequately with Zimbabwe's combined growth and distribution problems.

In its resettlement program, Zimbabwe already has what might be thought of as a distribution policy. As we have seen, economic and political considerations suggest that this program is unlikely to bring about a reduction of overpopulation in the communal areas except in the very long term. Lacking a demographic objective, moreover, the program does not take account of population growth which will continue at a rapid rate in the communal areas. It is already evident that the resettlement program needs to be coordinated with an upgraded job creation program if only to provide options for the children of current settlers other than to return to the communal areas.

This analysis suggests that Zimbabwe might find it productive to broaden its current conception of population policy so as to include both fertility-reducing and distribution-influencing measures in its population policy. This will be a difficult policy to design and coordinate. Nevertheless, Zimbabwe's natural wealth, strong economic base and, most of all, its active and increasingly well educated population, give assurance that a way can be found to correct the present imbalances.

Chapter VIII

Indonesia: Stresses and Reactions

Kartomo Wirosuhardjo

Introduction

Indonesia, like the rest of the developing world, currently faces a series of major problems arising from the interactions of population and environment. For the most part, these problems arise as a result of the proliferation of the human species. The current unprecedented rapid population growth represents both a triumph over high death rates, and increasing stress on the environment. Urbanization and industrialization also represent success in increasing human productivity, but at the same time they load the environment with wastes and toxins that threaten life. Agricultural development has increased food output and yields, but at the price of chemical inputs that damage the environment. Both population growth and the expansion of world-wide trade encroach upon our forests and threaten the great diversity of plant and animal species they support, and cutting out the trees that act as the lungs of the globe.

While all these positive developments wrought havoc with the environment, the world has also seen some attempts to alleviate these stresses. At the global, national and local levels, specific actions are being taken to reduce human fertility and to soften or reverse the resulting stresses on the environment. Indonesia has participated in this process with a number of new programs to address environmental stress.

The world as a whole has reacted rather quickly to the detrimental effects that come with progress. Non-governmental Organizations (NGOs), like the International Planned Parenthood Federation (IPPF) and the Ford and Rockefeller Foundations, began to express concern about rapid population growth in the early 1950s. In 1966, the United Nations formally recognized the population problem and established the United Nations Population Fund (UNFPA). On the environmental side, NGOs collaborated with governments to warn of the damage to the environment that comes from economic progress using vehicles such as the Club of Rome Report and the Stockholm

Conference. Today that concern is institutionalized in the United Nations Environment Program, and in a great variety of global, regional and national programs concerned with environmental protection.

Southeast Asia is a region that clearly reflects some of the more important effects of this progress with its attendant stresses, and reactions. The region is one of the world's fastest growing economic areas, with annual economic growth rates of 5 to 10 percent over the past three or four decades. Singapore has been transformed from a crowded, dirty city to a sparkling city-state that is one of the world's newly industrialized countries (NICs). Malaysia and Thailand are moving forward at the perimeter of this development. Although poorer, Indonesia, by far the largest country in the region, is not far behind. All four countries have experienced rapid population growth and economic development, and the powerful environmental stresses which accompany that development.

All have also reacted in various ways to these stresses. National family planning programs strive to reduce human fertility and many different programs have been created to promote environmental protection. It remains to be seen how successful these reactions will be in protecting the environment for future generations. Success depends in part on the character of the problem and the environment that is being affected. It also depends on what policies and programs a government can mount to address the problems. Finally, it depends on the actions and behavior of local communities.

In Indonesia, many aspects of the process of development, including population and environmental issues, may shed some light on those the world currently faces. Indonesia is almost two million square kilometers, stretching over 5700 kilometers along the equator. It is home to more than 170 million people. The Indonesian government's ambitious agenda calls for the adoption of mechanisms to balance the demands of economic development with environmental protection. Population is viewed as both a resource for and an impediment to this agenda.

Population

Like the rest of Southeast Asia, Indonesia's population grew very slowly until the 19th century. In 1800 its population is estimated to have been between fourteen and eighteen million. It had grown at about 0.2 percent per year for the past few centuries (Reid 1988). In the 19th century the growth rate probably rose to about one percent per year, increasing the population by forty million people in 1900. In the next half century the population doubled, to eighty million, with an overall average annual growth rate of 1.4 percent. Since then the growth rate rose steadily to a high of 2.41 in 1970-75, and subsequently began to decline. The United Nations estimate of the population for 1990 was 183 million. Although the growth rate is declining, the total population is projected to continue to grow, to an estimated 208 million by the year 2000.[1]

The total fertility rate (TFR) has shown a slightly different trajectory, perhaps reflecting some of the environmental conditions that affect population growth. The UN estimate shows the TFR rising to a high of 5.67 in 1955-60, then falling to 5.42 during the next five years, and rising again to 5.57 in 1965-70. From that point it declined steadily, projected to reach 2.1 or replacement level after 2005. The decline in the early 1960s probably reflects the growing economic pressure of the time and the great upheavals that came at the end of that period. With stability and economic growth following 1966, total fertility rose again.

The decline of fertility after 1970 is a result of what could be called the double impact of stress and reaction. Population growth and economic and social development produced considerable stress on families. This led to individual actions to reduce fertility. Urbanization and the growth of new employment opportunities, plus the expansion of education with the resulting increase in age at marriage has changed reproductive behavior. This can be considered a reaction to stress at the individual level.

As important, however, was the national level reaction to stress. The change of government in the mid 1960s dramatically altered the definition of population growth. Whereas the Sukarno government considered population growth a positive condition, the New Order government of President Suharto considered it a major obstacle to social and economic development. The new government also placed strong emphasis on economic development, thereby increasing the sense of urgency in reducing the rapid rate of population growth.

Indonesia initiated a strong family planning program in the early 1970's. Private associations had begun to provide contraceptive services in the mid 1950s, despite the government's then decidedly pro-natalist orientation. After 1966, these organizations began to receive active support from the government. In 1968, an Ad Hoc Committee was formed to advise government officials on population policy. Support from international agencies, such as the UNFPA, World Health Organization (WHO), the World Bank and NGO's like the Ford Foundation and the International Planned Parenthood Federation (IPPF), enabled Indonesia to take advantage of world community resources to address the population problem. First, it undertook a basic survey of population in 1968 with the support of the Ford Foundation. This showed both the dimensions of the problem, and the readiness for a national program to reduce fertility. In 1969, Indonesia invited a mission from UNFPA, WHO and The World Bank to advise it on population policy. That

1. Other projections range from a low of 196 million to a high of 239 million (Hugo, Hull, and Jones 1990, p.329).

mission urged the government to establish a strong family planning program, to be focused first on Java and Bali and subsequently to extend the program to the outer islands.

As a result a National Family Planning Coordinating Board (BKKBN) was set up in 1969. It was headed by a prominent gynecologist, Dr. Suwardjono, who eventually became Minister of Health. Under his leadership, world experts were invited to become consultants to BKKBN in fertility-related matters. Subsequently a new graduate from the University of Chicago, Dr. Haryono Suyono, took charge of BKKBN and continued the effective program to the present.

A two-pronged strategy was adopted in the early 1970s. First, reliance was to be placed heavily on information, education and communication (IEC). All aspects of the program were supported by research in the universities and the Central Bureau of Statistics. The object was to change reproductive norms. The second prong involved service delivery. Indonesia established a *village-based family planning program,* in the mid 1970s, which provided a family planning post in every village. The post was to be occupied by a prominent person in the village, and was to provide oral contraceptive pills and condoms and referral to clinics for other contraceptives, as well as information about family planning. The program emphasized both service delivery and normative change, promoting the idea of the "Small, Happy and Prosperous Family." The emphasis was thus on the welfare of the family through family planning.

The most recent study, *Indonesian Family Planning Prevalence Survey 1987*, estimates that more than 60 percent of eligible couples are now practicing family planning, 95 percent of couples had heard about family planning, 65 percent had ever practiced family planning and 46 percent were currently practicing. Indonesian demographers estimate that Indonesia will reach replacement level by the year 2006, ten years earlier than the World Bank projection. There are, to be sure, many problems for the future, but the success in the recent past provides some confidence that the problems will be effectively met.

The Indonesian Family Planning Program has been widely recognized as one of the world's most successful. There are several reasons for this success. Perhaps most important is the strong commitment from the President of the Republic of Indonesia. In his speeches, his organization of government and his selection of program leaders, he has provided strong and consistent support for family planning for the past twenty years and more.[2] Indonesia has the advantage, however, of having one national language, (Indonesian), which is based on a Malay dialect, a minority language. The Malay language has been

---

2. For this he has won numerous international awards. The latest was the prestigious United Nations Population Award, which he received in 1989.

a lingua franca for centuries. In a country with four hundred language dialects, signifying sub-cultural variations, it could be very difficult to introduce family planning. This makes communication easier, and the message of family planning can reach families even in the remotest area through a network of communications. Communications could also present a problem in a country as culturally and religiously diverse as Indonesia. While the majority of Indonesians, some 87 percent, are Moslems, almost all belonging to the Sunni branch, the remaining are Protestants, Catholics, Hindus and Buddhists. All leaders of the different religious groups have been invited to participate in the program. Thus, in addition to the officials of BKKBN, the religious leaders have been instrumental in getting the ideas across. It is not uncommon, for example, to find the village family planning post in the home of the local Imam (priest).

Finally, the program has been highly decentralized administratively. The National Coordinating Board has counterparts at the Province and District levels, with local committees and clubs mobilized at subdistrict and village levels. At each level there is considerable discretion in how to carry out the program, and there is also much local initiative in determining how the goals of normative and behavior change will be advanced in the context of local sensitivities and customs. Provincial governors, for example, are important in determining how much government support and pressure are given to family planning at the local level. Provinces differ greatly in economic conditions and in culture. This, decentralizatized administration provides the capacity of the national program to adapt to local conditions.

Despite the overall success and the support of the central government, there are important local variations that affect the course of fertility limitation. These variations derive from a mixture of political and administrative conditions, economic and logistical conditions, and local cultural conditions. East Java and Bali were the first provinces to register substantial success in family planning. Both have highly centralized political cultures, with strong capacities for organizing the villages. These political and administrative conditions apparently had an impact on fertility decline. West Java, on the other hand, has a flatter and more egalitarian political culture. It also probably lacked the strong political commitment to family planning that East Java had. Finally, it has been found that Sudanese tend to have higher fertility than Javanese, showing some deep cultural influences on fertility (Suparlan and Sigit 1980). Thus between East and West Java, we have a mixture of political, administrative and cultural differences that appear to have had an impact on the speed of fertility decline. In Sulawesi there are also striking differences with North and South Sulawesi experiencing more rapid fertility decline than Southeast Sulawesi. Again, some mix of political and cultural conditions appear to be at work, though we have no detailed examination of the elements of this mixture (ESCAP 1987).

A brief comparison with other Southeast Asian countries can help to identify important sources of the recent adaptation to the stress of rapid population growth. Structurally, the Indonesian program is almost identical to that in the Philippines. Both began at roughly the same time (1970) and both are directed by a central *coordinating board*, whose task it is to stimulate existing agencies to provide information and services for fertility limitation. The Philippines, however, has lacked the strong support of the president. Like the rest of the Philippines government, it has been highly centralized administratively. And, of course, the Philippines religious leadership opposes family planning and the use of modern contraceptives. The result has been a slower rate of fertility decline in the Philippines. This suggests that strong central political support, a decentralized administrative system, and the lack of strong religious opposition are important conditions for a more rapid adjustment to the stresses of rapid population growth.

Thailand, Malaysia and Singapore, on the other hand, have had a more rapid fertility decline than Indonesia. The Thai program began at the same time, in 1970, and was located in the Ministry of Health, under a specialized section for family planning. The program is well organized, with good administration and logistics, just as the Indonesian program. It is likely that the more rapid rate of fertility decline is due to both cultural and physical environmental factors. Buddhism is more tolerant of different individual behavior and thus poses no serious obstacle to family planning. Further, the country is more compact geographically, with a good transportation system. It has had strong economic and social development programs over the past forty years. The rapid fertility decline of Malaysia and Singapore can be attributed to high levels of economic development, urbanization and school attendance, and the penetration of the market economy. It is interesting to note that the Malaysian government program has slowed considerably, and now plays little role in reducing fertility. Central political conditions, and local social, cultural and economic conditions all play a role in determining both national and local effectiveness in responding to rapid population growth.

Finally, we must note that all of these countries have received generous amounts of international financial and technical assistance. This has come from the United Nations, bilateral agencies and non-governmental organizations. It is not too much of an exaggeration to say that these countries had as much foreign assistance as they wanted and could use. This assistance has been very important, especially for Thailand and Indonesia. It represents the quick reaction of the world community to the stress of rapid population growth. It is obvious, however, that if foreign assistance has been a necessary condition, it has not been a sufficient condition for fertility reduction, as is evidenced in the Philippines.[3] Something more than foreign assistance is necessary to effectively address the problem of rapid population growth. Foreign assistance can help to reduce fertility if, and only if, the receiving government and society are willing and able to act.

Agriculture and Forestry

Agriculture

Indonesia has long been a highly productive agricultural land. The rich volcanic soils of Java and Bali, and the well-watered valleys of so much of the country have provided an environment suited to wet rice in particular, but to cassava, corn and a wide variety of other crops as well.

Over the past two centuries food output has been driven largely by the demands that came from population growth. Until the decade of the 1960s, however, this increased output was attained primarily from an expansion of land under cultivation. Without a major change in technology, yields remained low, and at much the same level as they had been for centuries. As long as population growth was moderate, 0.5 to 1 percent per year, this level of increase was adequate. The rapid population growth during the post-1950 period, however, has greatly increased the demand for food, and especially for an increase in rice production. Between 1950 and 1965, rice output grew at an average annual rate of about 3.5 percent, just over one percentage point above the rate of population growth. Combined with increased per capita food consumption, this generated a need for substantial rice imports. Through the 1950s, Indonesia increased its rice imports from about three hundred thousand to almost one million tons. In the early 1960s, the country was importing about one million metric tons per year, making it the world's largest single importer, buying about one seventh of all the rice that entered international trade. Imports dipped in the late 1960s and early 1970s, then rose steadily to a peak in 1980 of two million tons or almost 7 percent of consumption. Imports fell drastically for two years then rose again in 1983 to over one million tons, or about 3 percent of total production.[4] The impact of rapid population growth was reflected directly in the fluctuation in rice production.

Since 1965, however, annual increases in rice output have been more than 2 percentage points *above rates of population growth.* From 1950 to 1965 wet rice output increased from nine to fifteen million tons, for an average annual growth rate of 3.6 percent. Furthermore, the increases before 1965 came

3. If one included China in this analysis, it is clear that foreign assistance is not always a necessary condition for fertility decline, since China achieved its most striking fertility reductions in the 1970s, before any foreign assistance was available to it.

4. Imports declined from their peak of 2 million tons in 1980, except in 1983, when they jumped from 300,000 to 1,168,800 tons. In 1984 imports were back down again to 414,000 tons and have continued to decline to inconsequential amounts since then.

largely from land extenuation, with yields growing only from 1.5 to 2 metric tons per hectare. Since 1965, yields have risen to nearly four metric tons per hectare, for an average annual increase of over 3 percent. Roughly the same has happened for maize, which rose to over five million tons, averaging 5 percent growth in the past two decades. Cassava, peanuts, soybeans and even dry rice have also shown impressive growth rates over the past two decades (IRRI 1985 and Government of Indonesia 1986).

This reaction to the stress of population growth has produced some impetus for the development of a new technology for rice and other food production. The change in rice production is the most dramatic. It began when the International Rice Research Institute (IRRI) began working on new high-yielding varieties of rice in 1960. These were adopted by the Indonesian government, especially after the new order government of President Suharto. There was a great deal of pragmatic experimentation in this adoption of the new rice technology. Foreign assistance was as heavily involved as in population, and again it came from the same broad range of donors and channels. Various campaigns were launched, some through large scale private corporations, others through local mobilization of the farmers. The ministry of agriculture now maintains a vast extension service through provincial and local governments. It subsidizes the distribution of fertilizer and pesticides and assists farmers in production and marketing with a complex set of government and private market arrangements. There was also experimentation in the expansion of irrigation for wet rice. Both large scale and many small scale irrigation projects were promoted and financed by government, many with foreign assistance. It appears that the small scale irrigation systems have been the most effective. They are cheaper to build, can draw on local labor and can be closely adapted to local hydrological and social and economic conditions.

Although this rise to self sufficiency in rice has been very welcome, it has also led to other negative impacts on the environment. Pesticide use is perhaps the most well known. By the early 1980s it was becoming evident that pesticide costs were rising too rapidly, and also that indiscriminate spraying was killing many natural predators of common pests and in fact led to epidemics of such insects as the brown plant hopper. The government responded in the late 1980s by banning many pesticides, and attempting to promote a new *Integrated Pest Management* system that was being developed both by Indonesian and IRRI scientists. Little is known, but much is feared from the health hazards of both pesticide and fertilizer use. These are new stresses that government is becoming aware of and attempting to address through scientific study and the development of more environmentally safe techniques.

There is considerable local variation in rice and agricultural production, just as there has been in fertility decline. It appears, however, that the causes of this variance lie more in ecological than in political conditions. Water is a critical ingredient, as are soils and topography. There has been little local cultural resistance to rice improvement in contrast to that in fertility decline.

Forests

Indonesia currently has about 144 million hectares of forest land. The forests are highly concentrated in only three provinces. Kalimantan (44 million hectares), Irian Jaya (40), and Sumatra (30) have 114 or about 80 percent of the total forests. Adding the 13 million hectares of Sulawesi brings the total to almost 90 percent in four Indonesian Islands. Java and Bali have very little forests, since their rich volcanic soils are so well suited to wet rice production.

Of the 144 million hectares, almost fifty million are designated protected areas, parks and forest reserves. Another sixty-four million hectares are designated as limited (thirty) or definitive (thirty-four) production areas. Finally thirty million hectares are designated for conversion to other uses, such as plantations, agricultural land or urban areas. The production areas have provided a rich source of timber, much of which provides foreign exchange through exports. For the past twenty years, Indonesia has harvested about fifteen million cubic meters of logs per year. This has produced twenty to thirty thousand tons of timber for export, earning up to US$ 20 million per year over the past decade. Much forest is also being cleared for additional agricultural land. This affects the outer islands, such as Kalimantan and Sumatra, where great forests and empty lands attract immigrants, both informally and under the government's transmigration program. Finally, many Indonesians still rely on fuelwood for cooking and heating. It is estimated that fully one half of the country's energy use is provided by wood (Worldwatch 1988). This reflects the overall poverty as well as the availability of wood for fuel.

The extent of forest exploitation has its own negative effects, however. Erosion, flooding, and species destruction are direct results of the cutting. To a certain extent stress is being addressed through an attempt to maintain some protected areas, and additional programs of reforestation. Beginning in the early 1970s about 100,000 hectares was being reforested yearly. That increased to almost one million hectares annually in 1980, and has since dropped to 300,000 hectares in the late 1980s. It is obvious that the central government must create more effective institutional arrangements to protect forests, and to continue sustained forest harvesting. This is one of the most difficult problems in the population-environment dynamic facing the Indonesian government.

Urbanization and Employment

Like most of the developing world, Indonesia is experiencing rapid urbanization. At Independence, almost half a century ago, about 10 percent of the population lived in urban areas, and most of these were small towns or large villages. The capital, Jakarta, was a city of 1.6 million in 1950, ranking forty-second among the world's large cities. Today about 27 percent of the population lives in urban areas. Jakarta is a sprawling metropolis of about twelve million, now ranking fifteenth in size (Hugo, Hull, Hull and Jones 1990). By the year 2000 Jakarta is expected to grow to seventeen million and to rank eleventh in size among the world's cities.

Urbanization proceeds in two ways. First there is the migration of people into cities and the population growth in specific cities. By this measure, Indonesia will probably have about 30 percent of its population in urban areas at the turn of the century. Second, cities emerge and grow as networks, with nodal points coming closer and closer to one another. This means that even people living in villages or suburban areas have access to urban services, become part of the urban life style, and also experience the same kind of pollution and environmental stress felt by people in the central cities. For example, although Jakarta is now about twelve million, there are an additional one million in Bogor about fifty miles to the southeast. Much of the population between Bogor and Jakarta can well be considered urbanized even though it resides in small villages. By this criterion, it is expected that more than half of the population could be considered urban by the year 2000.

While urbanization is generally considered one of the indicators of development or progress, when it comes as rapidly as it does in many Third World countries, it produces great stress on both the population and the environment. It is difficult for urban administrations to keep up with the demand for housing, and basic urban services like water and sewage. In addition, urbanization implies a great rise in transportation, especially by motor vehicles. It is well known that traffic congestion and air pollution are serious problems for rapidly growing cities in developing countries.

A recent survey of Asian urban administrators carried out by the Asian Urban Information Center of Kobe (1991) provides some interesting information on the nature of this urban environmental stress. As in other developing countries of Asia, the urban administrators surveyed in Indonesia perceived a rapid inflow of population into their cities, and saw this as a major problem. They also identified basic urban utilities, housing, traffic, and employment to be some of their most serious problems. The single most serious problem for them was the provision of sewage and safe water services. Weaknesses in these services, of course, represent a major health hazard. Pollution from vehicle exhaust, industrial effluents, and even simple human solid waste also pose problems in this urbiculture. Housing is difficult, but unlike India or the Philippines, it is housing for the poor not homelessness which is the critical issue. Indonesia is spared the great pressure of homeless people that can be found in India or in Manila.

Transportation represents still another challenge for urban administrators. In 1948 there were only 17,600 passenger cars in all of Indonesia (ECAFE 1959). By 1965 this had increased by a factor of ten, to over 170,000. Today there is an estimated 2.5 million passenger cars (Population Reference Bureau 1991). This dramatic increase in automobiles results in physical congestion and extensive pollution from exhausts.

The Indonesian government reacts to this environmental stress in a number of ways, usually in concert with a range of international organizations. For example, in the early 1980s the government organized a National Urban Development Strategy Project (NUDSP). This was a joint project including the Indonesian Department of Public Works and the United Nations Center for Human Settlements. This project attempts to chart the process of urbanization in Indonesia and to develop strategic plans for addressing the stress of rapid urbanization. Much of the work to be done involves construction of public infrastructure and the development of public utilities such as water, sewage, garbage, and power.

It is interesting to note that the urban administrator surveyed by the Kobe study did not see education and health and family planning services to be major problems. Shortly after independence, Indonesia launched a drive to expand its educational system. Schools were built and classes organized, using the new national language, Indonesian, as the language of instruction. This drive has been quite successful in getting most of the children into school. Even by 1965, 72 percent of all primary school-aged children were enrolled in schools; and the figure was 65 percent for secondary schools. Today both figures are over 110 percent.[5] Similar progress has been made in providing basic health and family planning. In education, the government has apparently responded adequately to the stress of population growth and urbanization.

Rapid population growth and urbanization have also brought stresses in the area of employment. The urban administrators considered this the second major problem after public utilities and infrastructure. The details of unemployment are obscured by the difficulties of agreeing on useful definitions and obtaining accurate counts. Nonetheless, the general picture is relatively clear, and is similar to that of other developing countries with rapid rates of population growth. There is more unemployment in urban than in rural areas. Unemployment is also negatively associated with age and education. The problem is especially acute among the urban young, fifteen to twenty-nine years of age. It is serious for those without education, or with only a primary

5. It is possible to have more than 100 percent of the children enrolled because the actual school age spreads beyond the 6-12 years of age that is used for the denominator of this ratio.

school education, but it is also a serious problem for those with lower secondary education. Slowing the rate of population growth will provide some relief, but not for the next fifteen years. The immediate response must be to expand economic activities, which requires capital and effective economic organization. While Indonesia's experience with economic growth has been positive, especially since 1965, the growth of the labor force continues to create substantial stress.

Conclusion

In these critical areas of life in Indonesia, we have seen that the relation between population and environment is complex and dynamic. Forces produce change in one aspect of life, which produce stresses that produce some reaction. Sometimes those reactions alleviate the stress, other times they may exacerbate it. What we have dealt with here are primarily public attempts to alleviate the stress.

As in many parts of the developing world, modern medical and public health technology reduced mortality drastically and quickly. The resulting rapid population growth produced great stress on the population-environment relationship. This stress produced a reaction both at national and international levels. The reaction was essentially a political one, as Indonesian and international political institutions came to define rapid growth as a problem that had to be addressed. To address the problem, a national family planning program was created to distribute the new contraceptives and to spread a new message about the value of fertility limitation. With strong support from the political center, and from international organizations, the program has been quite successful, and fertility is registering substantial declines. There are still local cultural forces that both support and impede the progress of the program, but overall there has been a good reaction to this environmental stress.

In basic food production and in the extent of forest cover, population growth placed specific stresses on the environment. The reaction again has come at national and international levels, and again political institutions have been important. New agricultural practices and technologies have increased rice production in particular and the production of other food crops as well. The forests have been extensively exploited both for more agricultural land and for timber exports. Now the state has entered into activities both to protect the forests and to replant those that have been lost. There are only partial successes here, and even these raise new problems. The process of stress and reaction is truly dynamic, and apparently, never ending.

Finally, population growth and economic development have brought increased urbanization. This is both a positive movement, and one that places great stress on the environment. Again government plans attempt to address the problem of stress, though it is too early to tell how successful they will be.

In effect, the population environment dynamic that we seek to understand is one of growth, stress, and reaction to that stress. All of these reactions have been more or less specific to the immediate problem at hand. This may be the only way to react to environmental stress, but it also raises the question of whether some more holistic or more integrated approach would not be more useful.

Chapter IX

The Phases of Agricultural Modernization in Brazil

George Martine

Introduction

Three different phases of agricultural modernization can be distinguished in Brazil during the recent period. As late as the mid 1960s, agricultural technology in Brazil was, for the most part, rudimentary. With the exception of the states of Sao Paulo and Rio Grande do Sul, few farmers in Brazil owned a tractor or other modern farm machinery. However, the situation was altered rapidly in the late 1960s and early 1970s due to the introduction of modern methods of production by the military government.

The trends set in motion during this period generated rapid change in productivity, as well as profound social and demographic transformations, which continued unabated until the late 1970s. The economic crisis of the early 1980s imposed severe limitations on subsidized credit thus forcing a new, tentative set of agricultural policies. During the latter half of the 1980s, partial recuperation of the economy and the onset of a democratic regime molded agricultural policy. To each of these phases of agricultural modernization corresponds a specific set of social, demographic and ecological consequences. This paper addresses itself to the description of these various 'phases' and to the analysis of their different social consequences or 'faces'.[1] Emphasis is placed on the first and longer period, not only because of its importance but also because of the greater availability of pertinent data.

---

1. This paper is largely based on George Martine's "*Fases Faces da Modernizacao Agricola*," Revista de Planejamento e Politicas Publicas 1(3):23-77, Brasilia, August, 1990; and "Changes in Agricultural Production and Rural Migration," unpublished paper presented at the Conference on The Demography of Inequality in Latin America, Gainesville, Florida, February 1988. Readers are referred to these papers for further documentation and data.

The Phases Of Modernization

The Conservative Modernization Era: 1965-1979

At the time of the military take over in Brazil (1964), a new technological package known as "The Green Revolution" was being widely disseminated among Third World countries. It promised to revolutionize agriculture through significant increases in production and productivity rather than through the redistribution of land. Basically, the new miracle package constituted an outgrowth of the farming system which had evolved gradually in North America since the 1920s; therein, mechanization and systematic research into higher yielding and more resistant crops had gradually multiplied the productive capacity of farm lands. However, in the 1960s, qualitative advances in the research of high-yielding-variety seeds promised the easy adaptation of this model to different soils and climates - providing that they were supported by fertilizers, pesticides, insecticides and mechanization.

This promising technological breakthrough had considerable appeal for developing countries. With the new method, the technological gap separating developed and developing nations might be breached; food supply would be assured, a large surplus would be exported and the other economic sectors would be stimulated by the need for industrial products in agriculture, as well as by the ensuing generalized prosperity. But the new package was particularly attractive in Brazil because of its snug fit within the "conservative modernization" style of development which was being designed at the time. This model basically aimed at updating the structure of production in all sectors and integrating them to achieve higher levels of productivity and greater wealth, without altering the existing class structure. To achieve this with the new technological package in agriculture required the proper manipulation of incentives.

The military government thus instituted a series of measures aimed at motivating farmers to adopt new technologies and new practices in the mid-1960s. The advantages for farmers in responding to proposed incentives were so great that the modernization process became, in a sense, 'compulsory'[2]. The most important stimulus, without a doubt, was the multiplication of subsidized agricultural credit, which deliberately favored the integration of agricultural, commercial, industrial and financial capital--thereby permitting the consolidation of the agroindustrial complex. Thus, subsidized credit was directed specifically to the purchase of high-yielding variety (HYV) seeds, farm implements, fertilizers, pesticides and insecticides. Other important measures were minimum price policies, crop insurance, indirect subsidies, and

---

2. For an excellent discussion of the modernization process in Brazil, Angela A. Kageyama et al. ii Novo Padrao Agricola Brasileiro: do Complexo Rural aos Complexos Agroindustrials, UNICAMP, Campinas 1987, p. 121 (Mineo).

a number of special programs directed to given crops, sectors or regions. The national systems of agricultural research and extension services were also expanded and streamlined to fit the new model.

All of these measures greatly benefitted from the sharp rise in international prices for commodities, during the 1960s. This coincidence was particularly fortunate for the emerging model since it occurred at a time when the growing internationalization of the Brazilian economy demanded that the export-oriented sectors increase their production so as to meet expanded external commitments.

In short, the takeoff towards a new model of organization for agricultural production occurred under particularly auspicious circumstances in Brazil. The type of transformation to be promoted in the agricultural sector fitted coherently within the chosen style of development at a given historical moment. Not only did it become possible to bypass reform in the structure of landholding but the growth of the agroindustrial complex guaranteed that the agricultural sector would consume an important share of industrial production at a time when industrial expansion constituted the nation's primary economic goal. Given the availability of open or under-utilized lands and the rising international prices for agricultural commodities, the new technologies stood to prosper and generate considerable wealth.

Results do show significant changes. For instance, the total land area incorporated into farming and ranching establishments grew from 250 to 365 million hectares between 1960 and 1980; meanwhile, the area under cultivation increased from 29 to 49 million hectares. In the same period, the number of tractors utilized on farms grew from 61 to 545 thousand units, the number of cattle from 57 to 118 million heads and the number of people occupied on agro-ranching establishments from 15.6 to 21.1 million. While productivity is difficult to synthesize into a single summary index, it is obvious that it grew significantly; moreover, the real value of agricultural production increased by 209 percent between 1970 and 1980.[3]

Such figures are undoubtedly impressive. Yet, some observers have looked critically upon these changes. The main point of contention between defenders and critics of the recent Brazilian agricultural transformation centers on the question "Who benefits?" Sympathizers of the modernization model argue that increased productivity inevitably favors society as a whole, although it may benefit some sectors more than others. Detractors argue that benefits have been so unequally distributed as to make it questionable whether society as a whole has gained from this process. The following discussion directly faces the issue of the inequalities resulting from agricultural modernization in Brazil.

---

3. Unreferred data cited throughout this paper are from Brazil's Agricultural and Demographic Censuses.

One of the major consequences of the new technological package was concentration of land, promoted by the social and spatial distribution of agricultural credit. The very logic of bank loans which requires a legal deed to the land or other collateral, and prefers larger commitments to piecemeal loans, favors concentration. Since agricultural loans were highly subsidized, ownership of larger farms guaranteed financial gains through mere application for loans. This fact, coupled with loose monitoring and control, stimulated land purchases for purely speculative purposes: the greater the land area owned, the greater the access to subsidized credit. In addition, the fact that loans were preferentially earmarked for the purchase of modernizing inputs and for export or agribusiness-oriented crops, tended to cluster the dispensation of available resources on certain types of farms. By contrast, small farmers, who make up the immense majority of all producers, received but a small share of the total volume of subsidized credit. As of 1980, only 21 percent of all farmers had access to credit; moreover, 10 percent of all establishments received 60 percent of the total value of all loans conceded.

Disparities in access to government subsidies increased the gap in access to modern farming equipment. As of 1980, 72 percent of all farming establishments did not have any kind of tractor, animal-driven plow, or similar tilling equipment. As a result of the disequilibrium induced by government incentives in the structure of production, one can perceive a growing duality and division of labor in Brazilian agriculture between 1965 and 1980. Therein, larger landholdings on more privileged land have access to credit, subsidies, research, and technological assistance. They produce basically for the export market or the agroindustrial sector. Meanwhile, smaller farmers are pushed off the land or relegated to less fertile and less accessible lands. They utilize traditional practices and exploit family labor to provide much of the foodstuffs consumed on the internal market, where prices are kept low because of depressed buying power of the urban masses.

The effects on the structure of land tenure are reflected in table 9.1. From 1920 to 1970, the size of agricultural establishments had been decreasing gradually, small farmers showed the largest growth while all other size classes progressively declined. With the intensification of the speculation in the 1970s, this trend was reversed; farms of one thousand or more hectares increased their share, while all others decreased.

These sweeping changes in the structure of agricultural production obviously affected employment. One important change has been to accentuate seasonal fluctuations in the demand for labor. Thus, the incidence of temporary wage labor has multiplied. Meanwhile, sharecroppers, tenants, small owners, and their families were ousted by the demand for land. The increasing use of machines also contributed to an overall reduction in the need for permanent manpower. The final result has been an increase in the instability of agricultural employment. With regard to trends in personal

TABLE 9.1 Distribution of Establishments and Total Area by Size-Class

Size-Class	Establishments							Areas						
	1920	1940	1950	1960	1970	1975	1980	1920	1940	1950	1960	1970	1975	1980
0-10HA	71.6	34.4	34.4	44.8	51.3	52.1	50.3	9.0	1.5	1.3	2.4	3.1	2.8	2.4
10-20HA	71.6	16.6	16.7	16.4	15.6	14.7	14.9	9.0	2.3	2.1	3.1	3.6	3.2	2.9
20-50HA	71.6	23.9	23.6	20.2	16.7	16.3	16.5	9.0	7.2	6.6	8.3	8.6	7.8	7.2
50-100HA	71.6	10.7	10.6	8.2	6.9	7.1	7.6	9.0	7.2	6.6	7.6	8.1	7.6	7.5
100-1000HA	24.4	12.8	13.0	9.4	8.4	8.9	9.5	27.6	33.5	32.5	34.4	37.0	35.8	34.8
1000HA	4.1	1.5	1.6	1.0	0.7	0.9	0.9	64.3	48.3	50.9	44.1	39.5	42.8	41.5
Total	648	1.90	2.06	3.34	4.92	4.99	5.16	175	197	232	249	294	323	364

Brazil 1920-1980 (in percent)
Source: IBGE, *Censos Agropecuarios*
Total includes undeclared (N=100%)

income, the issue is controversial but, overall, existing data present a sobering picture.

Since small farmers were most affected by such changes and since they make much more intensive use of manpower, their progressive elimination has produced a massive rural exodus: close to thirty million people left rural areas for the cities between 1960 and 1980. It is interesting to observe in table 9.2 that the more advanced states of the Southeast region, in which agricultural modernization occurred first (Sao Paulo, Minas Gerais, and Rio de Janeiro), show an earlier exodus than the rest of the country. During the 1970s, the rural exodus from the prosperous and relatively recent frontier states--Parana and Goias--is more notable in relative terms. Overall, rural out-migration has been less connected to relative levels of poverty and development than to the timing and rhythm of modernization. As a result, urbanization in Brazil showed a new face in the 1970s; for the first time rural areas showed an absolute decrease in population from 41 to 38.6 million.

Modernization has also caused serious environmental consequences. This is because the Green Revolution package is basically centered on HYV seeds, which require a supporting complement of chemical inputs - fertilizers, pesticides, insecticides, fungicides, etc. In many developing countries, the excessive and indiscriminate use of these inputs, particularly of weed and pest killers, is having serious environmental consequences. Between 1964 and 1984, the total consumption of agrotoxics increased by 279 percent (from 16,000 to 61,000 tons) in Brazil. Imports initially accounted for the large majority of this supply but gradually local industry has taken over most of the production. Given the highly polluting character of chemical plants and the loose controls exerted on them in Brazil, expansion of local production has undoubtedly contributed significantly to urban-industrial degradation in areas such as Cubatao. Unfortunately, data on the marginal contribution of industrial production for agricultural uses to ecological degradation are unavailable.

Inadequate controls over production, distribution and utilization of chemical inputs, coupled with the lack of training by farmers and agricultural workers in the use of chemical products, has caused serious damage to the environment and to human health. To a large extent, these problems are common to most developing countries and need only be touched upon here. Some of the main problems include the following: first, several high-pressure herbicides are highly volatile and can be carried by the wind to poison other food crops in other areas; some insecticides can also be carried great distances. In aerial spraying, between 10 to 70 percent of the products applied will contaminate unintended areas, including neighboring towns and cities.

TABLE 9.2. Net Migration in Rural Areas, by Region, and State. Brazil 1960-70 and 1970-80.

REGION & STATE	RATE	NET MIGRATION	
	1960-70 1970-80	1970-80	1960-70
NORTH	-447 +.006	-1	-0.279
Territory & Acre	-47 +0.491	+139	-0.218
Amazones	-195 0.446	-257	-0.409
Para	-205 +0.102	+117	-0.224
NORTHEAST	-4.373 -0.279	-4.990	-0.298
Maranhao	-598 -0.262	-586	-0.294
Piaul	-220 -0.279	-319	-.230
Ceara	-523 -0.348	-899	-0.238
Rio Gde. do Norte	-180 -0.224	-182	-0.251
Paraiba	-414 -0.338	-466	-0.317
Pernambuco	-846 -0.324	-760	-0.373
Alagoas	-223 -0.371	-355	-0.266
Sergipe	-180 -0.370	-180	-0.390
Bahia	-1.189 -0.282	1.243	-0.306

(continued)

TABLE 9.2 *Continued*

REGION & STATE	RATE	NET MIGRATION	
	1960-70 1970-80	1970-80	1960-70
SOUTHEAST	-6.801 -0.463	-5.038	-0.516
Minas Gerais	-2.933 -0.481	-2.611	-0.503
Espirito Santo	-273 -0.465	-408	-0.340
Rio de Janeiro	-641 -0.428	-467	-0.459
Sao Paulo	-2.954 -0.444	-1.552	-0.616
SOUTH	-1.079 -0.478	-4.395	+0.145
Parana	+166 -0.569	-2.516	+0.056
Santa Catarina	-391 -0.373	-617	-0.271
Rio Grande do Sul	-854 -0.406	-1.262	-0.286
CENTER WEST	-135 -0.455	-1.199	-0.070
Mato Grosso do Sul	+114 -0.399	-218	+0.209
Mato Grosso	 0.070	-25	
Goias	-249 -0.562	-956	-0.186
Distrito Federal	--- ---	---	---
BRAZIL	-12.835 -0.380	-15.611	-0.331

Source: Calculations based on IBGE, *Demographic Censuses.*
Rate of Migration = Net Migration, P1 to P2/P1

The second problem is the pollution of river and lake water in agricultural areas. Sources of this pollution include the following: residues from chemical preparations thrown into the river, spraying equipment washed in open water; chemical inputs in river side crops which filter down or are washed into the water, and the spraying of water bodies to control larvae and other hosts which poison the water supply. In addition to this, fish which remain at the river bottom have larger doses of contamination than those that live on the surface. Another serious problem is that many of the insecticides utilized in Brazil (DDT, BHC, Aldrin, dieldrin, chlordano, heptachlorex and mirex) remain in the soil for periods which can vary from a few years to several decades. Gradually they are transferred from the soil to other edible crops and to pasture; in this manner, residues are passed into vegetables, milk and meat. Tests have shown an alarming increase of organo-chlorides in the Brazilian population because of this.

Perhaps of even greater import is the well-known fact that greater and stronger quantities of toxic materials are necessary to control the increasingly more resistant and more diversified pests which have arisen since widespread use of insecticides and herbicides was initiated. Lastly, the number of cases in which farm workers and their families have been more or less seriously intoxicated in the manipulation of pesticides and insecticides has grown dramatically. Often illiterate and untrained workers, lacking special clothing or equipment, suffer permanent damage from fumes, spills or direct contact.

Farming practices linked to extensive monoculture have also been associated with widespread erosion of both rural and urban land. Mechanized tilling of relatively flat soils in the state of Parana, for instance, have combined with particular geological substrata and tropical rainfall to produce huge craters in previously high-yielding areas. Less visible, but much more common throughout the country, is surface erosion, whereby significant quantities of topsoil are lost yearly due to the impacts of wind and surface runoff, particularly during the planting season. Widespread deforestation for agricultural purposes has also been connected with increasing floods and other disequilibria in the hydrological cycle. Considerable concern is currently being expressed by environmentalists about the probable consequences of extensive monoculture in the newly opened Cerrados region.

In addition to these several interrelated but rather standard ecological impacts of the modernization process, it should also be mentioned that out-migration from traditional agricultural areas helped stimulate movement to the Amazon region. Although it would be wrong to attribute Amazonian deforestation primarily to incoming settlers (cattle-ranching fiscal incentives and major 'development' programs being much more significant) there is no question but that such flows, plus the public measures which they helped trigger, contributed to the large-scale devastation of tropical rain forests.

The Crisis Period: 1980-84

Between 1980 and 1985, the volume of credit applied to agriculture fell from 250 to 124 billion cruzeiros, in real terms. Surprisingly, overall agricultural production was not significantly affected. Although the industrial sector experienced drastic cutbacks during this crisis period, the agricultural sector apparently managed to hold its own. This feat was interpreted as a sign of a new maturity within a technologically consolidated production structure, capable of reacting not only to subsidies, but to market stimuli.

The appreciable performance of the agricultural sector during the crisis period was due to a combination of factors. First, although it is true that the total volume of subsidized credit was cut back drastically, the remainder was used more selectively to stimulate production of certain crops. Given Brazil's severe balance of payments problem, export crops and petroleum import substitutes such as sugar cane were given priority. The number of producers benefitting from subsidized credit was reduced drastically but those who did have access to it were subsidized at an even higher level than before. As a result, the production of export crops and import substitution crops actually increased, in some cases, significantly. Large and productive areas in the Center-West region were planted with soybeans, thus boosting the overall level of production. At the same time, the government instituted a more aggressive minimum price policy, manipulating prices on the basis of a closer monitoring of crop movements. International prices of some commodities suffered sharp increases thus further stimulating production. Finally, recessive tendencies in employment and income reduced internal consumption, thus freeing additional volumes of commodities for lucrative export.

The combination of these factors, rather than the incorporation of new technology, sustained agricultural production during most of the crisis period. Indeed, data from the 1985 Agricultural Census suggest a surprising reversal in many of the patterns which had characterized modernization during the 1970s (table 9.3). For the first time since the inception of the recent modernization process, the number of agricultural establishments showed a significant increase during the 1980-85 period (from 5.2 to 5.8 million), while the rate of increase in total land area showed an important reduction. Consequently, average land area per establishment actually decreased in the 1980-85 period. The smaller farms registered the largest growth in number of units during the 1980-85 period. Following a decade in which the total number of farms with 10 or less hectares had remained practically stable, this category showed an increase from 2.6 to 3.1 million establishments between 1980-85. The proportion of the total land area which accrued to these minifundios (small farms), however, only had a very slight increase (from 2.47 to 2.67 percent). In other words, the average land area of the

TABLE 9.3 Selected Indicators of Change in Brazilian Agriculture 1960-1985

Year	Establishments		Total Area		Average Size	Cultivated Area		Tractors		Cattle		Workers	
	#(1000s)	%chan.	#	%ch	Hectares	#	%ch	#	%ch.	#	%	#	%
1960	3.338	4.0	249	1.6	74.9	28.7	1.7	61	10.5	57.1	3	15	1.2
1970	4.924	0.3	294	1.9	59.7	33.9	3.3	165	14.3	78.5	5	17	3.0
1975	4.993	0.7	323	2.4	64.9	40	4.2	323	11.0	101	3	20	0.8
1980	5.160	2.5	364	0.6	70.7	49.1	1.3	545	3.6	118	2	21	1.9
1985	5.83	2.5	376	0.6	64.5	52.4	1.3	652	3.6	127	2	23	1.9

Source: IBGE, *Censos Agropecuarios, 1970,1975,1980; Sinopse Preliminar do Censo Agropecuario, 1985*
Workers = Total Number of Farm Workers

minifundios actually decreased in the interim. It is worth noting that much of the increase in the number of small establishments was registered in the poorer North Eastern region.

Other reversals from previous trends, noted during the 1980-85 period, involve a slower decrease in the incorporation of total land area and of land area under cultivation, while the increase in utilization of technology fell to one-third of previous-decade levels. Lastly, the growth of frontier establishments also suffered a severe decline during this time. What factors account for such changes? Although the data are only preliminary, it can be hypothesized that they reflect the impacts of a severe economic crisis. More specifically, the crisis seriously affected the availability of credit, incentives and subsidies and provoked the constriction of both the internal and external markets. Both of these phenomena reduced the attractiveness of the agricultural sector for capitalist involvement - whether as a productive activity or as an outlet for speculation. In this way, the crisis would seem to have re-opened interstitial spaces for small-scale producers.

The figures on occupied personnel would tend to bear this out. Indeed, the number of agricultural workers grew twice as fast in the 1980-85 period as in the previous quinquenium (five-year period). Moreover, most of this growth accrued to the smaller farms. Thus, of the 2.1 million overall increase in personnel, 64 percent was registered among farms with zero to ten hectares and another 25 percent in the ten to one hundred hectare category. Almost 50 percent of the total increase was registered in the northeast where 85 percent of the overall growth was located in minifundios.

It is interesting to note also that much of the increase in agricultural personnel has occurred in two categories of producers which had suffered severe setbacks in the 1970s--sharecroppers and squatters. The former had a 43 percent increase and the latter, a 22 percent increase. These are exactly the type of productive units which have the least-stable and least-formalized relations of production. Their ascendency would appear to imply a certain decline of capitalist production and/or a greater return to the search for rural-based survival strategies. In either case, the prevalent economic crisis are the root of these changes.

The various trends observed for the 1980-1985 period all point to a temporary regression of the speculative forces which had such a negative influence on employment and migration during the 1970s. During a period of severe employment crisis in the cities, the partial restoration of survival opportunities for the rural poor may have reduced the overall calamity. As a result of the processes described above and of the concomitant drastic reduction in fertility, it can be expected that the rural exodus will show a significant decrease during the first part of the 1980s, thereby relieving the rate of increase of the urban population. Nevertheless, there is no reason to believe that the short-term return to pre-capitalist forms of production, witnessed during the first half of the 1980s, constitutes a lasting trend.

As concerns the environmental impacts of these changes, it is plausible that the reduction of subsidies directed to the purchase of chemical products may have produced beneficial impacts during the crisis period. Indiscriminate utilization of fertilizers, herbicides and insecticides had to be curtailed in view of their cost. Overall, it is interesting that consumption of such products was significantly decreased without a noticeable decline in productivity levels.

The Era of "Super-Harvests", 1985-1990

Brazil's economy began to recover in late 1984, and the internal market expanded significantly in 1985-86, a period which coincided with the restoration of civilian government. A daring economic plan (the Plano Cruzado) quickly increased buying power and stimulated production across the board. Agricultural policy became more complex, utilizing a combination of both pre-crisis incentives and price mechanisms to stimulate both export production and the fulfillment of internal demand.

The combination of these various instruments soon produced a significant increase in the overall volume of agricultural production. The abrupt demise of the Plano Cruzado led to rapidly rising inflation as well as to reduced buying power which, in turn, freed a generous portion of production for export purposes. Since the balance of payments problem had once again become acute, the government strove to maintain production stimuli. As a result, the latter half of the 1980s repeatedly attained production levels of commodities in the vicinity of ten to fifteen million tons higher than the average of fifty to fifty-five million tons which prevailed during the 1975-85 period.

This fact has been repeatedly interpreted by politicians and researchers alike as proof of the growing maturity of Brazil's agricultural sector. Nevertheless, as will be seen in the next section, maintenance of production levels is precarious, dependent more on the ability to lobby for subsidies and incentives and less on actual productive capacity. As concerns the probable social and ecological consequences of recent trends, no reliable aggregate data are yet available.

Nevertheless, there are indications that the internal consumption of foodstuffs increased during the Plano Cruzado but subsequently fell again, generating part of the surplus which could later be exported. Meanwhile, the trend towards concentration of land resumed its previous course thereby reducing the number of options for small-scale subsistence farmers which had opened up during the crisis period. This, in turn, is likely to have forced a resumption of the rural exodus, although at lower levels than during the 1960s and 1970s. Available data indicates a significant reduction in the flow of migrants to Rondonia and other parts of the Amazon frontier during the second half of the 1980s. Overall, such patterns would suggest a new surge of urban concentration.

The Various Faces of Agricultural Modernization in Brazil

The transformations described above have generally prompted optimistic evaluations as to the current degree of maturity and integration of Brazil's agroindustrial complex. This position has, in turn, given support to more conservative political movements which, during the recent constitutional debate of 1988, managed to repeal any real attempts at agrarian reform, as well as to eliminate government agencies whose services were primarily aimed at small-scale producers. Some of the implicit and explicit assumptions underlying such political stances and their technical support should be examined more closely.

The growing strength of national and international conglomerates has helped to propagate the notion that without scale, economic salvation is impossible. To deduce from this that there is no longer any room for small-scale producers is but a short step. Without large land areas, access to large amounts of capital, ability to incorporate new technology and an entrepreneurial mentality, the argument goes, farmers cannot compete. Although it would be difficult to argue against the overall logic of this proposition, it does not necessarily follow, as intimated by the agricultural lobby, that: bigger is better, ad infinitum, in all sectors and on all occasions; the larger rural properties in Brazil adhere necessarily to the logic of capitalist production: small, unintegrated or informal producers must all disappear over the short and medium range.

Indeed, some farms in Brazil are larger than some European countries and there is no clear demonstration that they are inherently more competitive. "Large" and "small" in agriculture are concepts which permit considerable leeway. Moreover, many of the larger properties in Brazil continue to be unproductive, their main function being to serve as speculative chattel. As to the notion that modern capitalism necessarily eschews small-scale production, it simply does not correspond to the reality of economic production in the more advanced capitalist countries where family or informal-type enterprises of a small-scale variety continue to play an important role in many sectors.

In the specific case of agriculture, advantages of scale are decisive only in certain crops and activities, particularly where large-scale mechanization and uniformity is possible. But, in Brazil, much of the topography simply does not lend itself to the extensive use of machines, while several types of crops, particularly citrus fruits and horticultural products, still require considerable manpower. Overall, family properties continue to make more intensive use of available factors. This is why, in 1980, farms of under fifty hectares, with only 12.6 percent of the land area, accounted for 40 percent of total production and employed 69 percent of the agricultural labor force. In the case of certain crops and activities, there are additional factors which favor small farmers. Thus the nature of 'risk' which remains in much of agricultural production benefits smaller operations where daily monitoring can be a significant factor in productivity. Climatic factors can still be compensated only partially by modern technology. Various pests, plagues, fungi and bacteria are always evolving. The nature of the risk factor explains why, in some cases, large conglomerates prefer to contract small farmers for the provision of raw materials (poultry, hogs, tobacco, cotton, etc) rather than to take chances and produce it themselves.

The discussion of the relative efficiency of different scales of production cannot be divorced from the broader issue of the role of the State in the productive activities of modern society. To the extent that the State, and the basket of subsidies, incentives and benefits which it doles out, are not neutral entities but rather are apportioned according to the dictates of given power structures, certain segments of society are able to manipulate the State in the pursuit of their own interests. Currently, in Brazil, the role of the State is the subject of heated debate: one camp argues that the economy should be completely freed from State tutelage while another claims that the State is, and has been for some time, private property of those same forces which are now clamoring for 'liberalism'.

In any case, there is no question that productive activity, particularly in the agricultural domain, continues to be highly dependent on initiatives of the public sector which provide among other things, incentives, subsidies and preferential treatment. Indeed, the argument concerning the purported efficiency of the modern agricultural sector in Brazil tends to lose force when it is observed that this sector was dependent on public support in all phases of its modernization process, and that this dependency still continues in all stages of the current productive process. The central characteristic of the State's participation has been the unequal distribution of its favors by sectors, regions, social groups and economic groups. The current concentration of modern means of production reflect and perpetuate the profound inequity of the prevailing social structure.

Given the inherent limitations of public resources, their allocation is necessarily filtered by an auctioneering process among the more influential lobbies which act as intermediaries between government and producers in this sector. The power of the lobbies run by more integrated sectors is overwhelming and, thus, the struggle for resources is not between large/modern and small/traditional but among representatives of the different agroindustrial complexes.

In view of the efficacy of these lobbies in appropriating public resources and the fact that modernization often depends on government stimuli, it can be argued that the purported efficiency of the large-scale agroindustrial complexes is more political than economic. Obviously, this is not to argue that public resources should not be utilized to stimulate agricultural production; indeed, the experience of North America and Europe argues otherwise. But it is important to note that the power structure basically determines who will be benefitted by incentives and thus who will have a chance at being "efficient". Moreover, the voracity of the different lobbies has become somewhat counterproductive in that the proportion of all sector resources directed to research and technological development has declined significantly in recent years. As a result of this emphasis on immediate gain, current inferiority in Brazilian productivity levels (discussed in the next section) may tend to increase.

Another aspect which throws doubt on the current efficiency of the agroindustrial complex in Brazil is the limited increase in productivity which it has shown so far. Productivity of labor has undoubtedly increased due both to the significant growth of mechanization and to the rural exodus of almost thirty million people over two decades. But in a country where labor continues to be in large supply, productivity per cultivated area continues to be a more relevant indicator. In this respect, recent independent studies have shown that the increase in Brazilian agricultural production since the 1960s has been attributable, in large part, to an increase in land area. Moreover, the increase in productivity per hectare was basically limited to the 1970-75 period. However, the physical yield of Brazilian agriculture continues to be low in the majority of crops relative to international productivity standards.

The disadvantages of Brazilian productivity levels are particularly visible with respect to corn, rice and wheat but even in soybeans which, together with sugar cane passes for being entirely modern and competitive--yields per physical unit are about 20 percent lower than in the United States. Perhaps even more significant, productivity levels have practically stagnated. The inability of Brazilian agriculture to compete in the world market is particularly worrisome when viewed in light of the international farm crisis. Growing competition between the United States and the European Economic Community (EEC) has generated price wars--sustained by rich and powerful governments--which adversely affect the ability of Third World countries such as Brazil to compete in world markets. The possibility that current policy changes in the former Soviet Union will radically invert that country's traditional position as a large buyer of foodstuffs could also have unfavorable consequences for the Brazilian economy.

The events linked to the 1988-89 'superharvest' illustrate clearly the nature of the dilemma currently faced by Brazil. No sooner was the size of the record crop announced than the agricultural lobby organized a march of farmers on Brasilia to demand that the government provide price guarantees and incentives for the next planting. Unable to compete on the external market and with the

Brazilian internal market constricted by the poverty of a major segment of its population, super harvests generate little guarantee of continuity. In order to be repeated the following year, they require another liquid transfer of public resources to at least certain categories of producers. Given the technological, climatic, topographical, as well as socio-cultural advantages of countries in the Northern Hemisphere, it is improbable that Brazil will ever compete advantageously in cereals. Its development will depend on intensifying those crops in which it has comparative advantages--generally in the fruit and horticultural area.

Under such circumstances, expansion of the internal market could provide the necessary stimuli for continued growth and modernization of agricultural production. This would require not only sustained economic growth but also greater equity, so that through redistribution of the economic benefits , the population at large would have the means to acquire goods and services, including those produced by the agricultural sector. However, a country which systematically concentrates power, subsidies and incentives to production among elite groups has great difficulty in generating equity and this, in turn, leaves a large segment of the population outside the consumer market. This argument provides a new perspective on the question of who gains what from agricultural modernization. The main point is that concentrating resources for production in one limited segment of the producer spectrum tends to reinforce the concentration of wealth and thus the continued marginalization of a majority of the population. In this view, the integration of other types of producers and interests is necessary for the benefit of the market economy itself, perhaps even at the cost of a somewhat reduced productive efficiency--although this too has yet to be convincingly demonstrated.

Given the impact of modernization on the volume and forms of rural employment, the massive exodus registered in Brazil during recent decades was to be expected. Defenders of this model like to point to the history of developed Western nations where significant urbanization has occurred and where only a small proportion of the total population works in agriculture in order to justify the inevitability of the current situation in Brazil.

Such comparisons are misleading since the processes of agricultural modernization and rural-urban migration are occurring in an essentially different historical context in countries such as Brazil. First of all, the land tenure system, in North American countries at least, reflected a much more egalitarian system of social organization. Secondly, technological development in countries such as the United States or Canada was spread out over a period of fifty years. The Mechanical Revolution had begun in those countries back in the 1920s and, though it did promote rural-urban migration, its slow gradual nature, as well as its favorable impacts on industrial employment, helped ease the transition from rural to urban societies. The

Chemical Revolution, which occurred in the early 1960s, was based upon a capital-intensive, modern agricultural system and also led to a gradual increase in rural-to-urban migration.

In countries such as Brazil, on the other hand, the Mechanical and Chemical Revolutions both occurred at once, provoking unprecedented disturbances in the social organization of production. These were magnified by the fact that they took place in a country with a highly-skewed land tenure system and a system of social organization traditionally oriented to the benefit of a small segment of the total society. Moreover, such changes were imprinted on a society which was in the throes of a third major revolution--the Demographic Revolution--which was never experienced by developed countries. Finally, all of these processes are occurring in typical conditions of under-development which involve not only obstacles resulting from traditional and highly-stratified forms of social organization within the country as well as the dominance and dependency engendered by a highly-stratified international order.

In short, the conditioning factors of the recent rural exodus in Brazil cannot be compared to those which occurred historically in the developed regions. The massive exodus of some thirty million rural inhabitants during a twenty-year period is impressive. Particularly so in view of the fact that some 36 million people still reside in rural areas and that their rate of natural increase, though declining, is still substantial. Brazilian industry has had an exceptional record of growth from the 1950s to the present. Nevertheless, the supply of manpower has severely outstripped the creation of urban-industrial employment. Thus, a large part of the labor force which heads for the cities ends up in marginalized sectors having minimal productivity and eking out a precarious subsistence. This situation evidently has a deflatory impact on urban wages, even in the more advanced sectors.

The process of metropolization, whereby a large proportion of the country's total growth ends up being absorbed by the largest cities, provokes more sobering thoughts. (table 9.4) The current velocity of this process has forced the urgent examination of the functional limits of macro-city growth. A related concern stems from the observation that the majority of urban growth is occurring on the peripheries of the macro cities. A logical result of the unequal competition for urban space is that the more marginalized segments of the society systematically end up on the outskirts of existing cities. There, not only do living conditions tend to be even more unfavorable, but residents are additionally hampered by greater problems of distance to work and access to available social services. It is not surprising that increasing numbers of landless peasants have been willing to endure considerable hardship and violence in order to obtain land in frontier areas or in collective invasions of unproductive land in the South and Southeast.

TABLE 9.4 Selected Indicators of Urban Growth and Concentration in Brazil 1940-1980

Year	Urban Population	Number of Cities	Rate of Urban Growth	Number of Cities of 500,000	% of Country's Population in Cities of 500,000
1940	8041	50	4.6	3	10.8
1950	12622	82	6.2	4	14.2
1960	22913	148	5.4	8	21.3
1970	34586	246	4.9	11	26.7
1980	62238	393		14	32.3

Source: IBGE, *Censos Demograficos*

Another issue involves the supply and distribution of basic foodstuffs. Traditionally, small farms furnished much of the food consumed by the Brazilian population, both through auto-consumption in rural areas as well as production of a small surplus consumed by the poor urban masses. The exodus of some thirty million people, most of whom were food producers, and their transformation into low-income consumers, is cause for concern. Nevertheless, recent results of a national nutrition survey indicate that, overall, malnutrition levels have decreased between 1975 and 1989. This can be attributed to the general transformation of society which has affected information levels, fertility levels, per capita incomes within households, and specific measures in basic sanitation and general hygiene.

Conclusion

The foregoing comments regarding agricultural modernization in Brazil were not meant as a criticism of technical progress but intended to highlight how its impacts are molded by concrete socio-economic structures in particular historical contexts. Increases in production and productivity are a sine qua non for development. Yet there is no guarantee that the benefits of development thus attained will be distributed equitably. Brazil's history during the last twenty-five years shows that it is perfectly possible to maintain high economic growth rates without commensurate improvements in the living conditions of the masses. The patterns of expansion and the redistribution of the benefits of growth in a given society depends on its style of development which in turn is forged against the conflicting interests of different social sectors.

In this light, it is clear that agricultural modernization in Brazil transformed production structures and increased levels of output, but did little to modify one of the world's most skewed income distributions. In this perspective, continued attention to small-scale producers makes sense on two counts. First, because thirty-six million people still live in rural areas, the possibilities of productively absorbing a large proportion of this contingent in urban areas over the short run are small. In addition to this, access to a piece of land constitutes an immediate income gain and one of the cheapest forms of survival. Secondly, the potential comparative advantages of small farmers who are capable of incorporating technology, individually or in association with others, in certain types of crops and regions, has yet to be systematically explored. The political forces supporting large-scale production have effectively obstructed this alternative but, given their potential, the economic and social returns of investment in small farmers are still worthwhile.

Section III

The State as Environment: Population-Environment Dynamics in Small Communities

Introduction

The chapters in this section deal with population-environment dynamics in small scale social systems. They bring us close to the basic individual and small group behavior, to human adjustments to and human impacts on the immediate environment. *Human populations* are seen as collections of individuals and families with rich internal differentiations. Gender, family wealth, size, and power, as well as individual skills and mobility, become part of the fabric of that population, which has grown beyond the five billion mark. Similarly, the *environment* is seen in very immediate terms, with the same political and natural distinction seen in the previous section. An escarpment and a park boundary, a lagoon and a reef, a politically protected tropical rain forest being transformed into small coffee holdings, or a river valley of small farms--these are the small scale visions of the environment that emerge from ethnographic studies. It is at this level, that we become most sensitive to the issue of boundaries and to the balance of natural and human conditions that define a specific *environment*. Finally the *population-environment interaction* is one in which population growth clearly presses upon resources and forces changes in behavior. Thus population growth can be seen at least in part as deleterious to ecosystem health. It is a stress or perturbation from which a healthy system bounces back, but from which an ailing system only gets worse. Further, not only are technology and organization seen as mediators of the population-environment interaction, but both become more differentiated and elaborated. Close to individual behavior, these observations typically treat the political processes of the state as external and constraining, a powerful part of the environment that is usually perceived as given and beyond local influence. For all of the small groups the state intrudes, shaping the many transitions that can be observed. It is also a part of the environment that has a profound impact on the health of those transitions.

Population

At the local level, population in all four chapters is seen as families collected together in villages or other small geographic groupings. Family lines are the explicit connectors between population and environment. They also form the basic dimension of analysis for Low and Clarke, who examine fertility, migration, and resource control in nineteenth century Sweden. They use the long time-series on families made possible by the Swedish collection of parish records, which provides one of the most detailed historical descriptions available of a population over a long time period. Village or parish societies are local populations whose reproductive and migration patterns can be traced and related to environmental conditions. These populations can also be differentiated by land holding and by occupation, essentially by the power to control resources. The same-small scale society is found in the Stoffle et al. paper on the Dominican Republic, which focuses on a small fishing village in a relatively isolated bay. This population, too, is organized primarily around family lines. Individuals and families are differentiated by land and boat ownership, by skill and experience, as well as by occupational specialization. In effect, they are differentiated, as Low and Clarke have done, by resource control. Here, however, the elements of the distinction are more detailed, because they can be observed directly rather than through historical documents.

Brechin examines the population of a district in South Sumatra, Indonesia. In addition to being organized by families, it is also a population differentiated by geographic origins, language, and power. Indigenous peoples are different from the migrants--many of whom have been forced to migrate by population pressures in their home districts and have been attracted by the greater land availability in the research site. Although all can speak the national language, they are also distinguished by local dialects, which can be mutually unintelligible. There are also important differences in traditions, especially those that carry the power to protect the forests and sustain a known level of development from the past. Thus important non-economic social distinctions begin to appear as critical in defining the characteristics of a population.

The Benin population is similarly organized into families, in a subsistence agricultural community that is very much in the process of change. In addition to shifting agriculture and hunting and gathering, there is a movement toward the more exclusive planting of annual groundnuts on the same land, forsaking the older fallowing practices. The population is divided, as in Zimbabwe, by ethnicity and economic specialization. Conflicts now arise between the indigenous agriculturalists and the migrant Fulani herders. That is, the population is divided by ethnicity and language as well as by economic activity, and the latter divisions are changing.

Environment

The definition of an environment's boundaries emerges from these chapters as a critical variable. The four ethnographic studies all show boundaries that have a mixture of natural and human conditions. Further, the human element can be local or external to the specific group, and it can change radically, sometimes very quickly. The variance in the human dimension of the boundary definitions is thus identified as a critical element for understanding the character of the population environment relationship.

For Low and Clarke, the environment is defined by river valleys and parishes. The larger central political authority enters, granting the parish church the right and obligation to serve the local population, and relentlessly to record its life changes. But this is to a certain extent a political legitimation of a more natural locational pattern that emerged without much prior intrusion of the state. Like Rowe's *landscapes*, this environment is differentiated by soil fertility, crops, and different land uses. A critical distinction for the interaction we shall shortly examine is that between a more homogeneous and a more heterogeneous land use pattern. A parish-defined environment can be primarily agricultural and planted in one major crop, or it can be marked by more varied land use including agriculture, forestry, and fishing.

Brechin deals with a somewhat simpler environment, though it is also marked by both natural and human boundaries. A portion of a tropical rain forest is marked by the administrative boundaries of the district, subdistrict and village. Most importantly, new administrative boundaries define forest areas protected by the national government. There are both traditional and modern definitions of these boundaries, which are differentiated by the rights and obligations of the local population, or by the power of that population's political system to determine land use. Political institutions become an integral part of the definition and character of the environment. We also see more clearly here the growing encroachment of the modern state as it changes definitions of rights and obligations in a specific environment. If the parish in nineteenth century Sweden reflects a growing state's certification of a landscape ecosystem that emerged prior to the state intrusion, we see in Indonesia, especially in its transmigration program, a form of state intervention that will produce a new ecosystem, or a new set of boundaries and resources in the environment under study. There is another dynamic element in the environment as parts of the forest are now being illegally cleared and planted with coffee. What was previously an area protected by traditional authority systems is now being encroached upon because of the demise of that system, and its integration into a wider world ecosystem.

Stoffle et al. consciously use the ecosystem concept to examine the environment in Buen Hombre in the Dominican Republic. Natural elements appear more important here in defining the boundaries of the system within which local human actions mark out lines of differentiation. Here the larger external political system is less apparent than in South Sumatra or even in Sweden. Occupations in Buen Hombre define the environment by defining its land and water use. There is farming on plots of privately owned land, which itself reflects the impact of external political and legal institutions. The lagoon and the reef, distinguished from the land by purely natural conditions, have none of the internal boundaries that the institution of private property gives to the land. In the sea it is human skill and immediate activity rather than human institutions that determine boundaries and resource control.

Climate also becomes an important element of the environment in this isolated bay on the north shore of the Dominican republic. The arid climate marks agriculture as poor and especially risky. Fishing is less influenced by climatic conditions, but more by weather, which every day determines how much of what type of fishing can be done. This is the only chapter in which climate and weather become explicit elements of the environment (though Low and Clarke do note that some years of bad weather caused a reduction of resources, which pushed many families into the out-migration stream). Climate is thus introduced as an explicit condition defining the environment, and suggests the need for case studies to note the climatic conditions that implicitly mark environments as distinctive.

In Benin we see an environment defined by a sharply bifurcated set of natural and human conditions. A village, or collection of agricultural families, is pressed into a small space between a physical escarpment on one side and a national park boundary on the other. The escarpment provides a natural boundary beyond which agriculture is impossible. The park boundary represents an external political intrusion, beyond which neither agriculture nor traditional forms of hunting or gathering are *permitted*.

There are at least three aspects of the human, or socially constructed boundary that emerge as highly important in this micro-environmental analysis. First, the extent to which the boundary is *permeable and stable*, or to which it is crossed by humans, is determined in part by the control the external political forces can exert. It is also a boundary subject to negotiation and deliberate change. In fact, the boundary is changed to give more space to the village as government comes to understand that it is taking too much traditional village land in its original park boundary. Second, *boundary crossings by plants and animals* are not governed by the same external political forces, which produces some conflict around the boundaries. Experiences such as Chernobyl demonstrate that pollution does not respect human boundaries. In Benin, animals, as well, do not respect human boundaries. Baboons from the park feed on human crops, but are in part protected from human hunting by the law. If park law were fully enforced, farming could well become the endangered activity in this environment. As it is, the inability of the central political power to enforce the boundaries leads to some farmer protection against animal predators. Finally, the boundary is the product of a very *broad set of external forces*, including the national government and international organizations, especially those concerned with economic development and with environmental protection. This provides the opportunity for local versus national and international conflicts over environmental boundary definitions.[1]

All of these observations reinforce our earlier point that the "environment" is a messy conceptual bag. Not only can we not speak of the environment without also speaking of some form of technology, we also need to include that mix of human and natural conditions that define the boundaries of the environment. Because the *ecosystem* specifies boundary conditions, it emerges as a more useful concept by which the broad and messy *environment* can be specified. As we shall see in the summary, however, even the concept of the ecosystem does not resolve all boundary problems.

The Population-Environment Relationship

There are a number of propositions on the population-environment relationship that are either explicit or can be derived from the observations in these chapters. The most general proposition concerns population growth and migration. Population growth and resource control combine to affect patterns of migration. Population growth is a stress on an ecosystem, producing pressures toward migration, fertility limitation, or increased mortality. For Low and Clarke the original biological models of this relationship work fairly well in the direction of their predictions if not in the magnitude. Resource control directly affects the level of individual reproduction. Among other animals, however, the function is a step type, at least for males. Those without a substantial amount of resources simply do not reproduce. Most of the reproduction is done by males with great resource control. For humans the function is more linear, and lack of resource control can lead to individual migration in search of greater resources. Observation of the migrants is outside the purview of the Low and Clarke chapter, but it does not take much historical observation to know that many of the out-migrants from those nineteenth century Swedish parishes did indeed find greater resources elsewhere and did indeed reproduce, even if the resource control was only minimal. Among humans the poor also reproduce, though in pre-industrial systems they do so less than the rich.

In Buen Hombre and in Benin as well, population growth without increases in output from the environment led to out-migration. From Buen Hombre young people moved out to the towns in search of work, and floated back and

---

1. Africa appears highly susceptible to this type of international - local conflict over environmental boundaries. Elephant protection in Africa under pressure from international organizations often threatens local agriculture. (See Bell 1984)

forth between the village and the towns. The same was true in Benin, but the two cases differ somewhat in the constraints on resource mobilization from the environment. In Buen Hombre the constraints are those of nature, in Benin they come more from human boundary definitions. In all three cases, however, resource constraints led to migration rather than to a reduction of fertility. Here is a point at which the human experience appears to diverge substantially from that of the other animal forms on which Low and Clarke base their predictions.

In Brechin's South Sumatran forests and Agbo's Benin, the population-environment relationship is a multifaceted one that is changing while we observe it. Population growth in both cases leads to migration *into* the forests. The complexity arises from changing political systems. Inclusion into the larger Indonesian system brings transmigrators into the region as the external government deliberately sends people from the high densities of Java to the low densities of South Sumatra. Traditionally, there were also kin ties that gave people closer at hand a sense that they too had a claim upon the land. The traditional system mediated those claims, however, to protect a portion of the environment by refusing certain people the right to exercise what they considered to be their claims. With the demise of the traditional authority system, under the intrusion of larger Indonesia, there is no one to deny these kin-based claims, leading to further encroachments driven by population growth as well as by international economic conditions, such as the rise and fall of world coffee prices. In Benin, the political system intruded to create a new boundary of a protected area, reducing the land available to the local population. This along with the encroaching deserts in the North increased pressures of ethnically distinct Fulani herdsmen. Under lower population densities of the past, these herdsmen were permitted to move through the village to graze their animals. Now there are conflicts over these rights. Benin also experiences the transformation from subsistence agriculture to cash cropping, in part under increased population pressure, but in part under the impetus of government policy.

The Low and Clarke chapter provides yet another insight into the population-environment relationship. External impacts, such as the market declines that reduced income and also fertility in one parish had less impact on another. The critical difference appears to be the heterogeneity or homogeneity of land use. The more homogeneous agricultural land use in Locknevi also saw rises in food costs having a substantial depressing impact on fertility, since workers were more fully dependent on agricultural wages. In the more heterogeneous Tuna parish, fertility remained more constant, since resources from different parts of the environment could be substituted for one another when external market fluctuations hit the parish. This buffering was more substantial for those with resources. Heterogeneity of a resource base provides more insulation from external impacts for those who rely on that environment.

Many questions have arisen from social demography on the determinants of fertility, and especially on the long term and apparently irreversible fertility decline that is associated with economic development. There are only a few suggested answers in these chapters. Low and Clarke find economic impacts on fertility, but only under fairly well developed market conditions. They also find no evidence for the replacement hypothesis on the relation between infant mortality and fertility. Fertility in these nineteenth century Swedish parishes was apparently not driven by the desire to replace infants lost by deaths. None of the other chapters deals directly with fertility and its determinants.

The chapters in this set of micro-environmental observations have drawn us to a more complex and differentiated perception of population, of the environment and of the relationship between the two. This simply extends the increasingly elaborate perceptions developed in previous sections. In the concluding chapter, we attempt to draw all of these elaborations together into a new and more complex model of the population-environment dynamic.

Chapter X

Resource Control, Fertility and Migration

Bobbi S. Low
Alice Clarke

Introduction

Worldwide concern about the staggering growth of the human population and the proliferation of serious environmental problems has led to renewed interest in connections between the two key elements in the equation: population and environment (Teitelbaum and Winter 1989). This paper focuses on environment's influence on population. To predict population changes accurately, and to make effective policy influencing the direction and rate of such change, several levels of analysis are necessary: regional, local, social/economic sub-groups, and within-family comparisons. Appropriate scale of analysis has been discussed in general by Weins (1989) and, specifically in regard to population dynamics, by Lomnicki (1988). Observed changes in populations derive from the accumulation of behaviors of the individuals in the population; thus, what happens to a population depends on whether sub-groups behave differently, and on their representation in the population. Such study requires detailed data representing relatively long periods of time. Nineteenth century demographic data from Sweden is remarkably complete and appropriate.

Historical data may not appear to be relevant to populations in lesser-developed countries (LDCs) experiencing the highest population growth today, but there are important similarities between these high-growth countries and nineteenth century Sweden. Sweden then had primarily an agricultural economy, though industrialization was beginning (Eriksson and Rogers 1978). Health and hygiene advances led to decreased mortality rates (Brändström and Tedebrand 1986). All of these processes were geographically and temporally uneven. In Sweden the demographic transition also proceeded at locally different rates, sometimes reversed itself locally (Lockridge 1983), and appears to have been frequently related to resource availability, though demographers are still uncertain about causality. Migration was high, both to rural and urban areas within the country and to other countries (Eriksson and Rogers 1978; Norberg and Åkerman 1973; Norman and Runblom 1988; Ostergren 1988).

Some modern LDCs would fit this general description of agricultural/mixed industry economies or protoeconomies "awaiting" the demographic transition. However, there are real differences as well, such as the advent of modern birth control methods and the rapid spread of information regarding the "benefits" of small family size. Unfortunately, accurate long term demographic information does not generally exist for modern countries experiencing rapid population increases. The primary potential benefit of research on historical interactions between population and environment is that, while the particulars of the interactions may vary, the predominant causal links may well stay the same over time and across populations. Further, historical data may reveal non-obvious relationships, and may eventually help guide the collection of modern demographic data.

Population and Environment: The Family as Connector

Biologists, as well as demographers, are interested in lifetime reproductive patterns of humans. The questions asked are likely to diverge somewhat, because the underlying paradigms differ significantly. Demographic approaches tend to focus on population performance and specific proximate mechanisms (e.g., age at marriage, interbirth interval, etc.). "Why" questions of ultimate function are not raised, or are deferred. Evolutionary approaches converge on these questions. Questions on the behavioral ecology of resources and reproduction (Daly and Wilson 1983; Lumsden and Wilson 1981; Alexander 1979) arise from the simple, yet broad evolutionary paradigm. The organisms we see today are the descendents of those that successfully survived and reproduced in the past environments. Strategies for survival and reproduction are all-important, though their appropriate analysis may be very complicated.

It is important to remember that traits show intergenerational correlations, but are not "inherited" in the sense of being uninfluenced by environmental conditions. All phenotypic traits, behaviors as well as muscle and bone, for example, are part of this process. No trait can be determined to be completely independent of genotype, and no trait is "genetically determined" in the sense of being uninfluenced by environment, both physiological and external. Genes are chemicals that code for protein production or regulate the actions of other genes. But genes do not code directly for any trait--genes do not work in the absence of environment. Even traits we tend to think of as "genetically determined" are not. In fact, there are species in which males and females are genetically identical, but sex is environmentally determined (e.g., Bull 1980). When we consider complex behaviors, it is extremely important to remember this interdependence of genes and environment. In a very deep way, the old "nature-nurture" controversy is irrelevant (e.g., Daly and Wilson 1983), and the questions we ask are simply about behaviors in particular environmental conditions.

Perhaps, then, it is useful to explore, from the perspective of behavioral ecology, the role of selection in human life histories in complex societies. We do not know a one allele-one behavior link; we assume learning can be involved. A variety of proximate mechanisms might mediate any particular pattern. There is no doubt that human behavior is complex and exhibits amazing flexibility relative to most other animals, but here we will ask a simpler question: Under the described environmental conditions, what strategies will be favored--will result in enhanced lineage success, compared to alternate strategies? For those interested in patterns of human reproduction and family life, it seems worth exploring whether there are clear ecological patterns with the richness, controllability and predictability of important resources.

Available resource levels should influence fertility, mortality, and migration in specific ways. Abundant resources should be associated with high fertility and low out-migration from the region; reduction in resources should be accompanied by lowered fertility and greater out-migration. High fertility within families should frequently result in increased maternal and infant mortality. Migration may result in either an increase or decrease in the fertility of the migrants, depending on migrants' ages, the costs of migration (including increased mortality), and the relative resource base in the new area compared to the "home" region (Easterlin 1978; Rolén 1979). Out-migration may also influence the reproductive history of non-migrants and migrants' relatives. Demographers (e.g., Wrigley and Schofield 1981; Willigan and Lynch 1982) have constructed similar models. This approach does not, and need not, specify what constitutes a "resource," nor make explicit the mechanisms by which the proposed relationships might be accomplished. Fertility differences, for example, may arise because of differences in age at marriage, proportion of individuals marrying, interbirth intervals, etc. For our purposes, it is the net change in fertility that matters, while for many demographers, the mechanisms may be of more interest.

Predictions are tested at several levels: among regions, among population members within regions, and within families. Fertility, for example, should be higher in resource-rich areas than in resource-constricted areas; higher in richer (higher occupational status, or land-owning) families compared to poorer; and within families, higher for those with more access to resources (legitimate versus illegitimate; earlier-born rather than later-born) compared to those with less access to resources. Economic measures (prices, wages, and cost-of-living indices) are used to reflect available resources at the parish level. Within parishes, individuals, or families may vary in their circumstances. The study examines occupational status, land ownership (in Tuna Parish), and measures such as birth order (reflecting probability of inheritance). Using these measures, family patterns are compared for individuals in two parishes. Little affect of resource richness on women's reproduction is expected, except when conditions are so stringent as to affect infant survivorship and maternal mortality (Low 1989a, 1990a). This follows from the different ways in which resources affect the reproductive ecology of males versus females (Low 1990b). These predictions refer, therefore, primarily to men rather than women.

The Parishes, Sample, and Methods

Nineteenth Century Sweden

Nineteenth century Sweden was largely agricultural, with emerging protoindustrialization (Mendels 1981): geographically scattered market activity involving the transformation of raw materials into "made" commodities, but with a large part of the labor force working part-time or at home. The family could function as a form of economic enterprise (Flandrin 1979; Habakkuk 1955). Such protoindustrialization tended to develop in regions combining an underemployed, land-poor population, with access to urban markets (e.g., Tilley 1978). In Sweden it is probably related to land enclosure and inheritance changes during the nineteenth century.

Marriage followed the "European" pattern, with women marrying for the first time in their early to mid-twenties, and men in their late twenties. At marriage the new couple typically establishes its own independent household. A relatively high proportion of individuals never married (Low 1989a, 1990a).

From 1686 to 1809 or 1810, the nobility practiced "fideicommiss," or male primogeniture, with the constraint that the eldest son must continue the practice (Malmström 1981; Inger 1980). Until 1845, sons inherited twice as much as daughters. After this date, daughters had equal inheritance rights, although sons had first choice of the land and goods which were to be their inheritance, and sons could purchase their sisters' inheritance from them (Lo-Johansson 1981; Inger 1980). Thus, disputes occasionally arose over the value of the exchanged inheritance items. Purchasing needed land from a sibling could prove economically onerous, but also siblings sometimes complained that they did not receive fair value (not uncommon elsewhere in Europe; Habakkuk 1955). Even after the shift from fideicommiss and equal inheritance rights for both sons and daughters, inheritance biased by birth order was often evident (Gaunt 1987; Low 1989a, 1990a). A bias toward the first son was perhaps more evident in the northern areas. Legal agreements in which a father ceded his land to one (usually the eldest) of his sons before his death, typically in return for room, food, and certain other rights, were common. But as Gaunt (1987) notes, during the nineteenth century the payments delivered to the retiring father increased in size, and receiving a farm became an economic burden. Indeed, default was common (Gaunt 1977) and Gaunt (1983) cites contemporary jokes about arsenic as "retirement medicine." Thus, there probably existed some tension both within and between generations over resources.

Tuna Parish

The population of Tuna Parish, in Medelpad, rose from approximately twelve hundred in the early nineteenth century to approximately thirty-three hundred in the late nineteenth century (fig. 10.1, Demographic Database 1986, Basstatistik). Tuna was largely a farming parish, though forest and mining industries were also present in the early 1800s. Many men worked in the iron foundry (Matfors Bruk) as well as farmed. Tuna experienced rapid industrialization from 1850 onward (Norberg and Rolén 1979). Matfors Bruk closed in 1879, and reopened in the mid 1880s.

Locknevi Parish

In Locknevi Parish, in Småland, geography imposed constraints on farming (Gerger and Hoppe 1980). Only in the central valley were there fields fertile enough for farming. A small iron-works in the southwestern part of the parish provided supplemental income for some farmers until the 1880s. Population growth stagnated in the later part of the nineteenth century (fig. 10.1). At the beginning of this study, a few very large land-holdings existed in Locknevi, controlled by a small number of individuals. Agricultural day workers found employment here. The amount of cultivated land increased in the nineteenth century. However, large estates were divided and sold off (Gerger and Hoppe 1980), and their rich owners moved out of the parish, so that landholdings became progressively smaller. Some newly cultivated holdings were in agriculturally marginal land. Thus in Locknevi Parish during the period of this study, resource holdings shifted from being relatively uneven with some very large holdings to more even but more restricted holdings.

Aggregate Analyses

Yearly aggregate population data were available from 1821-1900 for Locknevi Parish, and for 1804-1896 for Tuna Parish, from the Swedish Demographic Database, Umeå University, Umeå, Sweden. From these crude birth, death, marriage and migration rates were calculated.

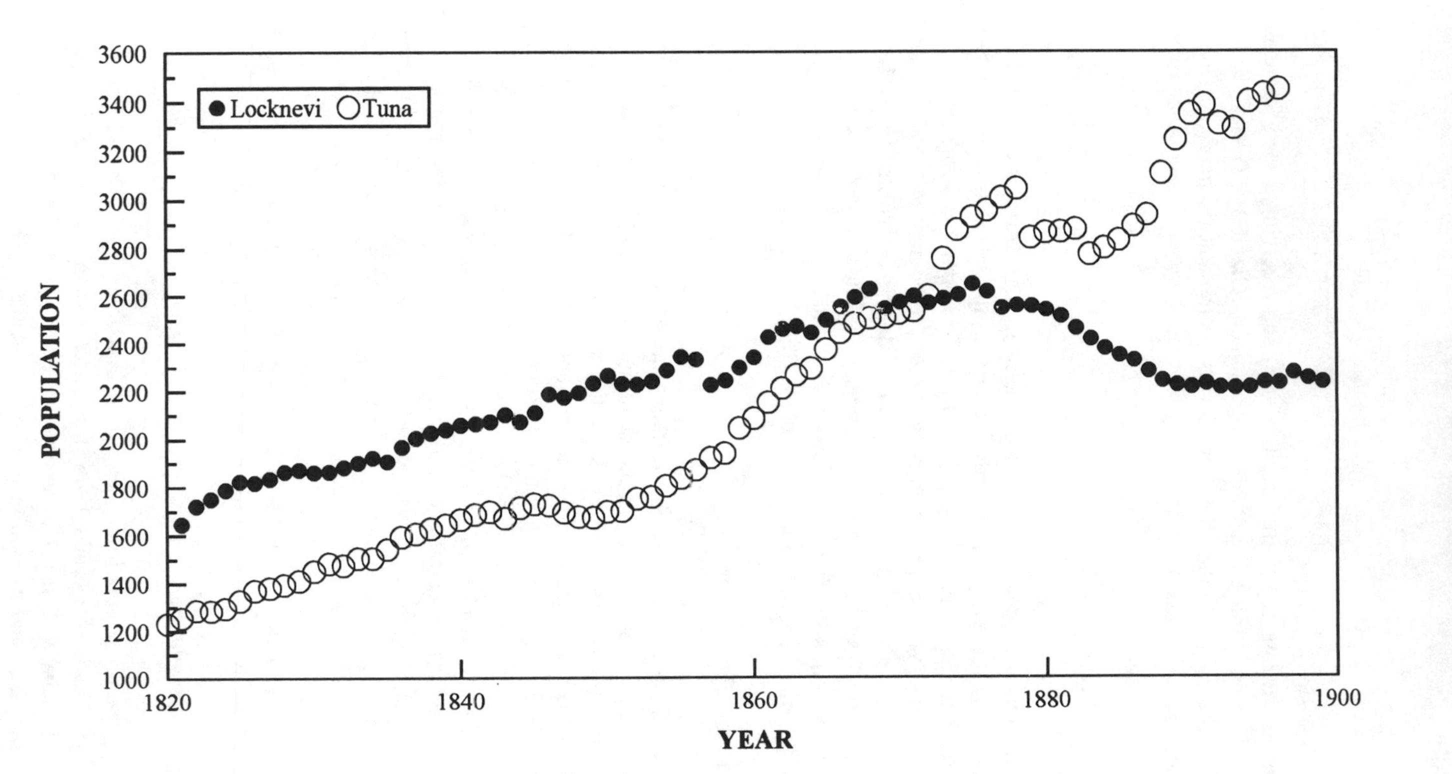

Fig. 10.1. Population growth, 1820-1900, in the two study parishes.

Individual Analyses

Analyses of family patterns were done for the period 1824-1896 using data from the Swedish Demographic Database, which contains computerized demographic and historical data (Low 1989a, 1990a). The structure of the data makes it possible to do longitudinal analysis, following family lineages, similar to the *Ortissippenbuch* used by Knodel (1988). This sample comprised an original cohort of men married between 1824 and 1840 (n=357), and all their male (n=1647) and female (n=1545) biological descendants in Locknevi and Tuna parishes. The original sample was structured this way for other analyses (Low 1989a, 1990a, Low and Clarke 1991). Biological descendants were defined as those whose parenthood, through birth or baptismal records, could be reliably established; the estimate is thus conservative (Low 1989a).

For all individuals in the sample, the following data were collected: birthdate, legitimacy, birth order, father's best occupation, number and dates of marriages, spouse's best occupation, dates of children's births and deaths, date of death or record loss, and type of record loss (e.g., death or migration out of the parish). Adults were defined as individuals surviving to age fifteen; sixteen was the youngest age at which any woman in this sample married. Comparisons were for all men or women born, all reaching age fifteen, or all marrying.

Total Fertility Rate (TFR), Net Reproductive Rate (NRR), and age-specific fertility, were calculated (Palmore and Gardner 1983). These standard measures are designed to allow predictions about populations. However, because they are aggregate measures, they can vary depending on the relative representation of sub-groups, and whether sub-groups behave differently. The strength of inter-group differences is difficult to assess, because intra-population variation is not reflected in the aggregate statistics. Because these measures are based on women, any extensions to men or families ignore death, divorce and remarriage.

This study measures actual completed lifetime fertility (NBC: number of children born to the individual, individual-level analogue to NRR), and number of children alive at ages ten (RS_{10}). RS_{10} is a reflection of "net success," an individual's net reproduction as a function of individual-specific fertility and survivorship. Of course, it would be ideal to measure children alive at marriage, but in this society many children left the home to work in other households (often outside the parish), as young as age fourteen. Thus, assessments were restricted to those children at age ten, before records became unreliable. NBC, RS_{10} and survivorship rates of children were calculated per individual, rather than as population averages. Such measures are important in making comparisons among individuals, family lineages, or reproductive strategies, because they allow statistical testing of inter-group differences (Hughes 1986, Low 1989a, 1990a).

NBC and RS_{10} were restricted to those adults living their lives within the parish. Age-specific fertility, NRR and TFR do not require this restriction. NBC and RS_{10} will tend to be somewhat lower than measures such as TFR, because some individuals die before completing their full potential reproductive lifespans. If there is any difference in the reproductive patterns of people who leave the parish and those who stay, NBC and RS_{10} will also differ from NRR.

Occupation

Though reproductive patterns may vary among individuals of different occupations (e.g., Eriksson and Rogers 1978), with aggregate measures the significance of these differences is difficult to assess. Individual analyses here follow earlier methods (Low 1989a, 1990a). Five occupational categories were used: upper middle class (those with several servants, owners of factories or large businesses, and/or large amounts of land); lower middle class (e.g., small businessmen, artisans, professional soldiers, with one or a very few servants), *bönder* (farmers who owned their land), *torpare* (farmed, had a long-term to lifetime interest in the land, but could not will it; see cottars); *statare* (hired farm laborers) and proletariat (household servants, retired servants, indigent poor) were lumped as a fifth category. Women in nineteenth century Sweden typically had no independent occupation after marriage, and their occupations prior to marriage (most commonly *piga,* household maid) rarely exceeded their father's occupation in status. The higher-status occupation, own-or-husband's, own-or-father's, was used. For some analyses, "richer" (upper middle class, lower middle class, and bönder) were compared to "poorer" (torpare, statare, and proletariat).

Land Ownership

Occupational status is only a crude reflection of the physical and social resources on which an individual can draw. Better tests will use specific resource information when data exist. Income was not taxed during the period of this study. Land ownership was reflected in *Mantalslängd*. A *mantal* was a measure of the amount of land required in any parish to make a family self-sufficient (Lext 1968). This is clearly complex, but within a parish, the mantal reflected a consistent land portion (Lext 1968; Wohlin 1912). Tuna Parish records were available for 1845-1875 from Dr. R. Ostergren, University of Wisconsin; 106 individuals owned land.

Commodity Costs

Climate and weather are obvious influences on agricultural resources (e.g., crop productivity), but their interpretation can be complex. LaDurie (1971), for example, criticizes Utterström (1955) for over-general and broad-sweeping assertion of causation--yet LaDurie also finds broad patterns. Local climate and crop production records are not reliably available for the period we consider. Jörberg (1972), however, compiled substantial data on crop and commodity market prices in Sweden, and found that crop prices co-varied with productivity. Jörberg (1972) also summarized yearly costs by county (=Lan; 24 in Sweden) for staple items, as well as day-wage units required to purchase important commodities. These offer a closer approximation to regional and temporal variation in living costs. Jörberg found that food costs comprised the majority of household budgets (83.3 percent; Jörberg 1972, 2:182), and also that different commodities fluctuated independently in price.

Several factors complicate the use of these data. Costs and availability were influenced by international import regulations and trade patterns within the country. Further, local crop failures which show correlations to people's reproductive lives at the local parish level (e.g., 1840s in Locknevi; Low 1989a) are not always reflected in regional crop prices or cost-of-living indices. Finally, the impact of cost changes may differ for richer versus poorer individuals, or for landowners versus day laborers. Thus, neither crop yields nor prices offer a complete measure of "better" versus "worse" periods and places for all individuals. Further, cost-of-living indices based upon a set "market basket" do not account for substitution of commodities. It is likely that people consumed locally fresh and cheap goods, so that no single cost index is appropriate.

Jörberg's (1972, 2:350) yearly cost-of-living indices are not available regionally for the period of our study, but he does provide five-year average regional comparisons of the cost of food relative to its cost in 1860. These costs represented 96 percent of the budget in Kalmar Lan (including Locknevi Parish) from 1805 to 1869, and 98 percent of the budget from 1865 to 1914 (Jörberg 1972, 2:353); in Västernorrland (including Tuna Parish) they represented 86 percent of the budget from 1805 to 1869 and 76 percent of the budget from 1865 to 1914 (Jörberg 1972, 2:355).

Even at the regional level, such cost-of-living measures require caution. In parishes such as Locknevi, in which migrant workers are an important part of the economy, the relevance of day-wage levels depends on how many days laborers worked in a year (Jörberg, 2:344). In Tuna, the diverse, mixed economy allowed many non-market alternatives to supplement day wages including barter, hunting and fishing. Hence food costs as measured by Jörberg comprised a smaller percentage of the budget. It is very difficult to measure all sources of income and resource acquisition in such environments.

Thus, stronger correlations of reproductive patterns with wages or purchasing power are expected in Locknevi than in Tuna. Finally, under the same prevailing economic conditions, some individuals may experience an economic improvement while others suffer hardships (e.g., as purchasing power declines).

Wages for day work were affected both by the number of available workers and the demand for laborers (Jörberg, 2:336). During the two periods when real wages fell, 1780 to 1800 and 1820 to 1840, there was a great increase in agricultural employment in Sweden. Number of farm hands and sons working for fathers increased from 180,000 in the 1770s to 235,000 in 1800, and to more than three hundred thousand in 1840. Production increased, real wages fell, and economic conditions for laborers declined, while conditions improved for farmers producing for the market (Jörberg, 2:343).

Given the above caveats, several measures are analyzed: day-wage units required to buy a hectoliter of rye, a kilogram of butter, and a hectoliter of dried fish (items not substituted for each other, though estimated by Jörberg 1972 to comprise only about 35 percent of his cost-of-living index); county-specific five-year average costs relative to 1860. The responses of different groups of individuals are also compared.

Statistics

Statistical analyses were done using the *MIDAS* and *BMDP* statistical packages on the Michigan Terminal System. For analysis of marital and occupational patterns, the sample was restricted to individuals surviving to age fifteen, in order to separate the effects of childhood mortality from effects of occupational status, age at marriage, and other adult measures. Analysis of variance was restricted to problems (e.g. survivorship rate within the family) for which data were continuous and normally distributed (Conover 1980). Although ANOVAs are typically used for analysis of number of children, the distribution is clearly non-normal (i.e., many more families have one child than have the maximum number), so the Kruskal-Wallis statistic (Conover 1980) was calculated.

Results

Resources and Fertility

Parish Differences

In Locknevi, several patterns suggest response to resource constriction. The population grew very slowly (fig. 10.1), then declined after 1870. From 1820, net migration was negative (out of the parish) in sixty out of the seventy-seven years (fig. 10.2). Migration out of the parish increased slightly over the study period. Large property owners and upper middle class families appear to have led the out-migration (Low 1989a), though there were few high-status families at any time. General fertility rates fluctuated (fig. 10.3), declining to lows in the 1830s and 1840s, rising in the 1850s and 1860s and declining to a low in the 1880s. Marital fertility and family size were reduced by the 1860s (Low 1989a). From 1841 to 1850, few couples in the individual-analysis sample married, and these couples had few children. This decade was a period of hard climatic conditions (Utterström 1957, 2:442-3), with several seasons having significant crop failures, followed by harsh winters. Further, while purchasing power had been low (fig. 10.4), it reached its lowest point in the 1840s, requiring thirty or more days' work to purchase a hectoliter of rye, a hectoliter of fish, and a kilogram of butter. Purchasing power remained stagnant until after 1860. Agricultural workers here as elsewhere were, until about the 1870s, paid at a considerably lower rate than industrial workers (1/2 to 2/3; Jörberg 1972, 2:345). From 1870 to 1880, out-migration increased, shrinking the labor pool (fig. 10.2). Perhaps as a result, the discrepancy declined during the 1870s. Farmers were forced to increase wages to keep or recruit labor, resulting in an increase in real wages during that period (fig. 10.5). Shifts in land ownership may also have contributed.

In Tuna, the population grew over the study period (fig. 10.1). Migration was net positive in forty out of the seventy-six years during the period (fig. 10.2). Migration contributed little to fluctuations in population numbers, except for the year 1879, when Matfors Bruk closed. Then Tuna experienced a net out-migration of over 250 people, well over twice the net out-migration in any other year (fig. 10.2). Crude marriage rates fluctuated over the period, and showed no temporal or economic pattern. Real wages declined somewhat from 1820 to the late 1830s, then improved steadily (fig. 10.4). The general fertility rate was relatively low until about 1850, then rose, peaking about 1890 (fig. 10.3); there was no consistent evidence of decrease in family size through the 1890s (fig. 10.6, Low 1990a).

Within-Parish Responses to Costs

In both Locknevi and Tuna parishes, relative costs changed over the time period (fig. 10.5). In Locknevi, as food costs relative to 1860 rose, completed family size fell. In Tuna, no relationship is obvious. However, food comprised more of the budget in Locknevi (Jörberg 1972), it is likely that more individuals were dependent on wages in Locknevi, and changes in commodity costs are likely to affect different individuals in quite different ways--as the day wages/hectoliter of rye increase, day laborers are certainly worse off, but landowners may be better off, as the rye they produce brings higher prices relative to the wages they pay their laborers.

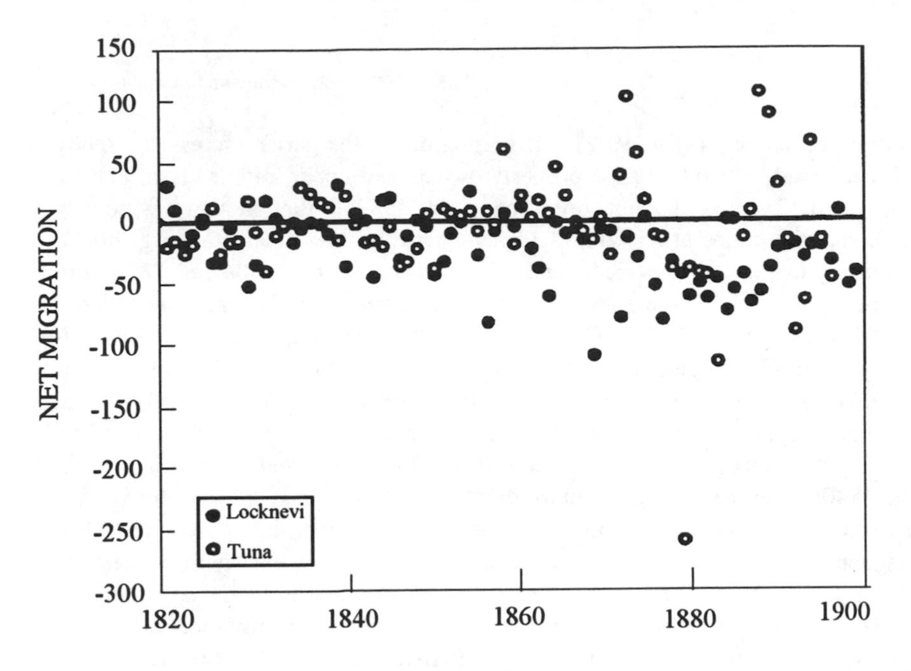

Fig. 10.2. Net migration patterns, 1829-1900, for the two study parishes

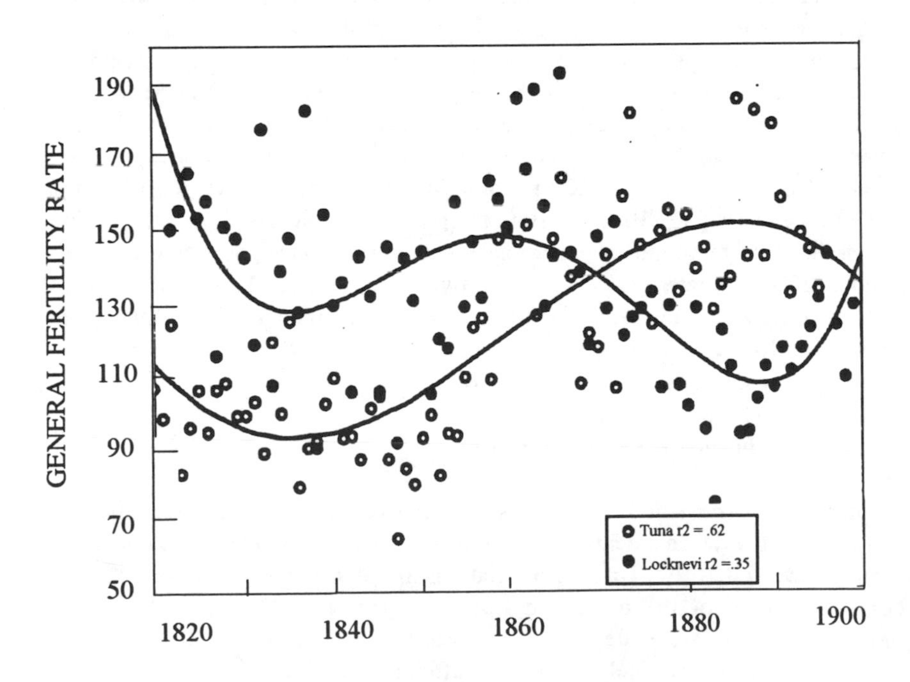

Fig. 10.3. General fertility rate, 1829-1900, for the two study parishes

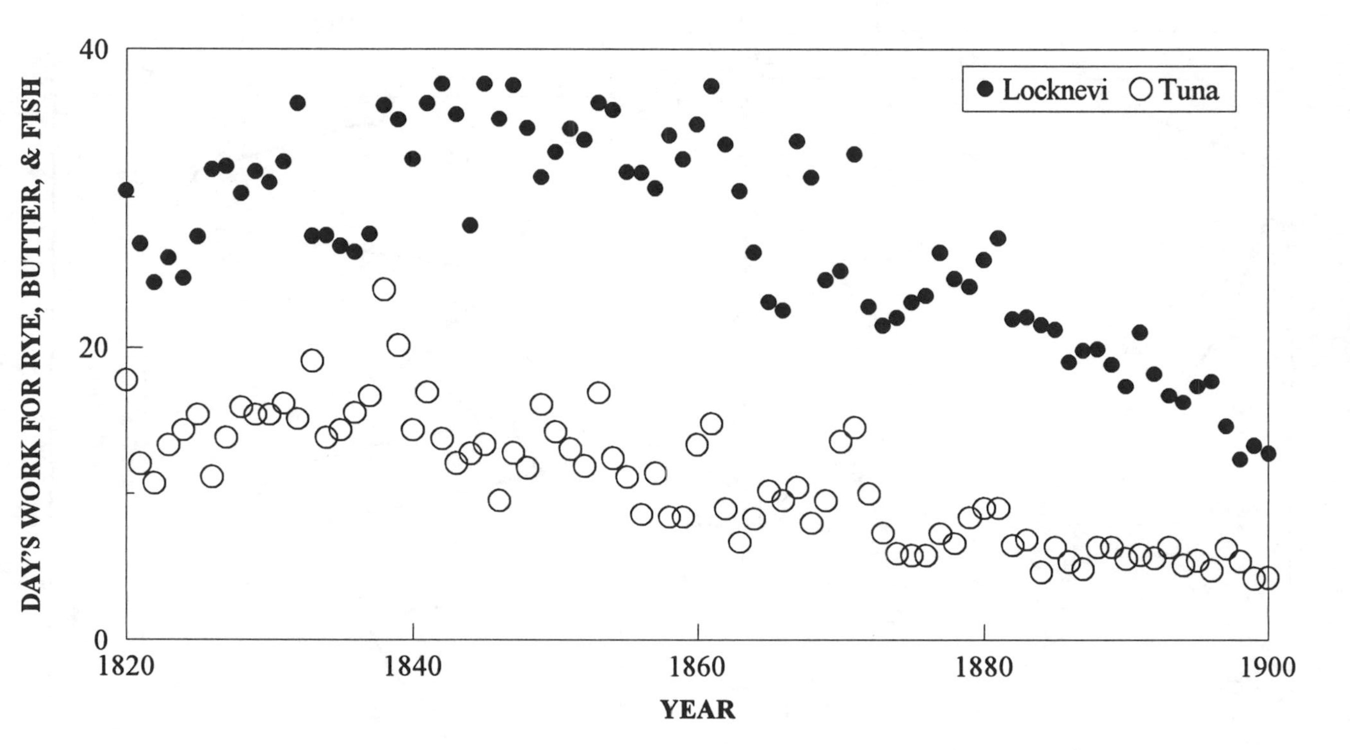

Fig. 10.4. Yearly fluctuations, 1820-1900, for the two study parishes, in the number of days' work required to buy a hectoliter of rye, a hectoliter of dried fish, and a kilogram of butter.

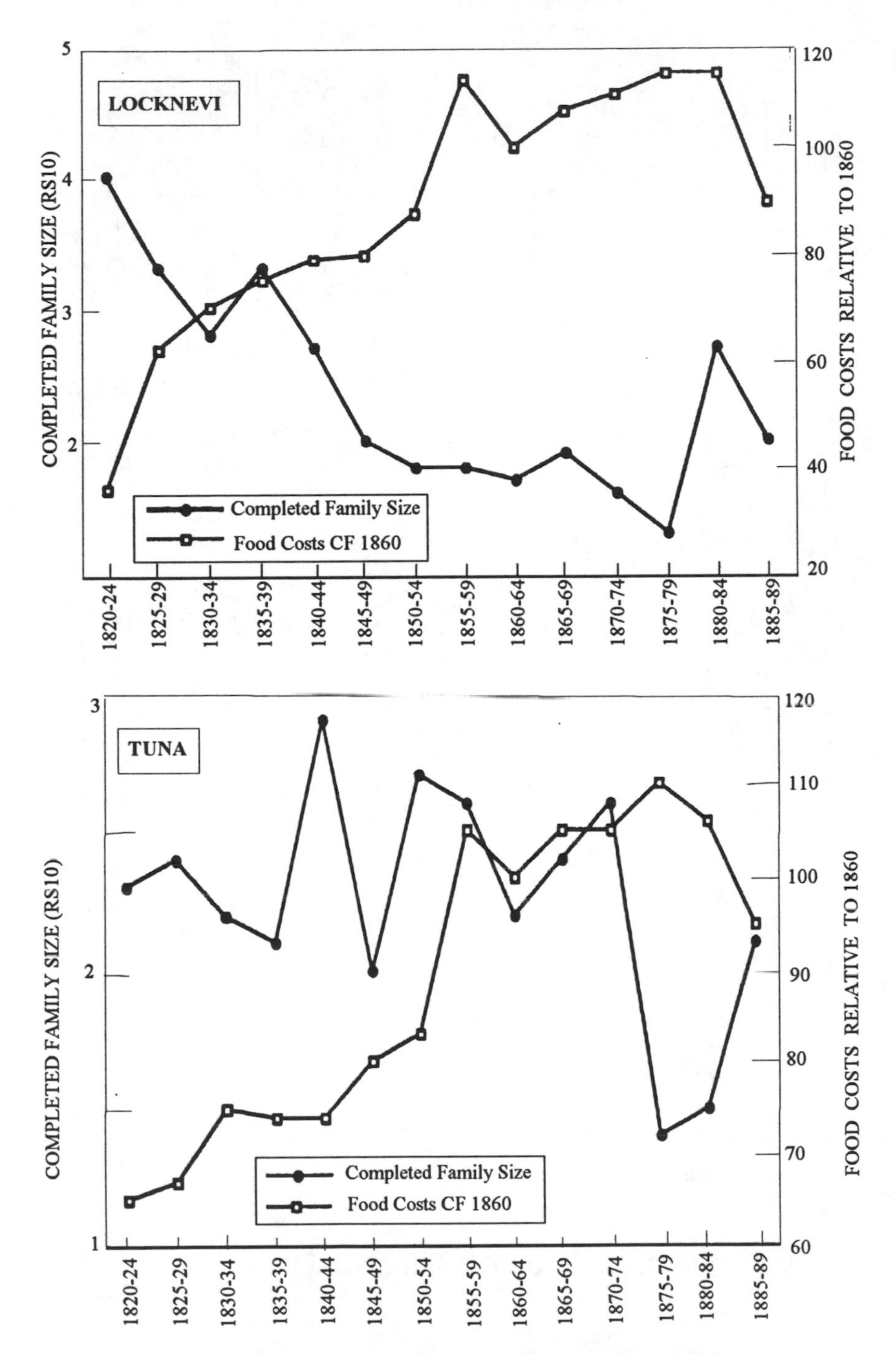

Fig. 10.5. Five-year averages in Jorberg's regional food costs relative to 1860, for Kalmar Lan (including Locknevi Parish) and Vasternorrland (including Tuna Parish), compared to five-year averages in completed family size (RS10) for individuals in the two parishes for whom we have lifetime fertility data.

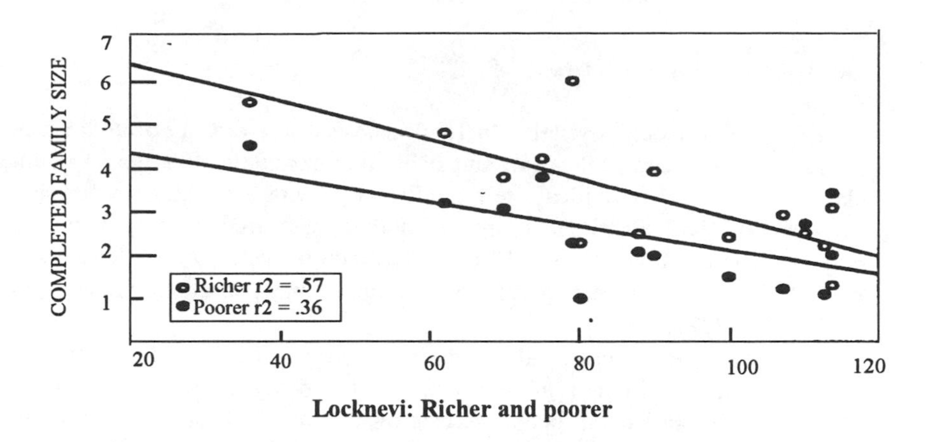

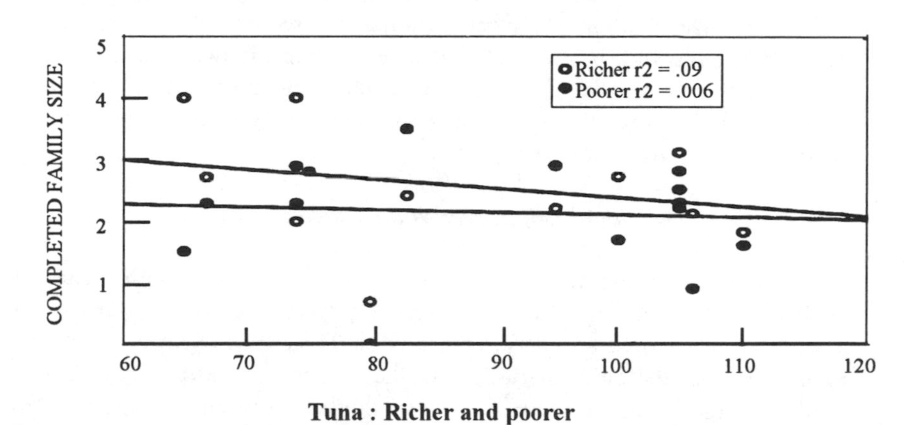

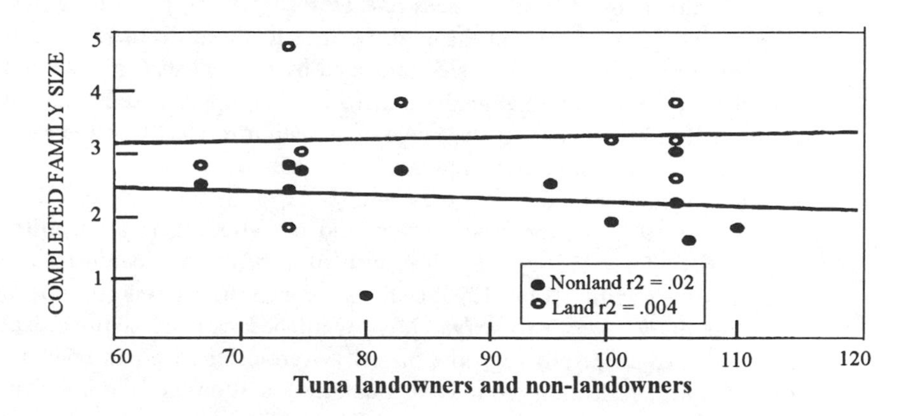

Fig. 10.6. Relationship, in the two study parishes, of the costs of food relative to 1860 and completed family size (RS10) for different categories of individuals.

In Locknevi, as costs relative to 1860 increased, the fertility of both poorer (torpare, statare, and proletariat) and richer (upper middle class, lower middle class, and bönder) individuals fell significantly, with a slightly steeper slope for richer individuals (in fact, upper middle class families literally left the parish, as noted above) (fig. 10.6). When costs were lower, there was a greater difference between family size of richer and poorer individuals (fig. 10.6; Low 1989a).

In Tuna, food costs, like the day wages required for important commodities, were not a good predictor of fertility behavior (fig. 10.6). There was no relationship between costs relative to 1860, or day wages units required for important commodities, and family size for either richer or poorer individuals (Low 1990a, who found no correlation between occupational status and family size). Men of all occupational statuses except proletariat owned land. When the fertility of landowners versus nonlandowners is compared to relative costs, there is a slightly negative, but non-significant relationship for nonlandowners--as relative costs increase, their family size decreased slightly; and a slightly positive for landowners.

Other Individual Differences Within Parish

In Locknevi, men's best occupation was associated with differential likelihood of marrying (Low 1989a). Sixty-four percent of statare and proletarian men failed to marry, compared to 34 percent of lower middle class men. Early in the study period, both the number of children born during a man's lifetime and the number of those children surviving to age ten varied significantly with a man's best occupational status: both decreased with decreasing status (Low 1989a). Prices rose (Jörberg 1972), purchasing power remained low (figure 10.4) and the fertility of richer men declined to that of poorer men (Low 1989a). This is complicated by the fact that in generations two and three of the individual analysis there were no upper middle class men remaining in Locknevi, so that the range of variation in wealth in the sample of richer men is not similar early and late in the sample.

In Tuna, men's survivorship varied with occupation (Low 1990a; Sundin and Tedebrand 1981). Men's occupations did not affect their probability of getting married (Low 1990a). Further, men of different occupations did not differ in lifetime fertility (Low 1990a). Sons of men of different occupations were differentially likely to marry. Men of all occupational statuses except proletariat owned land, although the majority were categorized as bönder, or farmers. *Mantal*, reflecting the relative amount of land owned, did not change between generations one and two (Low 1990a). Thus, there was no biasing correlation between age or generation and land accumulation (Gaunitz 1979; Easterlin, Alter & Condran 1978).

Land ownership, as reflected by the average *mantal* over the period, correlated with reproductive patterns. Landowners had higher fertility (NBC) and completed family size (RS_{10}) than non-landowners (Low 1990a). Landowning men were likelier to marry than non-landowning men (Low 1990a). Even when only marital fertility is considered, the difference persists; married landowners had more children and more children surviving to age ten than married non-landowners (Low 1990a). Landowning men married younger women, with approximately two more fertile years, than non-landowners. There was no difference in the interbirth intervals of children born to landowning versus non-landowning men (Cain 1985).

For women in both Locknevi and Tuna parishes, no patterns were apparent between their husband's or father's occupation and any measure influencing fertility: age at marriage, lifetime fertility, completed family size, interbirth interval (Low 1989a, 1990a). Only women's age at marriage was correlated with lifetime fertility and completed family size (Low 1989a, 1990a).

Within-Family Differences

Farmers (bönder), upper middle class, and lower middle class men had the potential to accumulate heritable resources (e.g., land and money) to a greater degree than torpare, statare and proletarian men. The larger the family size, the more constrained is the distribution of these resources. Until 1845, sons inherited twice as much as daughters in rural areas of Sweden (Lo-Johannsson 1981; Inger 1980). In 1845, sons and daughters were granted equal inheritance rights, but certain restrictions applied: sons had first choice of the land and goods that were to be their inheritance, and were given the right to buy their sisters' inheritance from them (Inger 1980). Thus, sons received relatively more resources directly from fathers than did daughters. But daughters could gain access to resources through their husbands.

Locknevi Parish

For resource-acquiring occupational categories, birth order was significantly correlated with reproductive behavior for men. Earlier-born men were more likely to marry, and had more children than later-born sons (Low 1989a). This pattern was not true for sons of torpare, statare, and proletariat, who had no land, and probably smaller monetary resources. Women's birth order showed no pattern with their reproductive lives.

Tuna Parish

Birth order was not associated with fertility differences overall, but earlier-born sons were more likely to own land (Low 1990a) and thus to have higher fertility. In the individual-analysis sample, nineteen families had land-owning fathers with at least two adult sons. In twelve out of nineteen families (63 percent), either the first-born son alone owned land (n=10), or the first-born son died in childhood, and the earliest-born surviving son owned land (n=2). In four families (21 percent), the first-born and later sons owned land. In three families (16 percent), only a later-born son with adult brothers owned land, but in one of these families, the oldest son died at age thirty-two, which, while adult, was before most men gained land. Thus land-owning men with more than one son distributed their resources unevenly, favoring first-born sons. Adult land-owning sons were far more likely to remain in the parish for their entire life, had more children and more children surviving to age ten than their non-landowning brothers (Low 1990a).

Resources and Migration

Parish Differences

There were very different levels of migration in Tuna and Locknevi. Fifty-three percent of men and 62 percent of women emigrated from the Locknevi Parish (Low 1989a). Migration appears to have played a key role in the population decline (Low and Clarke 1991) observed in this parish toward the end of the nineteenth century. In Tuna Parish approximately 28 percent of men migrated from the parish, compared to 39 percent of women (Low 1990a).

A greater percentage of women than men, regardless of marital status, migrated in both parishes. In both parishes, men and women were more likely to migrate if they were not married. Men may have migrated to acquire resources necessary for marriage, while women may have done so often for the sole purpose of marriage. Unfortunately, the study could not follow individuals.

Within-Parish Differences

In Tuna Parish, men of different occupational statuses were differentially likely to emigrate (Low 1990a). Farmers were most likely to stay, proletariat most likely to leave (Low 1990a). Sons of men of different occupations were also differentially likely to migrate (Low 1990a). Sons of upper middle class, lower middle class, and farmers were most likely to stay; sons of torpare were most likely to leave. Landowners were more likely to remain in the parish than non-landowners. Men who did not own land were more likely than other non-landowners to remain in the parish if their father owned land. For these men, own occupation was not important in the decision to stay or leave. For those men who neither owned land, nor were the sons of landowners, their own occupation was important in staying or leaving. Men of lower occupational status were more likely to out-migrate.

In Locknevi, men of different occupations were not differentially likely to migrate. Possibly, as resources become increasingly limiting, previous occupational advantages disappear (Low 1989a; Gerger and Hoppe 1980). The sample size was not sufficient to allow analysis of this pattern over a series of time intervals. Migration was related to father's occupation. Both men and women were more likely to migrate if their father belonged to the lower classes (torpare, statare and proletariat), suggesting that an individual's access to resources is to some degree related to their father's resources.

For women in Locknevi, birth order made no difference in probability of migrating. Birth order was important for men. Earlier-born men (first or second born) were more likely to stay in the parish (59 percent) than men who were third or fourth born (51.7 percent), or later-born (33.6 percent) (Low 1989a). In Tuna, non-landowning sons were later in the birth order and more likely to leave the parish than their earlier born, landowning brothers (Low 1990a). In Locknevi, married women were equally likely to stay in the parish whether they had children or not; unmarried women with children were more likely to leave the parish than married women or unmarried women without children (Low 1989a).

Thus, within-family differences may also influence probability of migration, perhaps through differences in access to resources (primogeniture, sanctions against illegitimate births). Differences at the family level may not be apparent at higher aggregate levels of analysis. When population-level analyses mask diversity in sub-group behaviors, differences and similarities between and within populations may be overlooked. Further, inappropriate theories of causality may be derived (see Hawkes and Charnov 1988).

Direction of migration (to rural or urban areas), may influence individuals' dependence on wages, versus wage-independent subsistence. Wage-dependent economies may offer greater opportunities for larger numbers of people to increase their resource holdings. Wage-independent economies, if sufficiently diverse, may provide greater insurance against extreme economic fluctuations, especially for low income people.

Resources and Mortality

In Locknevi, men's survivorship did not vary with occupational status (Low 1989a). In Tuna, there was no pattern early in life to men's mortality with occupation, but by age fifty-five occupational differences in mortality were apparent (Low 1990a). This pattern is also reflected in the average age of death of men of different occupations: proletariat (47.8 years) tended to die earlier than lower middle class (56.0), bönder (58.8), and torpare (55.2) ($n=176$, $F=2.49$, $p=.04$). The pattern for torpare and bönder parallels that found by Sundin and Tedebrand (1981) in Swedish iron foundry parishes. Torpare and bönder tended not to not take part in iron foundry production, and had lower mortality than proletariat men and some middle class men, who worked as smiths and production workers. Women's survivorship did not vary with either father's or husband's occupational status (Low 1989a, 1990a). These are results for individuals who spent their entire life in the parish. Mortality patterns in relation to resources and occupations might look quite different for migrants.

Migration, Fertility, and Mortality

How migration influences fertility and reproductive success is important, but little examined, perhaps because there are so few appropriate data on individuals, their movements, their fertility, and the survival of their children. This study was not able to follow the fate of migrants, and thus could not test either mortality risks or possible changes in fertility. In the few studies that have tackled this problem, there is evidence that migrant reproductive success can differ from that of non-migrants in ways related to the relative richness of the new, versus the old, area (e.g., Easterlin 1978; Pernia 1980). Greater understanding of migrant family patterns is vital to accurate population predictions.

Discussion

Tuna Parish possessed a more diverse economy (forestry, ironwork, mixed crop agriculture) than Locknevi. Perhaps, as a result of this diversity and the availability of non-market alternatives (hunting, fishing), population measures did not correspond with measured economic fluctuations in Tuna Parish, while they did co-vary in Locknevi (Jörberg 1972; Sundin 1976). In Tuna, there was also no difference in response between richer and poorer individuals, or landowners versus non-landowners. Landowners, however, had larger families at all times. Perhaps land ownership provided a buffer against hard times, over and above the non-market alternatives. Interestingly, landowners' families not only were larger, but showed less variance (NBC: d.f.=1,481, F=11.07, p=.0009; RS10: d.f.=1,481, F=8.27, p=.004) than those of non-landowners.

In Locknevi, there was a substantial wage-dependent agricultural working class, the *statare*. Purchasing power, as reflected by the number of days' work required to purchase a hectoliter of rye, a hectoliter of fish, and a kilogram of butter, varied more than in Tuna. Perhaps, except for the few richest families, people's economic lives were more uncertain in Locknevi. Family patterns in Locknevi also showed more variance than in Tuna. As economic times got worse, then better, family sizes fell, then rose (figure 10.5).

These differences may be related to ecological and resource differences. In many areas in Southern Sweden, single crops dominated the economy. Rye and corn were major crops, the former was particularly labor-intensive and influenced broad sectors of the population. Bad harvests created real hardships and were reflected in prices and purchasing power. In the North, barley was an important grain crop, but agriculture was more mixed and was consistently supplemented by fishing and forestry. Thus, failure of any particular crop was likely to have less impact on people's lives. Harvest and price information alone are insufficient reflections of conditions in some areas. Sundin (1976), in an analysis of theft and penury in Sweden, found that famines and high food prices were good predictors of theft in the counties relying on one principal agricultural crop. However, in "mixed" counties (e.g., Kalmar, which includes Locknevi Parish), the correlation was weaker. In the northern "forest" counties (e.g., Västernorrland, which includes Tuna Parish), forestry and ironworks as well as hunting and fishing supplemented agriculture, yielding a diversified economy. In these counties, Sundin found no significant relationship among crop yields, cost-of-living indices, and theft rates. Thus, the weaker patterns in our aggregate data from Tuna Parish, compared to Locknevi Parish, are not surprising.

Resource Decline and the Demographic Transition

Lesthaeghe and Wilson (1986) argue that while secularization of attitudes may be an important proximate (Williams 1966) trigger to changes in fertility, only measures that treat the variation in both resource control and fertility across populations will answer whether constriction of resources is a necessary correlate of fertility decline. Men's fertility, in both Locknevi and Tuna parishes, was related to measures of resources, though different measures were important in the two parishes. Lockridge (1983), in a broader study of the fertility transition in Sweden, found that trends in family size, in areas like Åsunda, and Hälsinglands Norra Härad, were reversible and appeared linked to economic conditions. In other areas, like Hedemora, Lockridge found evidence of fertility decline with no evidence of resource constriction, and no later increase. It may be worthwhile examining the interface of population growth and environmental conditions through analysis of individual family patterns.

Mortality and Fertility Patterns in the Demographic Transition

As Knodel (1986) has pointed out, the classic demographic transition model gives central importance to correlations in fertility and mortality; the question is whether fertility decline occurs in the context of mortality decline/survivorship increase (Livi-Bacci 1986). In Tuna Parish, there was no decline in fertility during the study period (Low 1990a). This is of particular interest since economic conditions appeared to improve during the period, in contrast to Locknevi Parish (Lockridge 1983, on the local temporal variation in the demographic transition in Sweden). In Locknevi, overall survivorship of infants and children up to age ten actually decreased slightly during the study. Fertility first decreased, then increased. Further, survivorship increased markedly among children of proletariat, whose fertility did not decrease, while both fertility and survivorship decreased among the lower middle class (Low 1989a). The sample sizes in this study are small, but the results suggest that breaking down samples appropriately may shed light on the actual relationship at the micro level.

Because fertility and infant mortality often co-vary and interbirth interval is typically shorter after miscarriage or infant death, the question of whether individuals "try" to replace infants who die (e.g., Knodel 1988) is sometimes raised. These correlations give no clue as to cause, but individual data allow another test. If parents were in fact trying to replace lost infants, then those individuals with lower within-family rates of survivorship to age ten should also show the highest fertility (NBC and survivorship should co-vary). If, however, there is no such attempt, there should be instead a relationship between survivorship rate and number of children alive at age ten (RS_{10} and survivorship should co-vary). Low (in press) has found that for a sample including these two parishes, fertility is unrelated to survivorship. Rather, there is a strong, positive relationship between survivorship and the number of children alive at age ten. There was no evidence of replacement attempts.

Livi-Bacci (1986) and others have argued that fertility declines are typically led by households of prominent or high-status individuals, while Lesthaeghe and Wilson (1986) proposed that economic factors influence fertility shifts. Though they restricted their argument to considerations of mode of production (i.e., labor-intensive farm production), their argument may be extended to the question of resource level and type (e.g., Low 1989a). In Locknevi, high-status individuals had larger families in the first generation (although the sample size is small); in the second generation, family sizes were more even across occupational status. Further, while migration probability showed no significant statistical pattern with occupation for adults of either sex, children of the three upper middle class men in the initial cohort had mostly (eleven out of thirteen) left the parish by the time they were twenty and before marriage. This out-migration coincided with the breaking up of land parcels into more numerous parcels of smaller size.

Resources, Fertility, and Investment: The Ecology of the Family

Considerable attention has been paid in demographic and sociological literature to the proximate correlates of fertility and mortality patterns: how was some pattern, or shift, achieved? Was marriage age, or interbirth interval, involved? This paper concentrates on another level of the problem. Not only humans, but all living organisms, must solve the problems of resource acquisition and use for survival and reproduction. Contrary to what many non-biologists might imagine, maximization of fertility seldom results in maximum net lifetime reproduction (lifetime family size). Rather, there are correlations between external factors such as richness of available resources, predictability of resources, variance among individuals in ability to acquire resources and patterns of fertility, mortality, and migration.

These results (Turke 1989; Low 1989a, 1990a, 1990b) suggest that, at least in some situations, when resource differentials are great, men can use them to increase their lifetime fertility to a much greater extent than can women. When resources become constricted, as in Locknevi, reproductive differentials are likely to disappear. The extent to which we can predict such shifts as a result of economic conditions or purchasing power will depend on a number of factors: how much parental investment assists individual children, whether individuals are wage- or market-dependent, etc. For example, if there is considerable migration from urban to rural areas, with subsistence farming (lowered dependence on market economies), there may not be a fertility decline in rural areas. Both resource differentials and migration patterns in lesser-developed nations are important. In LDCs, as resources begin to decrease, the risks of migration may be perceived as less onerous. Migration is expected to increase if it can offer individuals an alternative to decreasing fertility. As noted above, little is known about the fertility patterns of migrants.

However, resource constriction is not the only force that can favor fertility decline and in some cases an apparent increase in available resources correlates with fertility decline (e.g., Taiwan: Hermalin 1978). Is there a more general pattern? Perhaps the relative costs and benefits of children themselves are influential (e.g., Easterlin 1978; Becker and Barro 1988; Turke 1989). Figures 10.7 and 10.8 summarize one set of possible relationships. In figure 10.7, the ecological externalities are such that resources may allow enhanced fertility (for men), but resources are relatively ineffective in lowering children's mortality or enhancing their competitive success (e.g., in getting married and starting a family). Such conditions obtain in many traditional societies, and, we suspect, in a good many pre- and proto-industrial societies. Among the Turkmen, for example, Irons (1979) found that richer men had significantly higher lifetime fertility than poorer men. Among the Meru of

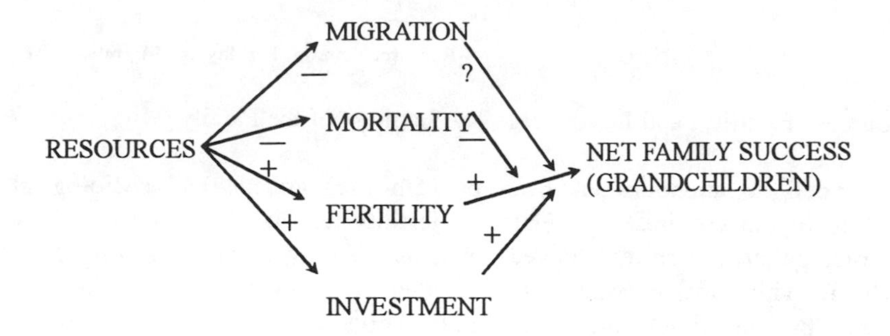

Fig. 10.7. Proposed impact of resources on fertility, mortality, and migration, and investment in children when investment has little impact on children's survivorship or eventual success.

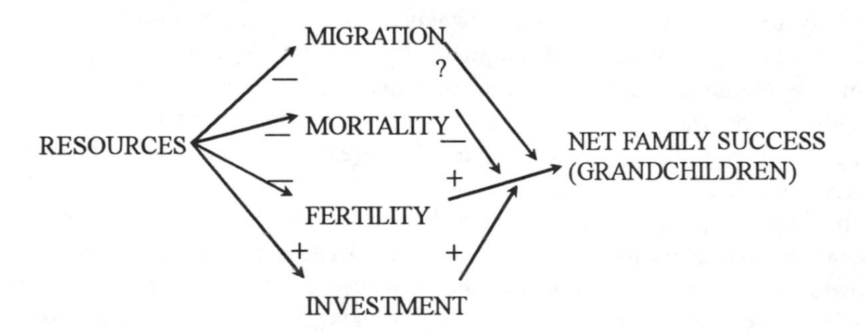

Fig. 10.8. Proposed impact of resources on fertility, mortality, and migration, and investment in children when investment in a child results in a higher probability of that child's surviving and successfully startinga family.

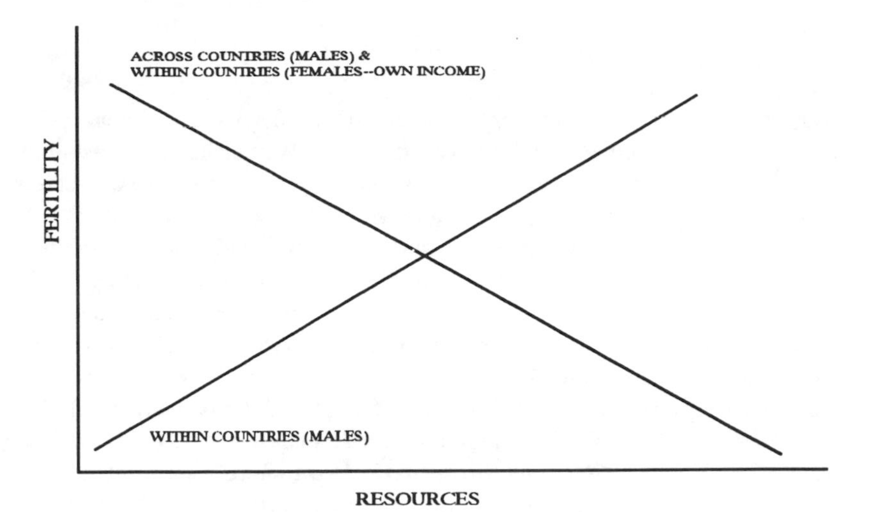

Fig. 10.9. As resource base increases, different fertility patterns are predicted for women within a society, and cross-national comparisons,versus men within a society.

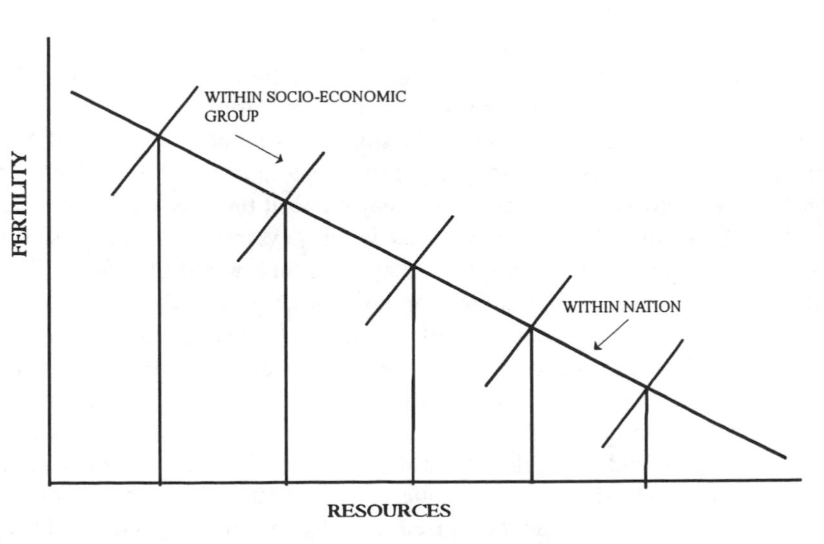

Fig. 10.10. The patterns are further complicated by fertility-resource relationships within socio-economic groups.

Kenya, men with more resources could afford more wives (Fadiman 1982). Even among the Yanomamö, who have little in the way of heritable resources, and who have been widely viewed as an extremely egalitarian society, reproductive success varies greatly, and is correlated with a man's status (Chagnon, 1979,1982). Borgerhoff Mulder (1988) found that the brideprice required for a woman was directly related to her reproductive value, so that richer men could afford younger wives with more reproductive potential. However, in none of these cases was there evidence of a correlation between children's survival and father's resources.

This is really an ecological approach to fertility, and has many similarities both to "individual decision" demographic models (Easterlin 1978; Becker and Lewis 1973; Tilley 1978), and to "proximate variables" approaches (e.g., Bongaarts 1978, 1982) (review by Farooq and DeGraff 1988). As such, it is in contrast to many demographers' view that the main contribution from the field of biology to an understanding of fertility is a physiological one. Thus, while using Hutterite fertility as a benchmark against which fertility of other societies may be measured (Coale and Trussel's m, for example) can be extremely useful, it makes no sense whatever to call it "natural (biological) fertility". It may be as close as humans can come to maximum possible physiological fertility when a variety of ecological constraints are removed. However, most traditional societies probably fit better the term "natural", most such societies have far lower fertility than the Hutterites, and they also show interesting variation in fertility with ecological conditions.

Among the !Kung, for example, Blurton Jones (1986, 1987, 1989) found that maternal strategies are extremely well-tuned and very responsive to the constraint that investment in one child may come at the expense of a mother's ability to invest in other children. That is, closely spaced pregnancies, when nutrition or other factors are limiting, may result in lowered lifetime fertilities. !Kung women have birth intervals of about four years (Blurton Jones and Sibley 1978), far longer than Hutterite women. But !Kung women depend on bush foods, which means that in their harsh environment, women must walk far, carrying her child on her back. Because predators are prevalent, she may have to do so for up to six years. Blurton Jones (1986, 1987) found that modelling "backload" (weight of child plus foraged food), he could predict inter-birth intervals and mortality patterns quite accurately. Thus number of successful descendants was optimized not by maximizing births, but by responding to the interaction of producing a new child versus the costs of such production on the success of existing children. The !Kung represent a well-studied case, but as we learn more about more societies, the pattern appears to be quite general.

Under conditions in which increased investment in individual children enhances their ability to survive and reproduce (fig. 10.8), net lineage success can be enhanced by shifting more resources into investment in children (education, savings, health insurance, resource gifts, etc.). Unless there is a net increase in total resources, the allocation of available resources must be into fewer children (Becker and Lewis 1973, Farooq 1988).

An analogous body of theory exists in behavioral and evolutionary ecology. MacArthur and Wilson (1967), in examining the role of conspecific competition (in the context of island biogeography), noted a potent change in the direction of natural selection, depending on the density of conspecific competitors. They argued that, when the density of conspecific competitors was low, selection favored "productivity," and that competitive efficiency of offspring was relatively unimportant to their eventual success. Under such circumstances, parents should simply maximize number of offspring produced. As the environment filled up with competitors (a comparatively dramatic process on small islands), selection favored the production of more competitive (better nourished, better taught) offspring--at the cost of number of offspring. Parents should shunt resources into offspring investment, even at the expense of offspring numbers--net lifetime reproduction was enhanced not by high fertility, but by lower fertility but better-invested offspring. They named these two conditions "r" and "K" selection, after the areas of the logistic curve at which we would predict these conditions to be manifested. Pianka (1970) noted some correlates of r- and K-selection, and something of an ecological fad followed, with wide misuse of the theory, so that whole species were sometimes described as r- or K-selected, muddling the important central question: *how does the advantage to parents of better invested, competitive, versus more numerous, offspring compare?*

In other species, the ecological and life-history responses to this selection are relatively simple. For example, small organisms, with high or uncertain adult mortality (and thus uncertain ability to give effective parental care), are likely to produce larger number of offspring than those in which adults are relatively safe and long-lived. Those with "safe" parents and uncertain food supplies, in which parental teaching affects offspring effectiveness (e.g., vertebrate predators), have fewer offspring than those in which parental investment is relatively ineffective.

These simple patterns are of relatively little help to us. The considerable diversity among human societies, is in fact closer to MacArthur and Wilson's original argument than across-species comparisons. In many traditional societies, with little or no medical facilities, fertility is by far the strongest correlate of net lifetime reproductive success, and in such societies, men typically strive for resources and status, using these directly in reproductive ways (above, Low 1989b). In fact, the training of boys in these societies is related to their potential reproductive gains from striving (Low 1990b). On the other hand, *complexities in either the ecological or social environment which result in increased effectiveness of parental investment, should result in more investment, at the expense of fertility itself (figure 10.8).* The degree of this shift should correlate with resource richness--if resource richness also increases as the importance of investment increases, the negative impact on fertility will be moderate; if resource richness decreases as the importance of investment increases, the negative impact on fertility will be severe.

For women, the conflict between investment capability and fertility may be even sharper than for men, even in traditional societies (e.g. the !Kung women above). When monetary resources become central to children's success, women may shift from traditional maternal investment patterns to market employment, typically with additional negative impact on fertility (Farooq 1988).

How robust are such patterns? General trends suggest that across modern populations, as resource levels increase (e.g., as measured by GNP), fertility levels decrease. It seems important to ask, in each case, whether effectiveness in getting resources requires more (especially monetary) investment by parents in individual children; when it does, these authors predict that fertility will decline (Knodel et al. 1990) but when it does not (protoindustrialization and labor markets using unskilled labor), fertility decline is not expected. Further, studies of populations at a less aggregate level have shown that fertility may actually increase with income within socioeconomic groups. These correlations highlight an important complication. Because perception of resources may vary for families of different status, and even of different background (do current resources represent an increase or a decrease?), there may not always be well-tuned responses that agree with external assessment of resources. Johnson and Lean (1985) review relevant studies, which suggest that couples assess their income relative both to their parents income in the previous generation and to others in their social-economic group. Similarly, Freedman and Thornton (1982) have shown that in the United States, families make deliberate decisions about family size in response to their judgment of available resources, and that, when deliberately chosen family sizes are considered, there is a correlation between income and family size. When accidental pregnancies are considered, the picture becomes less clear. Studies suggest that when income is judged as favorable relative to others, fertility is increased.

These findings suggest that the investment level required to produce successful offspring varies with environment. If poorer parents cannot substantially enhance their children's' success, then one might expect a more "Mediterranean" form of investment, with larger families, concentration of resources in one or a few children and with others living with the family or leaving early. Couples at the high end of the socioeconomic ladder may do better by investing more per child to allow them to be competitive with their peers (e.g., education, clothing, status acquisitions). The required investment may limit the number of children they can afford. Within sub-groups, however, those with more than sufficient resources may be able to support additional children and still have all be adequately invested.

From this broader perspective, perhaps re-examination of existing data is useful. For example, industrialization may not be, of itself, a force driving toward lower fertility, unless success in an industrial environment requires greater training or monetary investment by parents, resulting in later marriage ages, and often, fewer children marrying (e.g., the "European" marriage pattern; Hajnal 1965). This data also suggest that if high fertility is not a response to infant loss, medical and public health measures leading to increased infant survivorship, though warranted in their own right, will not, in themselves, lead to lowered fertility. Perhaps, any force enhancing the effectiveness of increased investment in individual children should be looked for: ability to purchase medical services, ability to will reproductively useful resources such as land or status (e.g., the heritable status of the British peerage; Hollingsworth 1957). As parents' ability to influence their children's eventual success via investment increases, fertility is expected to decline and resources to be routed through investment, decreasing mortality and increasing success of children (figure 10.8).

Measuring "resources" is, of course, a difficult task. Important resources can differ significantly between and within regions. These differences result not only from physical differences in the environment, but also from the social structure of the population. A "resource" to one occupational group may be a simultaneous cost to another. Relationships between resources and reproductive patterns can be easily clouded by aggregate data and inadequate measures of total resources. Thus, data required for predictive analysis of population-environment interactions in LDCs will depend on careful evaluation of variation in both physical and social environments.

In sum, this review of two Swedish parishes leads to the suggestion that an additional focus will help in understanding part of the population-environment dynamic--the influence of environmental resource richness on population level and distribution, with the family as mediator of population change. What happens to populations is the statistical sum of what happens to family fertility, survivorship, and mobility, as a result of familial resources and effectiveness of investment by parents in individual children. If various kinds of families respond differently to external shifts in resources (perhaps because they have differential access to those resources, or because the shifts profit some while costing others), then what happens to population numbers, and ultimately how the environment in its turn is affected by the population, depend on what proportion of the population comprise different kinds of families. Failure to measure the appropriate resources, or if we look simply at aggregate measures, we may make the wrong predictions and set inappropriate policies.

Relevance to Modern LDCs

In less developed countries today, in which considerable differences exist in familial access to resources, such micro-level analyses might prove especially useful. Within populations, resource-fertility patterns need to be identified and examined. Also, consideration should be given to how developments will affect the effectiveness of parents' investment in children; for example, industrialization, from this perspective, will not necessarily lead to a fertility decline--the important factor is whether unskilled labor is enough, or whether, to be successful, entrants into the labor market will require more training, resulting in more parental investment to afford children's training, and very likely later age at marriage.

The demand for children has been suggested to be driven in large part by a perceived flow of wealth from children to parents and in particular by a desire for old age insurance. Often, the suggestion is that in agricultural areas, children have higher value than in urban or industrialized areas (e.g., Cain 1982, 1983; Hammel et al. 1983). Actual data do not always support this hypothesis (e.g., Gaunt 1977, 1983). Turke (1989) reviews the literature on this subject and provides evidence, including his study of the populations of Ifaluk and Yap, that suggests that children are more likely to represent a net economic cost to parents throughout their lives, in societies with all sorts of resource bases (e.g., hunter-gatherers, agriculturalists). Thus true demand for children is unlikely to arise from a view of children as a source of income.

When children can aid in their own support, however, parents may be able use children's efforts to offset some of the costs of reproduction. In the United States, for instance, industrialization provided a large employment source for children, prior to the establishment of child labor laws (Zelizer 1985). Parents' perceived valuation of children as workers was still high (Zelizer 1985), yet family sizes fell well before the child labor laws prohibited children working. Such evidence weakens the argument that it was a reduction in need for children as laborers with growing modernization that initiated the drop in family size. A better analysis of the societal changes which shift individual skill requirements may provide an understanding of changes with potential to lower family sizes in LDCs.

Chapter XI

Protected Area Deforestation in South Sumatra, Indonesia

Steven R. Brechin
Surya Chandra Surapaty
Laurel Heydir
Eddy Roflin

Introduction

The purpose of this chapter[1] is to explore the relationships between population and the environment found at the local level. Empirically, we attempt to determine why small-scale coffee farmers have deforested large portions of established protected areas (i.e., designated as protection forests and wildlife reserves) within the district of Lahat, South Sumatra, Indonesia. The relationships between population and the environment tend to be complex, fluid, and mediated by a number of additional factors. The relationships between farmers and protected forests found in Lahat are no different.

This more complicated notion of the relationship between population and the environment, however, is frequently overlooked in the literature.[2] Since these are forest areas under a managerial regime, it follows that politico-administrative factors must have contributed to their deforestation. But how important are other factors? And, are they related to changes in population?

---

1. The information presented here is from an on-going collaborate research effort between The University of Michigan, Ann Arbor, Michigan, USA; Princeton University, Princeton, New Jersey, USA; and Sriwijaya University, Palembang, South Sumatra, Indonesia. Portions of this chapter were presented in a paper at the Third Symposium on Social Science in Resource Management, Texas A&M University, College Station Texas, May 16-19, 1990. This chapter has been substantially revised from a paper presented at the International Symposium on Population-Environment Dynamics, The University of Michigan, Ann Arbor, October 1-3, 1990. The authors would like to thank Profs. Stephen Siebert, Jill Belsky, and the Press' reviewers for their useful comments.

2. In his review of the population-environment literature related to development concerns, Myers (1991) is amazed how little actual research has been conducted on the interrelationships of these two important variables. In applying this same observation to a specific resource, the Overseas Development Administration of the United Kingdom makes note strikingly few systematic studies on the links between population and tropical deforestation, hindering our understanding of this important environmental problem (ODA 1991).

Finally, from a policy perspective, can we gain any insight from our research into how to best correct local population-environment imbalances?

The chapter is divided into several sections. It begins with a brief discussion of the literature on tropical deforestation and the status of protected areas. The main body of the chapter contains research findings on the probable causes of protected area deforestation. It also includes a more conceptual analysis of population-environment relationships in general and a review of future policy alternatives.

Tropical Deforestation and Protected Area Status

Deforestation of the world's tropical forests is a major international issue that needs little introduction. Environmentalists and others are concerned about the loss of biological diversity, possible climatic change, the replacement of forests with unsustainable agricultural activities, flooding, erosion, loss of hydrological functions, and more (World Resources Institute 1990-91; Global Coalition 1990; Gradwohl and Greenberg 1988). The pace of tropical deforestation is alarming. The World Resource Institute (WRI) has estimated the rate of tropical deforestation at approximately 20.4 million hectares per year (WRI 1990: 102). This latest estimate almost doubles Food and Agricultural Organization's 1980 estimate of 11.4 million hectares per year (WRI 1988).

From the latest figures on tropical deforestation, Indonesia is ranked third among all countries in annual forest loss, losing an estimated 900,000 hectares of tropical forests each year, (a rate of 0.8 percent per year) (WRI 1990:102). Throughout the tropical countries, including Indonesia, the principal forces behind the deforestation of tropical forests are said to be agricultural expansion (due largely to increasing population), and unsustainable commercial logging. However, both these factors can usually be traced to governmental policies of one form or another (WRI 1988;1990; Repetto 1988); as well as other issues such as the technologies being employed and cultural practices. Specifically, in Indonesia, slash and burn farmers cause about 50 percent of the country's deforestation; the government's resettlement program creates 40 percent; and commercial loggers, 10 percent (Repetto & Gillis 1988). Consequently, when attempting to control tropical deforestation in Indonesia, understanding the behavior of rural people and the pressures they face become essential tasks.

Parks and other protected areas throughout the world, likewise, are seriously affected by events originating outside their borders. They include: industrial pollution, excessive tourism, shrinking or nonexistent budgets, land fragmentation, economic development pressures, growing rural populations seeking arable land, and angry residents (Machlis and Tichnell 1985; Meganck and Goebel 1979; Brechin and West 1990; West and Brechin 1991a). Once again the problems facing our world's parks and protected areas can certainly be traced to a number of causes including population growth pressures, economic development activities, changes in lifestyles, poverty, lack of economic alternatives, and short-sighted governmental policies.

In Indonesia, it has been estimated that 17 percent of the country's protected forest areas (i.e., forests that are not to be cut) have either been logged or cultivated by farmers (Vatikiotis 1989). Although there has been some international work to investigate the effect of population factors such as growth and migration on tropical deforestation in general, it has not generally extended to their effects on specific protected areas (Allen and Barnes 1985; Vayda and Sahur 1985; Whitten 1987; Potter 1988; Rudel 1989; and Cruz and Cruz 1990). Likewise, in the study of protected areas, numerous publications have noted the problem of farmers and others encroaching on protected areas (Vogt 1946; Wetterberg 1974; Eckholm 1976; 1978; Meganck and Goebel 1979; Machlis and Tichnell 1985; West and Brechin 1991a). Few, however, have looked at the population-environment dynamics of farmer encroachment in any detail. This case study investigates the socio-political causes of farmer-based tropical deforestation and its effects on conservation management efforts.

Description of Study Site and Methods

South Sumatra is a vast (109,254 square km) province of Indonesia, on Sumatra, one of the country's major outer islands (Sriwijaya University) (figure 11.1). The province was home to about 6.3 million people in 1990, and contains a variety of ethnic groups, 80 percent of whom live in rural areas (1990 Census, and Sriwijaya University). South Sumatra is blessed with natural resources such as forests, oil, gas, coal, and other minerals, and produces many agricultural products. Its capital, Palembang, is a national center for the chemical and cement industries. Ecologically, the province consists mostly of lowlands and coastal wetlands. The exception is a mountainous region in the extreme western portion of the province, including its highest point, the volcano Mt. Dempo, at 3159 meters (10,425 feet). Mt. Dempo is the climax of a larger mountain range, known as *Bukit Barisan*, which runs north-to-south along the western edge of Sumatra (Dalton 1988).

Geographically, the study area is located in the Kabupaten (district) of Lahat, which is in the western highlands (figure 11.2). A rich agricultural region, Lahat is a major center for the coffee which is cultivated throughout the higher elevations. Within Lahat there are protected areas under several different management categories (conservation/national parks, protection forests, limited production forests, and regular production forests), which, after their expansion in 1982, cover about 290,600 hectares. Consequently, about 41.4 percent of Lahat is technically under forest management

Fig. 11.1. Lahat District of South Sumatra
Source: Imagepro, Inc. Ann Arbor, MI

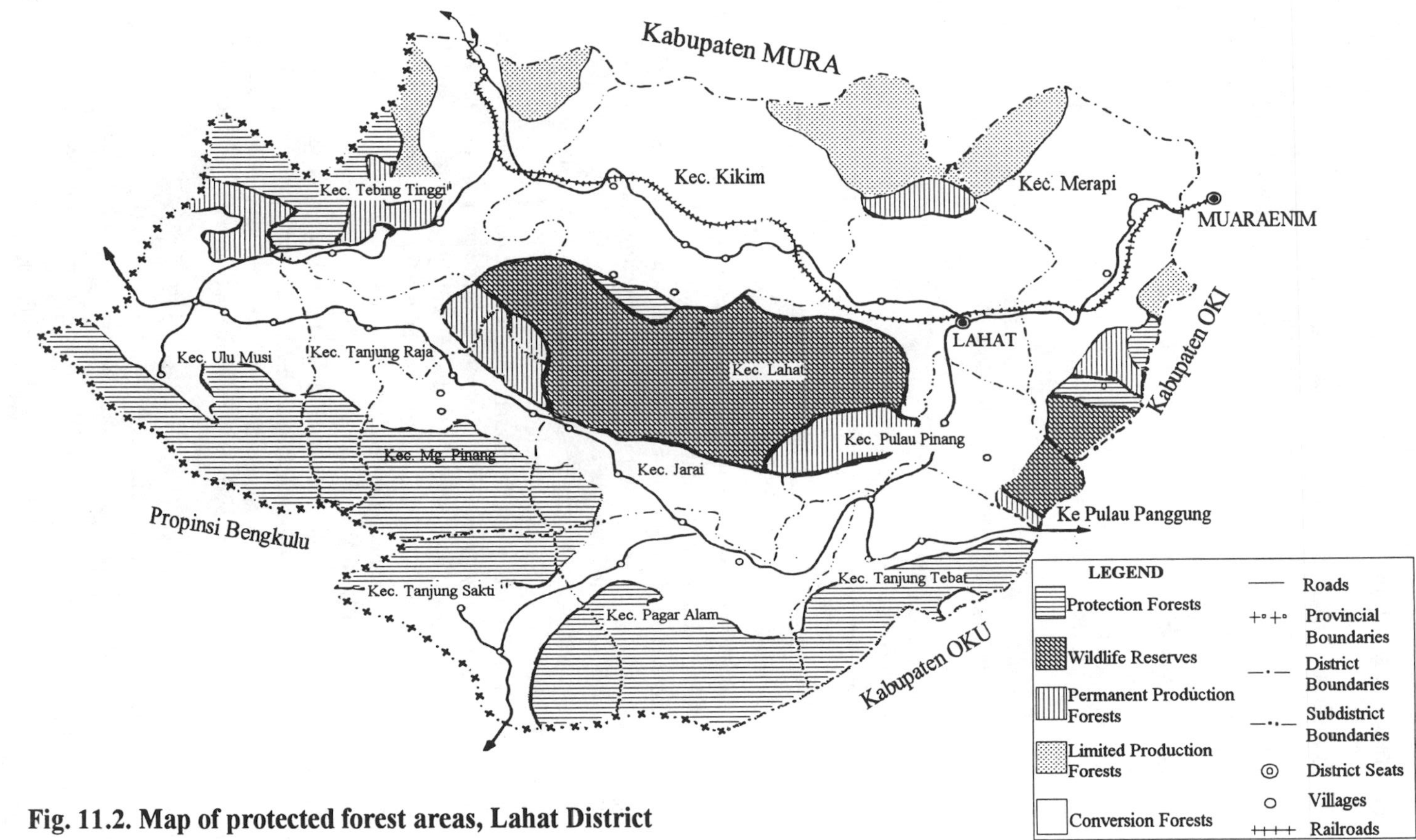

Fig. 11.2. Map of protected forest areas, Lahat District

TABLE 11.1 Protected Forests in Lahat

TYPE	SIZE (HA)	PERCENT TOTAL
Conservation/National Parks (Hutan Suaka Alam)	79,500	27.3
Protection Forests (Hutan Lindung)	149,600	51.5
Limited Production Forests (Hutan Produksi Terbatas)	21,750	7.5
Regular Production Forests (Hutan Produksi Tetap)	39,750	13.7
TOTAL (ALL TYPES)	290,600	100

Source: Lahat Forestry Department; Surapaty et al. 1991

(Surapaty et al. 1991)[3] (table 11.1). Such a large protected area has placed considerable pressure on available land resources. Nearly 80 percent of Lahat's protected areas are non-commercial conservation areas (protected forests and wildlife reserves), not meant for harvesting (table 11.1).

The study, to date, has concentrated only on five southern kecamatans (subdistricts) within Lahat: Pagar Alam, Jarai, Kota Agung, Pulau Pinang, and Tanjung Sakti. These were selected because: (1) their deforestation of protected areas is relatively high and thus they have received considerable attention from the government; (2) the people found in these kecamatans share a common language, Pasemah[4]; (3) coffee is widely grown there; and (4) it is home to many of the illegal farmers. Data is still being collected for this study, but research teams have made field visits in July 1989, August 1990, January 1991, and May 1991. Information has been collected from a number of sources. In depth, conversational style, interviews with farmers, political officials, and former traditional leaders have been conducted. Secondary data on population, economics, agriculture, and forestry have been obtained from subdistrict, district, and provincial governmental agencies, including forestry

3. For forest management purposes, Indonesia uses a classification system of: Conservation/National Parks; Protection Forest; Limited Production Forest; Continuous Production Forest; and Conversion Forest. Percentage of forest protection area is based on Lahat land area of 7,014.23 Km2 - Table 11.4.

4. There are three other languages in Lahat area: Lematang, Kikim, and Lintang (1991 Field Notes).

departments, development planning boards, statistical offices, trade associations; and from published literature.[5]

Probable Causes of Deforestation

The problem of illegal farming in the protected forests of Lahat appears to have begun in the mid-1970s. The movement of farmers, which began as a trickle became a steady stream by the mid 1980's. Satellite images clearly show dramatic loss of forest cover between 1982 and 1989 and even between 1982 and 1985.[6] In 1988, a government report concluded that illegal farmers in Lahat were responsible for deforesting 29,399 hectares (or about 18 percent) of the district's protected areas (see Tempo 1990; Surapaty et al 1991)[7] (table 11.2)

The environmental consequences have been locally and provincially significant. Complaints from villagers at the foot of these mountains have grown in recent years. In particular, villagers are noting formerly uncommon problems such as floods during the rainy season and the lack of water in the dry season. Irregular water flow has disrupted village life, bringing increased health problems, silting of the traditional irrigation systems used for rice cultivation, and even some deaths (1991 Field Notes, Tempo 1990).[8] Soil erosion throughout the region appears to be clogging important natural waterways. For example, provincial authorities noted that the Musi River, the area's largest, is rapidly silting-up, affecting both commercial water traffic, and the river's fisheries (1989 Field Notes, Donner 1987).

In a month-long operation from August 1990, government personnel with police escorts used helicopters to forcibly remove all illegal farmers from these protected forest areas including: Mt. Dempo, Gumai Pasemah, Mt. Patah, Isau-Isau Pasemah, Isau-Isau Lematang Ulu, and at the mountains near Kota Agung. A number of the farmers, along with a local official, were jailed. At a few locations coffee trees and farmers' temporary houses were burned (Sriwijaya Post 1990ab; Tempo 1990; 1991 Field Notes). This has frightened

---

5. We must note that due to the formality of the Indonesian government, all data collection activities have to be formally approved and are monitored. Before field research begins,colleagues at Sriwijaya University obtain written approval from the Provincial Governor in Palembang. This written approval is hand-delivered to the Bupati (the administrative head of District) in Lahat. After his approval, a representative of the district office, usually a planning officer, accompanies the research team to the field. The accompying official serves essential functions as local guide and is a formal point of contact to local-level officials, which is essential for obtaining their cooperation. Still, this official is present during all interviews and may even participate in the discussion. Our sense is that the official did not significantly influence the answers we obtained. Still, this possibility needs to be considered.

6. We are working with the Environmental Research Institute of Michigan (ERIM) on remote sensing applications to deforestation issues in South Sumatra. Much of this work is under the Population-Environment Monitoring System (PEMS) program of the Population-Environment Dynamics Project, The University of Michigan.

7. The 18% figure is based on the amount of protected area that existed prior to the 1982 expansion, 165,900 ha.

8. The Environment Research Center at Sriwijaya University with funding from the Ford Foundation has been studying some of these consequences. See also Naning, M.I. et al. 1988.

illegal farmers sufficiently to keep them out of the mountains so far. In so far as the government is concerned, the era of illegal coffee farming in Lahat has come to a close.[9]

TABLE 11.2 Protected Area Deforestation by Forest Name and Subdistrict (Unexpanded Area, 1988)

FOREST NAME	SUBDISTRICT	AREA (HA)	DEFORESTATION AREA (HA)
Gunung Dempo I	Pagar Alam, Ulu Musi	3,750	100
Gunung Dempo II	Pagar Alam, Tanjung Sakti	8,500	200
Bukit Dingin	Tanjung Sakti, Pagar Alam, Ulu Musi	34,300	1,000
Bukit Runcing	Merapi	8,640	500
Gunung Patah	Pagar Alam, Tanjung Sakti, Kota Agung	33,775	3,000
Bukit Raja Mendaro	Pagar Alam, Jarai	7,450	1,200
Bukit Hitam	Pagar Alam, Jarai	4,460	400
Isau-Isau Pasemah	Pagar Alam, Jarai	3,276	50
Gumai Pasemah	Lahat, Jarai, Tebing Tinggi, Ulu Musi	12,810	4,000
Isau-Isau Lematang Ulu	Kota Agung	2,286	714.5
Gumai Lematang Ulu	Tebing Tinggi, Ulu Musi, Pagar Alam, Lahat	29,210	17,526
Gumai Tebing Tinggi	Lahat, Tebing Tinggi, Ulu Musi, Pagar Alam	3,863	650
Bukit Balai	Tebing Tinggi	13,585	58.5
TOTAL		165,905	29,399

Source: Lahat Regional Forestry Department

From this research there appear to be several factors that have encouraged certain farmers to continue to illegally cultivate coffee in the protection forests. They include: population density pressures from natural growth and in-migration; the coffee production cycle, including the traditional shifting cultivation practices of local coffee farmers; inadequate protected area

9. These matters are delicate indeed. In 1989, violence broke out in which a number of people were killed when governmental authorities and farmers clashed over, among other issues, the removal of illegal farms from protected areas in Lampung Province (south of South Sumatra) (1989, 1990 Field Notes).

management practices; and economic incentives created by changes in the international coffee market.

Population Pressure Factors

Numbers and Density

A critical issue of this research concerns the possible impact population pressures may have had on the area's land resources, especially the forests. Increases in Lahat's population size and density over time may have resulted in farmers eventually overwhelming the available arable land for coffee and other types of cultivation. Deforestation of the area's protected forests and nature reserves could be, then, the result of farmers being forced to cultivate the steeper slopes of the protected areas, possibly the only available lands left for cultivation. Indeed, support for this notion comes from population data which compares Lahat to South Sumatra (table 11.3).

Table 11.3 clearly demonstrates the level of population increases for Lahat and South Sumatra. From 1961 to 1990 Lahat increased from 310,035 individuals to 599,347, an increase of about 93 percent. During the same period, the South Sumatra province grew from approximately 2.8 million to about 6.3 million, an increase of about 125 percent. Although Lahat's population has nearly doubled over the last thirty years, it rose considerably less than the average of all the districts. It is not understood why this is the case.

Although population size is commonly linked with discussion about environmental impacts, population density is a more useful indicator for gauging land pressure (table 11.4). It is quite clear that, after excluding the urban areas of Palembang and Pangkal Pinang, Lahat is the second most densely populated district within South Sumatra. This tends to support the possibility that farmers in search of new land were forced up the mountains. A number of those interviewed indicated that farm land began becoming noticeably scarce in the 1970s.[10]

Table 11.5 lists the population density of our study area in 1990, adjusted for the land area under protection status. When the protected areas are subtracted from the total land area, the population density of Lahat and our study area increased significantly.

Table 11.5 demonstrates that the subdistricts of Jarai and Pagar Alam are relatively much more densely settled than the other subdistricts of the study area. Jarai, at 315 people/square km (1990), is near the density of Bali in 1961 (320 people/square km) (Biro Pusat Statistik 1987), which is not an

10. It should be noted that there are no government or private plantations or other large land holdings in the study area. Of course, the establishment of these types of landuses would place greater strain on remaining lands.

TABLE 11.3 Population Figures in South Sumatra by District: 1961, 1971, 1980, 1990

No.	District	Numbers 1961	1971	1980	1990	%Population Change 61-71	71-80	80-90
1.	Kodya Palembang	474,971	582,581	786,607	1,139,926	2.27	3.50	4.49
2.	Kodya Pangkal Pinang	60,283	74,733	90,068	113,163	2.40	2.05	2.56
3.	Musi Banyuasin	296,226	374,876	591,074	883,719	2.66	5.77	4.95
4.	Ogan Komering Ilir	378,260	445,788	564,031	771,463	1.79	2.65	3.68
5.	Ogan Komering Ulu	381,524	538,575	750,763	963,794	4.12	3.94	2.84
6.	Muara Enim	332,456	363,769	430,827	586,075	.94	1.84	3.60
7.	Lahat	310,035	372,821	484,814	599,347	2.03	3.00	2.36
8.	Musi Rawas	185,693	252,420	366,081	512,077	3.59	4.50	3.99
9.	Bangka	251,639	303,804	399,855	513,946	2.07	3.16	2.85
10.	Belitung	102,375	128,694	163,599	192,972	2.57	2.71	1.80
	Total	2,773,462	3,438,061	4,672,719	6,276,482	2.40	3.46	3.56

Source: 1990 Census
Population Change (61-71) defined by (population 71-population 61)/(population 61)*100 %)/10 years

TABLE 11.4 Population Density in South Sumatra: 1961, 1971, 1980, and 1990

NO. DISTRICT	AREA (KM^2)	1961	1971	1980	1990
1. Kodya Palembang	244	1,946.60	2,387.63	3.223.80	4.671.83
2. Kodya Pangkal Pinang	32	1,883.84	2,335.41	2,814.62	3,536.34
3. Musi Banyuasin	25,669	11.54	14.60	23.03	34.43
4. Ogan Komering Ilir	21,658	17.47	20.58	26.04	35.62
5. Ogan Komering Ulu	10,408	36.66	51.75	72.13	92.60
6. Muara Enim	9,575	34.72	37.99	44.99	61.21
7. Lahat	7,014	44.20	53.15	69.12	85.45
8. Musi Rawas	21,513	8.63	11.73	17.02	23.80
9. Bangka	11,614	21.67	26.16	34.43	44.25
10. Belitung	4.532	22.59	28.40	36.10	42.58

Source: 1990 Census
Population density defined by population in the year divided by land area (People/Square Kilometers)

TABLE 11.5 Population Density of Study Area (and Other Subdistricts in Lahat), 1990

NO.	DISTRICT/ SUBDIST.	AREA (KM^2)	PROTECTED FOREST (KM^2)	POPULATION 1990	DENSITY (POP/KM^2)
	LAHAT DISTRICT	7,014.23	2,906.00	599,347	146
1.	Pulau Pinang	344.57	169.65	21,612	124
2.	Kota Agung	436.82	167.20	31,179	116
3.	Pagar Alam	586.79	161.50	106,075	249
4.	Jarai	391.86	250.00	44,686	315
5.	Tanjung Sakti	482.71	193.50	24,865	86
6.	Merapi	677.18	228.75	32,659	73
7.	Pendopo	269.83	100.00	41,538	245
8.	Ulu Musi	750.68	399.15	45,360	129
9.	Muara Pinang	441.91	287.85	51,256	333
10.	Tebing Tinggi	703.05	233.95	52,813	113
11.	Kikim	1,215.05	426.65	53,657	68
12.	Kota Lahat	713.78	287.80	93,627	220

Note: Population density= Size of land area of district/Subdistrict minus amount of forest protection area, divided by population size, 1990.
Source: Office of Statistics, South Sumatra and 1990 Census

TABLE 11.6 Relationships Between the Amount of Deforestation, Number of Illegal Farmers, and Population Density by Study Area Subdistricts

Subdistrict (STUDY AREA)	Amount of Deforestation (HECTARES)	Ilegal Farmers (NUMBERS)	Density (POP/KM^2)
Jarai	722.25	1782	315
Pagar Alam	559.50	1142	249
Kota Agung	295.25	1034	116
Pulau Pinang	256.50	744	124

Source: BAPPEDA Office, Lahat District and Table 11.5

insignificant level. It is important to point out, as well, that most of the protected area deforestation within this study area is found in these two subdistricts.

In comparing the amount of deforestation, with the number of illegal farmers, with the density levels (all by subdistrict), a series of striking correlations emerge. When evaluating subdistricts, those with low to high amounts of deforestation correspond exactly to those with low to high numbers of illegal farmers, and again with those with low to high population densities. These numbers are presented in Table 11.6.[11] The figures obviously suggest a strong relationship between population density and deforestation.

Migration: Transmigrants or Local Migrants

The existence of a relatively high population density, however, tells little about how the area became that way or where the illegal farmers come from. In-migration is a likely possibility. In addition to the natural rate of increase (i.e., population growth resulting from number of births exceeding number of deaths) of 2.38 percent per year, South Sumatra, including Lahat, has experienced significant in-migration. The province has, for some time, been a major designation site for the government's transmigration program (Romsan 1989; Whitten 1987). Romsan (1989:54) estimates that between 1934 and 1988, 741,425 persons were relocated to South Sumatra from Java and the other densely populated inner islands of Indonesia. Although most of these families were sent to lowland areas, a number of transmigrants were relocated to areas within Kabupaten Lahat as well.

This influx of migrants may well be a possible source of illegal farmers. There are stories throughout Indonesia of failed relocation projects, forcing the transmigrants to seek livelihoods elsewhere (Secrett 1986; Whitten 1987; Hanson 1981). There have also been cases of transmigrants invading the protected forests of Indonesia (Whitten 1987 suggests it is a minor problem, while Secrett 1986 suggests it is major). Romsan (1989; Romsan, per. comm 1991) has found transmigrants to be important sources of forest destruction in some parts of South Sumatra. Table 11.7 shows the transmigration numbers for South Sumatra and Lahat from 1980 to 1987.

Table 11.7 shows that 31,928 or about 11 percent of the transmigrants to South Sumatra settled in Lahat. Although it only represents about 5.3 percent

11. The only minor exception concerns the inverse order of density between the two lowest subdistricts Pulau Pinang and Kota Agung. This slight anomaly, however, seems to have an explanation. In Kota Agung residents and officials alike said that many farmers have not been cultivating all of their land holdings. Instead, many have been "saving" parcels for future use. Although physically more land exists, socially it is unavailable, as some farmers are withholding parcels of land from production. Those interviewed considered the practice to be selfish and inequitable, noting that some people didn't have land to farm. The result has been a "defacto" increase in density, but the physical availability of the land would tend to decrease the density in the actual figures.

of Lahat's total population in 1990, it is not an insignificant number. The arrival of thousands of people needing land could have directly or indirectly encouraged the deforestation of the area's protected forests. In addition, the greatest deforestation appears to have occurred during this same time in the mid to late 1980's.

TABLE 11.7 Transmigration in South Sumatra and Lahat. 1980 - 1987 (Number of People)

Year	South Sumatra	Lahat	Percent Within Lahat
1980-1981	67,167	9.014	13.42
1981-1982	103,472	6,851	6.62
1982-1983	50,896	3,600	7.07
1983-1984	17,847	2,012	11.27
1984-1985	20,039	2,764	13.79
1985-1986	4,844	2,872	59.29
1986-1987	32,510	4,815	14.81
TOTAL	296,775	31,928	10.76

Source: Statistical Office of South Sumatra Province

Based upon the interviews with officials and farmers, however, it appears that illegal farmers are not from ill-fated transmigration projects. Rather, the illegal farmers tend to be local migrants, i.e., from other local areas (kecamatans or subdistricts) within Lahat, from an adjacent Kabupaten, or from Bengkulu, a neighboring province. This finding tends to support the conclusion of Whitten (1987) rather than those of Secrett (1986) and Romsan (1989). It is likely, however, that in-migration has indirectly encouraged protected area deforestation by reducing the amount of unused arable land, as reflected in the relatively high density rates.

According to the field research, there appear to be four different groups of illegal coffee farmers in Lahat's protected areas:

1. *Tanjung Sakti*. Many illegal farmers are from this area. They are local people from the Lahat District (from the Kecamatan Tanjung Sakti).
2. *Semendo*. These illegal farmers are from an adjacent kabupaten, Muara Enim. They are then outsiders to the Lahat, but not to South Sumatra. They have their own native

land, but arable land is very limited. Many young families are in search of new farms.

3. *Manna*. Outsiders to Lahat from Bengkulu, an adjacent province directly west of South Sumatra. They share a common ancestry with the Pasemah peoples centered in Pagar Alam and believe they have some claim to the land there.

4. *Javanese/Suhdahese*. Only a relatively small number of illegal farmers are from Java. Those that are here are not from failed transmigration projects, but have come in search of adventure or for employment. They tend to serve as laborers for the more wealthy local illegal farmers, such as the Tanjung Sakti.

The Semendo are traditionally rice farmers from the low lying subdistrict of Muara Enim. Their system of inheritance is "tunggu tubang", in which the oldest daughter when married acquires the parent's property. This arrangement forces the remaining family members to find new agricultural land elsewhere. Some have found themselves growing coffee in highland areas. The Manna are more traditional coffee farmers (i.e., practicing farming as a way of life) and generally farm a one to two hectare plot. The third and apparently largest group, the Tanjung Sakti, are very aggressive farmers who cultivate coffee as a short-term means to acquire wealth. Their goal is to save enough money to move to the urban areas to pursue other occupations while maintaining coffee farms in the hills. The Tanjung Sakti frequently establish several farms and hire Javanese "interns" as tenant farmers to occupy one site while they move on to establish another (Heydir et al. 1990; 1991 Field Notes).

There also appears to be a unique combination of illegal farmers in each subdistrict of the study area. Table 11.8 shows estimated breakdown (by percentage) of illegal farmers by ethnic group (or home area) in each kecamatan of the study area. From the interviews, it appears that the arriving individuals sent home news of their success which encouraged others to come (1991 Field Notes). This appears to be particularly true of the Tanjung Sakti who tended to illegally farm the wildlife reserve (Gumai Pasemah) north of the towns of Jarai and Pagar Alam.

TABLE 11.8 Percentage of Illegal Farmers by Ethnic Group or Origin for Each Study Area Subdistrict

SUBDISTRICT	GROUP	PERCENT
Pulau Pinang	Tanjung Sakti	50
	Pagar Alam	15
	Jarai	10
	Javanese	5
	Locals/Others	20
Pagar Alam	Manna	90
	Semendo	10
Jarai	Tanjung Sakti	90
	Javanese	10
Kota Agung	Semendo	40
	Pagar Alam	20
	Javanese	10
	Locals	30

Source: Field Notes 1991

Coffee Production Cycle: Shifting Cultivation

Coffee is not native to Indonesia. It was introduced by the Dutch colonialists around 1699 as a cash crop (Heydir et al 1990) and in South Sumatra some time later. Today, coffee is produced in 13, of Indonesia's 27 provinces. In 1989, 369,667 tons of coffee were produced nationwide with approximately two-thirds of it exported, mostly to Japan (26 percent), Germany (23 percent), Netherlands (16 percent), and the United States (11 percent) (Biro Pusat Statistik 1989). Twenty-five percent of all Indonesian coffee comes from South Sumatra alone, the most of any one province (Biro Pusat Statistik 1989). Within South Sumatra, the district of Lahat supplies nearly sixty percent of the province's coffee production (Coffee Export Association, Palembang; 1991 Field Notes; Heydir et al 1990:4). In short, coffee is clearly an important crop in our study area.

Protected area deforestation is likely caused, in part, by the way coffee is produced. In Lahat, at least, coffee farmers have traditionally been shifting cultivators. New ground is broken and coffee trees planted. Fruit is not harvested until usually the third year. Harvesting takes place once a year, extended over about a four month period, usually May through August. At their peak in productivity, trees yield an average of two to three tons per hectare. After about eight years, the coffee yield declines significantly. In anticipation of the decline, the farmers move on to seek new land, thereby restating the cycle only after the third or fourth year (1991 Field Notes; Heydir 1990:34)

The shifting cultivation cycle of coffee farmers is significant for at least four reasons. First, established tradition makes opening up new land for cultivation an understandable practice. Shifting cultivation can, of course, be a sustainable practice under conditions of low population density (Dove 1985). It is also a behavior that might not be easily changed. Second, because of the long lead time required to establish new coffee crops, new land is opened ideally while other land is in production. This type of cultivation practice obviously doubles the strain on land resources. Third, farmers who practice shifting cultivation have traditionally had little incentive to cultivate intensively which would ease the pressure on the land. Finally, under conditions of growing population density, local farmers as well as newcomers looking for land would most likely be pushed farther up the mountains in the direction of protected forests, the only unoccupied lands left.

Protected Forest Management

Because these forests are under a managerial regime, their invasion by farmers obviously suggests an administrative failure of one sort or another. Of some interest is the history of these forests. Far from being products of modern conservation efforts, a significant part of these areas were established centuries ago by local authority structures (marga) as forests to serve a combination of woodstock reserves and watershed protection functions (Ayek Tulung) (1989, 1990, and 1991 Field Notes; Heydir et al. 1990; Brechin et al. 1990).

The Dutch Colonialists made their way to South Sumatra in 1859. In 1874 they initiated "Domein Verklaring" in which all unclaimed land came under state rule. Traditional marga systems, while under Dutch control, managed their own lands, including forests. Although marga officials still actively helped regulate their use, the Dutch in 1916 formally incorporated the marga forests with their forest areas and collectively called them "Bosch Wezen" or registered forests. In 1967, after independence, the Indonesian government continued this arrangement under Forestry Principle #5. As under the Dutch, the Pasirah or marga head, with his council, regulated their forest use through traditional law or "adat." This arrangement ended in 1983 when the marga system in South Sumatra was completely dismantled by the central government and replaced with the "desa" or village system. Presently the country's forests are under the jurisdiction of the Ministry of Forestry and are administered in an hierarchical manner from the central government to province to district level.

From the research, it seems that under national government control, the managerial regime existed mostly on paper, lines on maps with little actual initial enforcement. Whitten 1987 found the same for other parts of Indonesia as well. Government control of protected area boundaries became a post-hoc matter, years after they were initially invaded. It appears the forests were more tightly controlled under Dutch rule. There are reports that illegal farmers were shot occasionally (Heydir et al. 1990). With national independence, after World War II, the level of supervision of forests fell dramatically due to tight budgets and limited personnel. It was reported that during the 1970's the level of forest supervision became even weaker. Even today there is also, on average, only one forester for every three kecamatans (1991 Field Notes).

Unlike the Dutch foresters, the Indonesian foresters in Lahat today, except for a special police force, are unarmed. They also have no vehicles, i.e., they are completely on foot, and walk alone through the forests. Their tasks in the protection and other non-commercial forests are to observe local situations and report boundary violations to their Forestry superiors at the district level (1991 Field Notes). Under this system, subdistrict administrative officials, including the head (Camat), have no direct authority over the local forestry officials or their activities[12] (1991 Field Notes). There are obvious drawbacks to this supervisory system, including the lack of coverage, but also the creation of an atmosphere of intimidation and corruption which is discussed later.

Economic Factors: Coffee Prices

For farmers, coffee has been a relatively lucrative cash crop. And most coffee farmers are considered fairly wealthy by local standards. Until recently, coffee generally held a 7:1 to 13:1 domestic price advantage over paddy rice, a major staple crop.[13] In 1976 and 1977, however, the price of coffee skyrocketed due to coffee crop failures in Brazil (1991 Field Notes; per. comm. National Coffee Association (New York) 1991; and deGraaff 1986). This created a price shock waves throughout the international coffee markets. For Indonesian coffee farmers the domestic price differential between coffee and rice rose to 53:1 in 1977 (1991 Field Notes).[14] Figure 11.3 shows coffee and rice prices, as well as coffee production levels over time. The decline in

12. One Camat we interviewed complained about his lack of control over forestry officials. He noted that subdistrict officials have nothing to say about where they go or what they do. He complained that it was 8 months after he arrived as the new Camat before he met the local forestry official (1991 Field Notes).

13. This information is from South Sumatra Commerce Department, the Lahat Statistical Office; and Coffee Export Association, Palembang. Prices were in Rupiah per Kilogram.

14. The numbers presented above are based upon national-level data collected on coffee and rice prices noted above from the South Sumatra Commerce Department. Although no hard figures were collected from the field, local farmers and officals consistently noted a 10:1 coffee price advantage over rice. This probably reflects local prices paid to farmers as opposed to the numbers presented in Figure ll.1 (1991 Field Notes; Heydir et al. 1990).

coffee's advantage over rice reached its lowest mark in at least fifteen years in 1987.

Discussion

In summary, the protected area deforestation within the study area appears to have been the result of a complicated set of factors, including population density pressures, the coffee production cycle, inadequate protected area enforcement, and a rise in international coffee prices.

The illegal farmers responsible for this deforestation tended to be local migrants, who were lured to the protected forests by the usually high price for coffee, caused by a series of severe frosts in Brazil during the mid 1970s. They were not members of unsuccessful transmigration projects. As local lands were occupied, the protected forests were in effect the only lands available for cultivation. The farmers' entry into the protected forests was facilitated by the initial lack of boundary enforcement from forestry officials, a little corruption, and some confusion as to the precise location of the boundaries.

This study found two groups of illegal farmers. In fact there are many more families within the second group than the first. The first group is those who more or less purposefully invaded the protected forests to cultivate coffee; these have been the focus of the study. The second and much larger group is the farmers whose holdings became illegal as a consequence of the government's 1982 decision to substantially expand the size of many protected areas by redrawing boundaries. Thus, a distant governmental decision has transformed many rural families into illegal occupants of state owned protected areas.

The second group is noteworthy for several reasons. Most important, the government, by its efforts to correct perceived deficiencies in its conservation program, has unwittingly but significantly increased the population density of the region by decreasing the amount of available land. This has greatly complicated the situation and will make solutions that much more difficult to achieve. Second, in its treatment of the matter, the government is making little distinction between the two groups of illegal farmers.

Of considerable interest is the fact that the protected areas under central government control were deforested first. Although all protected forest areas were technically under the control of the Ministry of Forestry, many of the areas included former marga forests which effectively remained under the local control of the marga head, Pasirah and regulated by "adat" or customary law. Local control of the forest areas seems to have been quite effective up until the traditional marga system was dismantled entirely in 1983 (1989, 1990, 1991 Field Notes; Romsan 1989; Heydir et al. 1990; Poffenberger 1990a). After

1983, farmers began to invade these parts of the protected forests as well (1991 Field Notes).

From a farmer's perspective, the uncertainty regarding the precise location of the areas' boundaries has further complicated the situation. Many markers are missing or have been moved numerous times, both legally and illegally, to the point that no one is certain of the boundaries' correct location. In some cases, it was noted that certain forestry officials had changed boundary markers for a price. Even more honest forestry officials, however, would be powerless to stop a large influx of farmers into the forests. In short, an unarmed, solitarily forester on foot is no match for a group of matchet-wielding farmers. In one area there are reports of collusion among local government officials who sold protected land to unsuspecting farmers eager to grow coffee. This greatly complicated the situation with illegal farmers being able to provide documents of ownership (Tempo 1990; 1991 Field Notes). Also of interest, several officials commented that enforcement of the protected area seemed to lessen precisely at the time the coffee prices rose dramatically (1991 Field Notes). This may only be coincidence or the result of more conscious action by powerful figures in more central positions with economic ties to coffee markets. In a similar vein, corrective action is presently taking place at time when coffee's price advantage over other crops such as rice is at a fifteen year low point (fig. 11.3).

It seems that since independence in 1945, the protected forests of South Sumatra have undergone three expansions: in 1971, 1975, and 1982. A fourth change took place in 1986, but it only reorganized the classification of existing protected areas, new areas were not added. Significant change occurred in 1982. This was the result of a decision to change the criteria used for defining protected areas and determining their classification (1991 Field Notes). The former criteria consisted of forests with elevation greater than 700 meters and slope of 45 percent or greater. The new criteria was a formula which took into account slope, soil type, and rainfall.[15] The result was nearly a 350 percent increase (from 1,562,783 to 5,214,700 hectares) in the size of protected areas in South Sumatra. Many villages and residents are now technically illegal occupation and are expected to be relocated. In Lahat, it appears the amount of protected forests increased from approximately 165,000 to 290,600 hectares, an increase of about 76 percent (Lahat Forestry Department; Surapaty et al. 1991).

---

15. The formula was (Slope x 20 + Soil x 15 + Rainfall x 10). Total score determined type of area. For example a total score of 175 + = protection forests; 124-174 = limited production forests; <124 = production forests (South Sumatra Provincial Forestry Department, Palembang).

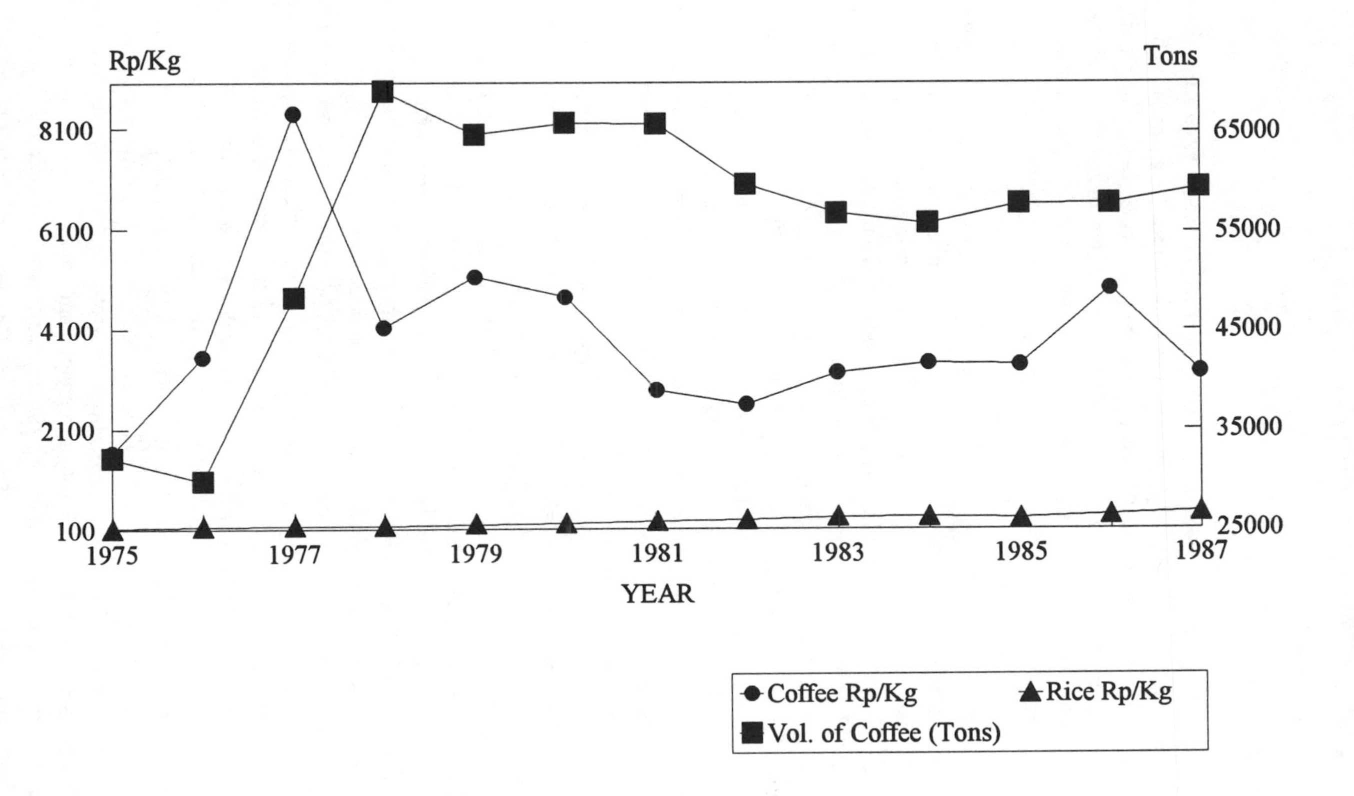

Fig. 11.3. South Sumatra 1975-1987: Coffee & rice prices with coffee export volume
Source: South Sumatra Commerce Department

In attempting to understand the relationships between population and environment from the case study, each of the four factors (population density; coffee production cycle; conservation management practices; and rising coffee prices) appear to have collectively contributed to the deforestation. The most powerful factor, however, in determining the amount of deforestation seems to be population density. As was noted above in table 11.6, the most densely populated areas were the sites of greatest deforestation.

The four factors, however, are interrelated. For example, the high population density surrounding the protected region encouraged farmers to seek out the protected forests as the only remaining unoccupied lands. However, these farmers could have been stopped from entering the forests given better resource management efforts. The lack of a substantive conservation management regime allowed farmers unimpeded access to the forests, at least initially. This may help explain why population density appears to be highly correlated with the amount of deforestation. More effective boundary enforcement may have forced some other dynamic. The fact the marga forests remained intact when the marga system was operating while the state-regulated protected forests were invaded suggests that certain control mechanisms might have worked.[16]

Similarily, the dramatic rise in coffee prices alone is an insufficient cause. Coffee prices created tremendous incentives for cultivation. Again, this became a factor due to the lack of alternative arable lands, and was compounded by the tradition of shifting cultivation among the coffee growers and the weak enforcement structures. The increase in demand for more coffee cultivation might have been met by utilizing unused agricultural lands or intensified use of existing lands. A host of other likely scenarios could be conjured up using different dimensions of these same factors.

From the case study, it is obvious that the relationships between population and environment must consider the impact of other variables. The dynamic is not unilaterally determined. Rather, it is actually the result of the confluence of a number of factors occurring at different scales and at different times. For example, the poor weather that destroyed much of the coffee crop in several high production states in Brazil contributed to the deforestation of specific protected areas in Southeast Asia. The obvious link is international market mechanisms. Other factors may be more controllable, such as the character and effectiveness of state conservation management policy. Others factors may depend on local customs, such as the shifting cultivation of coffee farmers and the effectiveness of the marga system. These may be so ingrained in everyday life as to be extremely difficult to change without creating other problems.

16. In many situations within developing countries, resource management problems seem to develop when the resource control responsibilities are shifted from local to state levels. The effectiveness of some local institutions in regulating forestry use in developing countries has become well documented (Uphoff 1986; Brokensha and Riley 1989).

Clearly, though, effective state policies and implementation could have greatly reduced the impact of the exogenous influence of market forces and the movement of people. But from a population perspective, given the uniform lack of enforcement across the study area, the increased land pressure through population increases, as reflected in density, certainly appears to be the single most powerful factor in determining the amount of deforestation in each subdistrict of the study site (table 11.6).[17]

Possible Policy Directions

It is difficult to predict what will happen to the farmers and forests of Lahat. The future will be determined, in large part, by the implementation of specific governmental policies.

The main policy currently being pursued by the government is the relocation of illegal farmers. This includes both types of illegal farmers discussed above. Here illegal farmers include those individuals who were the subject of our investigation and those villagers who are now considered illegal because of the government's decision to expand the boundaries of protected areas. 1,167 families (or 4,720 individuals) are in this group (Surapaty et al 1991; 1989, 1990, 1991 Field Notes). Because of the number of families involved and the lack of suitable relocation sites, however, it is unlikely this program will be very successful. Little concrete action has been taken so far due to the lack of capital and alternative lands.[18]

In the summer of 1991 most of these illegal farmers were biding their time in the local towns such as Pagar Alam, and harvesting existing crops. The government has agreed to allow illegal farmers to harvest the 1991 crop only if they don't clear any additional land. After this harvest, they are not to return to the protected areas (Tempo 1990; 1991 Field Notes). They are also waiting to see if the whole affair will blow over so they can return to their lucrative enterprise. Thus, this may be only a hiatus in the deforestation of Lahat's protected areas. Relocation by itself is not the answer to the problem. The government has yet to institute any changes in regard to its resource management policies. It appears content with using dramatic means when it becomes necessary to enforce protected area boundaries, some time after they have been violated.

---

17. This is supported by the data collected so far from four subdistricts within the study site. Obviously more data from similiar additional sites, which would allow for statistical tests, are required before we could confidently substantiate this claim.

18. One exception is the village of Semidang Alas (Kecamatan Pagar Alam). Villagers are presently being relocated to a site at lower elevation, called Padang Muara Dua. The site is one of only a handful of unoccupied lands left in the District (government owned). At 650 hectares the site will provide land for about 200 - 250 families, which is slightly more than the present size of Semidang Alas. At a lower elevation within Lahat, the soil and climate are not ideal for coffee. They will be required to cultivate rubber trees, a crop with which they have no experience, nor is it a crop as financially lucrative as coffee (Heydir et al. 1990).

Clearly if the government is to respond effectively to issues involving population-environment relationships, a more intergrated, or at least comprehensive, multi-sectoral approach is required. Piecemeal solutions to complex, interrelated problems will likely succeed only in creating more problems. Likewise a more integrated monitoring system is required to observe the many varied connections that compose this affair. Some elements of a comprehensive policy could consist of the following:

Revised Conservation Protection Policies and Administration

As was discussed in the chapter above, one of the main reasons for the invasion of protected forests was the lack of immediate control over their boundaries. Consequently, tighter control over important forest areas is desperately needed to eliminate similar problems in the future. This is especially true for those farmers who illegally invaded the forests at the higher elevations to plant coffee on the steeper slopes, and, as a result, caused most of the environmental damage. More personnel, better equipped and supervised, would be an important first step to implementing such a policy. Another option would be to return control of forests to more traditional governmental systems. Before they were dismantled, the margas were fairly effective in regulating forest use. Empowering traditional governmental authorities with local resource management responsibilities is an option that is gaining some support in the recent resource management literature (Poffenberger 1990a; Brokensha & Riley 1989; and Uphoff 1986). Although this type of action would presently contradict existing government policies, we believe it could be quite fruitful for the government environmental protection efforts. By finding ways to resurrect selected traditional enformcement structures and integration them with the new national governmental structures could possible create more effective regulatory mechanisms.

There may be a need to simultaneously revise existing conservation policies that require the automatic removal of resident people from protected areas. This refers specifically to the class of farmers declared illegal due to the expansion of protected area boundaries. This concern ties in with the relocation option discussed below. Instead of automatically removing residents, perhaps other options could be initiated that would help to achieve the conservation objective but not require moving large numbers of people. Various alternatives that regulate certain land uses or initiate preventative measures may be far more appropriate, especially when alternative lands are scarce (West and Brechin 1991b.) Conservation zones, for example, are widely used throughout the world. In addition, perhaps certain types of agroforestry practices could be established to help encourage more sound and sustainable agricultural activities. This would require substantial changes in the way the Ministry of Forestry is presently pursuing forestry practices in South Sumatra.[19]

Reforestation Activities

As of yet, the government has failed to initiate any program to reforest the damaged protected areas. Flooding and silting of irrigation systems will undoubtedly continue in some form for some time to come, especially in those areas where coffee trees have been destroyed. The government should take active steps to replant trees where needed and to stabilize soil and water resources of the region. An opportunity exists to constructively include local people in these useful conservation activities (see Dani and Campbell 1986).

Population Control

Although Indonesia in general remains a model of effective population control through voluntary family planning programs, rural South Sumatra's fertility rate still remains relatively high. Presently South Sumatra has a growth rate of 3.09 compared with an average of 1.98 for all of Indonesia (Biro Pusat Statistik 1990). More active population programs in this region can be beneficial in reducing the population dimension of future population-environment relationships in the region. This suggests policies geared toward: (1) limiting fertility through family planning programs; (2) limiting in-migration to the area; or (3) relocating some farmers to less densely populated areas if appropriate areas can be found. If relocation is to be pursued and equity maintained, effort will be required to make important distinctions among the two types of illegal farmers.

Agricultural Intensification

On a positive note the government, as part of its general development program, is promoting intensive cultivation practices throughout Indonesia. In Lahat, there appears to be some limited success with coffee. Of course, with little in the way of alternative land resources, most coffee farmers have been forced to stop their more traditional practice of shifting cultivation. In one community, the village head has been actively working with other local farmers and encouraging them to cultivate intensively by using coffee plant waste as fertilizer (1991 Field Notes). To be more effective, however, agricultural intensification needs greater local emphasis, with special attention to coffee cultivation.

---

19. See Poffenberger 1990bc; and Peluso and Poffenberger 1989 for examples of alternative approaches.

Creation of Economic Alternatives

Given the relatively high density of the rural highlands, another option would be to create greater economic opportunities in the urban areas and sectors. Urban pull may help to draw excess populations from the hinterlands where they practice unsustainable agriculture because they are forced to cultivate the more marginal lands. Another option would be to pursue the development of alternative but equally lucrative crops that could be grown in the less-densely-populated lowland areas. This last option usually requires the development of infrastructure, such as roads, as well as markets. Both options are difficult and would have to be included as part of larger development agendas.

Relocation

West and Brechin (1991b) in their review of parks and people issues note that relocating residents from protected areas should be an option of last resort. In many countries relocation tends to be the first and only option considered. In locations where population density is relatively high and pressure on existing land severe, relocation is likely to only substitute one set of problems for another. This would probably be the case in Lahat. Unless the authorities are prepared to move the illegal residents outside the district, relocating several thousand farm families successfully to alternative sites nearby will be extremely difficult because of the lack of available land.

If it becomes necessary to determine who should remain and who should go, authorities may want to review carefully the characteristics of the various groups of illegal farmers may be classified as "intentional" and "inadvertent." In addition, there are important differences among the intentional group. Some are impoverished people who out of necessity farm one to two hectare plots for subsistence. By contrast, most of the environmental destruction caused by the intentional group came from commercially oriented farmers who frequently cultivated several plots of two to three hectares for profit. Greater compassion should also be directed toward those farmers who are inadvertent victims of changes in resource management regimes and for those who are truly impoverished.

Finally, several important governmental officials expressed the need to more strongly regulate the movement of local migrants (i.e., that by individuals and families within the same District which is not recorded presently) (1991 Field Notes). It was their feeling that the problem of illegal farmers stemmed largely from the government's inability to control the movement of its citizens. Although there is a logic to their thinking, the problem of illegal farmers could have been managed without reducing further the personal liberties of its citizens through, among other things, more sophisticated resource management personnel and practices. In addition to the preservation of personal freedoms, a stronger resource management administrative system could provide other benefits as well. Such a system would be in a better position to re-weave conservation practices into everyday village life, sustaining productive livelihoods for future generations. It would also reduce the occurrence of serious environmental problems and destruction. In addition, if conservation measures could be adopted by more rural people, the need for drastic measures, including arrests and relocation and the expense (social, fiscal, administrative, and environmental) that it entails could be avoided for the greater benefit of all.

Chapter XII

An Ecosystem Approach to the Study of Coastal Areas: A Case Study from the Dominican Republic

Richard W. Stoffle
David B. Halmo
Brent W. Stoffle
Andrew L. Williams
C. Gaye Burpee

Introduction

The world-wide degradation of fragile ecosystems has been the focus of a great deal of recent research in the social sciences (Browder 1989; Little and Horowitz 1987). For example, scientists have been monitoring the potential effects of deforestation on global climate (Graedel and Crutzen 1990; Postel 1988) using sophisticated remote sensing technology (Green and Sussman 1990). Natural resources and fragile ecosystems, such as steep, high altitude hillslopes (Ives and Messerli 1989) and tropical forests (Browder 1989; Clay 1988; Denslow and Padoch 1988) have become the targets of intensive efforts for conservation.

Coastal areas, especially those containing coral reef ecozones, also are undergoing degradation and potential destruction (Bunkley-Williams and Williams 1990; Britton and Morton 1989; Clark 1985; Robben 1985). Despite containing some of the most biologically productive ecozones in the world, coastal areas are extremely fragile. Studies have traced the deterioration of coastal ecozones to a number of environmental factors such as climate, turbidity, hypersalinity, disease, sedimentation, and changes in water temperature. Other studies trace the deterioration of coastal ecozones to the human extraction of natural resources for subsistence and the commercial market, and the intrusion of pollutants from human activities. The combination of environmental and human factors creates a synergistic effect that compounds the degree and intensity of environmental change in coastal ecozones.

It is argued here that the study of population-environment dynamics in coastal areas must address changes in both terrestrial and marine ecozones because they comprise a single ecosystem. The concept of "ecosystem" has scientific and policy implications for coastal area studies. From a scientific perspective, it means developing descriptive and explanatory models that include human and environmental variables as these operate on the land and sea. From a policy perspective, it means the holistic management of the human and natural resources that scientific studies demonstrate exist and are functionally integrated at the junction of the sea and the land.

Environmental research clearly demonstrates that in coastal areas the marine and terrestrial ecozones are inextricably linked by natural processes (Mosher 1986,244), but the role of human use patterns is less well understood. Smith (1977,7-8) notes that social science research on what are called "fishing communities," tends to focus on the dangerous and exciting aspects of fishing while ignoring other adaptive activities associated with farming and tourism. Even when a study considers both marine and terrestrial use strategies, it rarely describes how these strategies function as a single adaptive system on both a personal and community basis, and almost never considers the environmental implications of these adaptive strategies.

This paper is based on a series of studies of population-environment dynamics involving the residents of a small community located on the north coast of the Dominican Republic. These studies were conducted with the permission of the people of Buen Hombre. The community is located between the port cities of Monte Cristi to the west and Puerto Plata to the east (see fig. 12.1). For over a hundred years, the people of Buen Hombre have used four ecozones associated with their community. The two maritime ecozones include: (1) a tidal shore ecozone composed of beach, lagoon and mangrove swamp microzones that are used for gathering plants and seafood and (2) a coral reef ecozone used for fishing. In addition, two terrestial ecozones, (1) hillslopes used for mixed crop agriculture and (2) upland mountain forests used for plant collecting and hunting, form the land-based component of the local economy. Currently, local strategies for using these ecozones and the ecozones themselves are threatened by drought, inflation, nonsustainable use by outsiders, and by government interventions to protect the ecozones by restricting access.

The research has both scientific and applied goals. The research explores the hypothesis that fragile coastal marine and terrestrial ecozones tend to be deteriorated by the traditional adaptive strategies of local people. Evidence that apparently supports this hypothesis has led to natural resource management strategies that focus on the modification or elimination of traditional adaptive strategies. The research findings presented in this paper, however, argue that the traditional adaptive strategies of the people of Buen Hombre have not deteriorated the fragile coastal ecosystem beyond what is

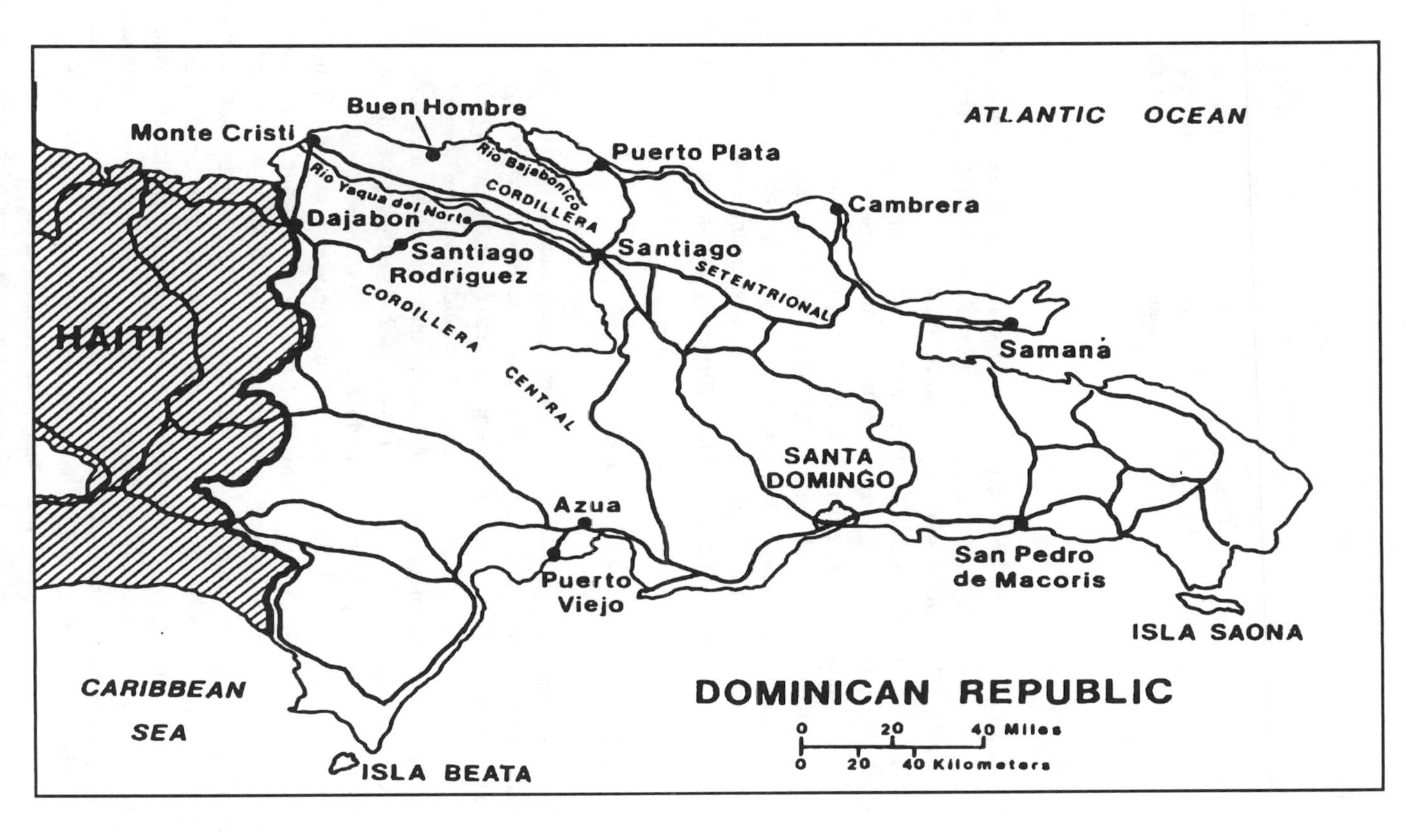

Fig. 12.1. Map of study area

normal, given the population size and living standards of these people. These data argue that the people of Buen Hombre have sustainable marine and terrestial use strategies. The analysis concludes that natural resource management strategies should be designed to encourage the persistence of traditional use strategies and modify or eliminate outside intrusions into the coastal ecosystem.

Methodology

This analysis derives from three social and ecological assessment studies carried out in the village of Buen Hombre (Stoffle 1986; Stoffle, Halmo, and Stoffle 1990). The studies were conducted in the summers of 1985, 1989, and 1990. Several research methodologies were used in these studies including informal interviews, focus group interviews, survey interviews, oral history interviews with community elders and participant observation. A total of 284 interviews were conducted with local fishermen, farmers, women, and government administrators. Additional data was derived from analysis of local fish catch records and 174 person days of participant observation. Soil samples were tested at Michigan State University as part of the agricultural practices study.

Historic Demogragphy of Hispaniola

The dynamics of population-environment interaction on the island of Hispaniola are not recent in origin. The island has been occupied by humans for thousands of years. Native Americans arrived on the island by at least 5000 B.C., and its pristine environment would never be the same. As American Indian people settled the island, they expanded in numbers and modified the natural environment. Figure 12.2 visually represents the direct relationship between population growth and environmental disruption on the island. The American Indian population probably rose steadily from 5000 B.C. until about 800 A.D. During this time, they increasingly modified the environment as they became increasingly sophisticated at hunting and gathering. With the adoption of corn, beans, and squash as well as other tropical cultigens, horticulture became the economic base upon which the population expanded until it was among the densest and most socially complex in the Caribbean. Extensive environmental use by dense native populations probably reached the apex in terms of environmental alteration well before 1492. Based on analogy with other American Indian populations in the New World, it can be assumed that American Indian people in Hispaniola had recognized the limits of the natural environment to support their people and had developed a wide range of conservation measures long before 1492.

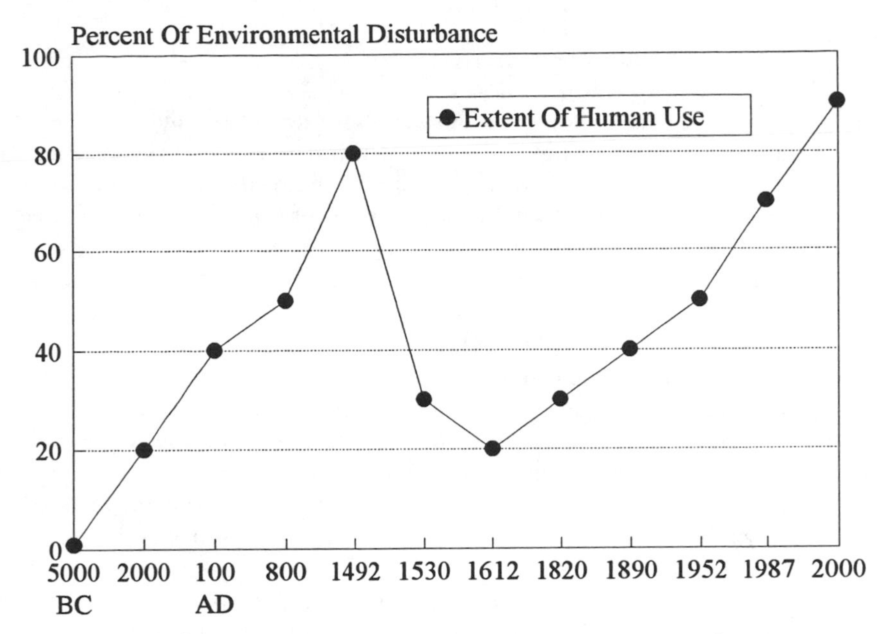

Fig. 12.2. Environmental disturbance by population & adaptive strategies

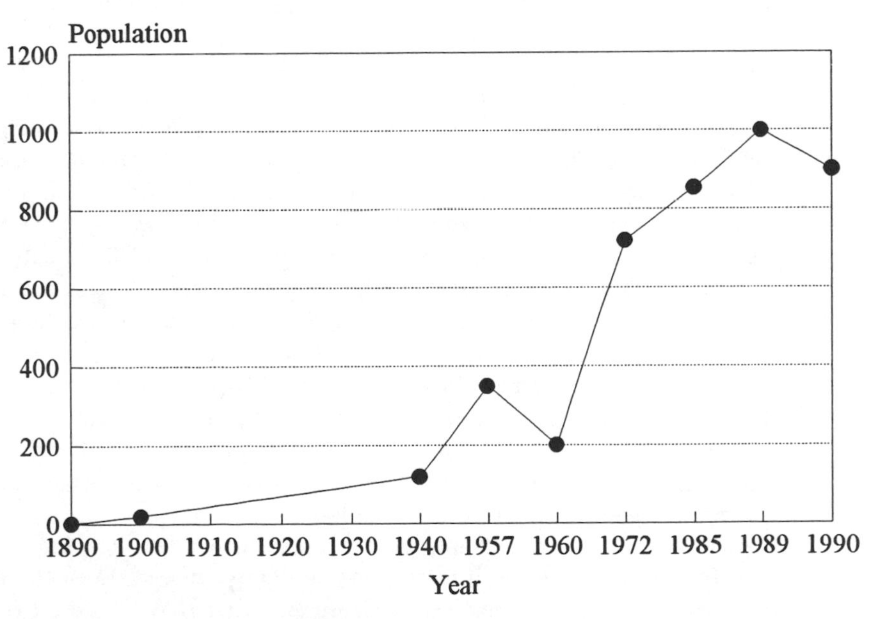

Fig. 12.3. Buen Hombre population changes
Note: These figures are based on data of varying quality so they should only be used to generally represent trends.

On the north coast of Hispaniola, American Indian people combined horticulture with the intensive use of marine resources. In the Buen Hombre area, for example, numerous sherds of stylized, decorated ceramics and heavy concentrations of conch shells in middens and mounds observed in the hillslope fields surrounding the village attest to the presence of a permanent American Indian community. The thick layers of black soil are likely anthropogenic black soils (terra preta) like those described for the Amazon basin (Smith 1980) and elsewhere.

Pre-conquest population estimates for the island of Hispaniola vary. Angel Rosenblat (1976) made a conservative estimate of over one hundred thousand. Eyewitness accounts led Bartolome de las Casas to estimate the population at three million (Thornton 1987,16). Based on a detailed examination of documentary sources, two distinguished historical demographers (Cook and Borah 1971,1:379-410) estimated the indigenous Native American population to be as high as eight million in 1492.

The Columbian discovery drastically altered the population density and land use practices on the island. Indian populations suffered heavy mortality from warfare and slavery, but it was Old World pathogens such as smallpox that decimated Native American inhabitants (Dobyns 1983; McNeill 1976; Purdy 1988). Aboriginal American populations had no immunity to this and other Old World diseases.

Cook and Borah calculated that the native population collapsed from an estimated eight million in 1492 to 3,770,000 in 1496 (cf. Dobyns 1983,257). The first recorded outbreak of smallpox originated on Hispaniola in December of 1518 (Dobyns 1983,259), and from there began the first hemisphere-wide pandemic. Between 1496 and 1518, a span of just twenty-two years, the population of Hispaniola fell from an estimated 3,770,000 to only 15,600 (Cook and Borah 1971,1:401). Only thirty-five hundred remained in 1538 (Cripps). Vazquez de Espinosa (1942,39) noted that all American Indian people were gone from the island before he arrived in 1612.

Because of American Indian population collapse and Spanish failure to repopulate the island, environmental recovery probably began in 1492 and lasted until at least 1612. After this time, Hispanic people slowly began to reestablish a population that would be as dense and as extensive as that of the American Indians before 1492.

American Indian inhabitants of the area surrounding present day Buen Hombre surely must have suffered heavy mortality. Indeed, Buen Hombre lies forty-five kilometers to the west of Columbus' first New World settlement of La Isabela and about seventy-five kilometers east of the settlement established at La Navidad (in contemporary Haiti). These settlements contained small Hispanic populations. On the north coast of Hispaniola, the Hispanic population did not begin to establish itself until the late 1800s. Consequently, the Buen Hombre area, the natural environment may have undergone as much as 250 years of recovery before being redisturbed by humans.

In summary, the Spaniards did not conquer a "virgin land" comprised of pristine ecosystems. Rather, in the words of two eminent ethnohistorians (Jennings 1975,15; Dobyns 1976; 1983,8), they "widowed" an already occupied and extensively altered natural environment. Because of aboriginal depopulation by the early seventeenth century and slow Hispanic repopulation, the natural landscapes and seascapes underwent hundred of years of regeneration before being extensively disrupted again in the late nineteenth and early twentieth centuries.

The Community of Buen Hombre

The village of Buen Hombre is situated along the arid northwest coast of the Dominican Republic. The people of Buen Hombre look north to the sea and south to the mountains. The community is located in a cove between an extensive coral reef zone and the flanks of the Cordillera Septentrional mountain range. A deep break in the coral reef and a shallow lagoon are two more of the community's natural resource assets.

The history of the community of Buen Hombre began with the settlement of the area by immigrants from Cuba. According to elders interviewed in 1990, the community of Buen Hombre was founded in the late 1890s by a family of thirteen Cuban refugees who made landfall on the north coast of the Dominican Republic after fleeing their homeland during the second Cuban War of Independence (1895-1899). The early settlers found an uninhabited ecosystem characterized by fertile soils, regular rainfall, and dense secondary forest, reflecting a two to three hundred year "fallow" period following the extinction of the indigenous people living in the area.

Historic data on the early use of marine resources is still somewhat sketchy. The early settlers must have used the beach and mangrove ecozones to collect shellfish, but oral accounts suggest that they focused their subsistence activities on agriculture. The settlers cleared tracts of land near the lagoon and established a diversified agricultural system that included the cultivation of such crops as plaintains, cassava, maize, beans, potatoes, peas, tomatoes, bananas, and rice. The cultivation of both rice and bananas on the leeward side of the mountains suggests that precipitation was more regular and reliable at that time than it is today. People also raised a variety of animals, including donkeys, horses, pigs, goats, sheep, chickens, and cattle, for transportation, traction, and food. Their initial stock of animals was likely acquired from the town of Villa Vasquez and other villages over the mountain slopes, thus initiating a network of trading relations with surrounding communities that continues today.

Village elders indicated that by 1937 there was a large fish market on the beach front. The early economy of Buen Hombre eventually expanded to consist of two interdependent economic systems, one centered on fishing and the other on agriculture. Such a relationship fits a common pattern of coastal fishing communities supplying interior agricultural communities and towns with marine products. Village elders stated that there was a substantial increase in the use of marine resources by the mid 1940s. They recalled the period as being one in which the coral reef and tidal shore provided an abundance of fish, large lobster, conch, octopus, and manatee. Fishing in these earlier periods was characterized by the use of lines, nets, and harpoons. Women were also important contributors to the fishing economy at this time. Between 1890 and 1940, then, the economy of Buen Hombre could be characterized as a subsistence economy based on a mixture of agriculture and fishing that supported an expanding population of approximately 120 persons.

The acquisition of water has historically been a central challenge for the inhabitants of Buen Hombre. Initially drinking water was obtained through a combination of household water catchment systems and earthen cisterns that captured runoff rainwater from hillside crevices and small gorges. During the dry season and times of drought, community members were forced to travel over the mountains by foot, mule and horse to collect water from the Rio Yaque del Norte. Thirsty individuals also traveled to the surrounding communities of Villa Vasquez, Las Aguitas, and Las Canas to obtain water. Domestic animals were able to take advantage of standing lagoons or water holes in the village which, during times of drought, were also used as sources of water for agricultural production.

Prolonged periods of drought, however, did not seem to have been a significant problem for the early inhabitants of Buen Hombre. Village elders recall this period as a time of abundance in terms of agriculture, animal production, and fishing. The productivity of this coastal ecosystem seems to have encouraged migration into the region. The population of Buen Hombre steadily increased until 1957 (see figure 12.3). During the late 1930s, President Rafael Trujillo had initiated a large-scale road building program in order to increase mechanized transportation to the central-northern frontier (Georges 1990,61, 63-64). Thus immigration was further facilitated by the clearing of a small road from Villa Vasquez to Buen Hombre in 1952 as part of the national program. Although the road was rugged and became impassable after any substantial rainfall, it provided access for the first motorized vehicles to enter Buen Hombre and thus facilitated the importation of food and water into the community. The road probably also stimulated exportation of marine and terrestrial surpluses.

The road proved to be an invaluable asset when a three-year drought struck the region in 1957. The drought was catastrophic for the community. Village elders recalled that there was widespread crop failure, loss of animals and hunger. What precipitation did occur was not fully accessible because the condition of the earthen catchment cisterns and canals on the mountain slopes had deteriorated due to deforestation and resulting erosion. Village elders mentioned that without the government's daily shipments of food and water, which were made possible by the road, everyone in the community would have died or been forced to migrate. Despite government assistance, village elders estimate that the village population declined to approximately two hundred people (see fig. 12.3)

During the early 1960s, the drought subsided and the road into the community was improved once again. As a result, the area experienced a resurgence in population growth. Whereas early muleteers served to establish connections between nodes of production on either side of the mountains, the improved road opened up the community to buyers, intermediaries and merchants to an unprecedented extent. The village economy expanded beyond subsistence production to include commercial production. Development of new infrastructure in terms of transportation networks and expansion of port towns seems to have been a significant development in the human ecology of the region. Growth led not only to a significant increase in the population on the north coast, but also to increased exploitation of marine and terrestrial resources. Key consultants who assisted in conducting a census of the community for the Dominican government estimated that the population of Buen Hombre had reached 721 in 1960.

While local population data are scant, ebbs and flows of population continued between 1960 and the present. Georges (1990,176) notes that the Central Valley and Sierra region of the Dominican Republic, just south over the mountains from the Buen Hombre coast, suffered severe droughts in 1966-1967 and 1975-1976, with dire consequences for small-scale farmers. Buen Hombre farmers must also have experienced the effects of these drought episodes.

Today Buen Hombre still lacks potable water. The community remains relatively isolated because its only transportation link with interior communities is the poorly maintained road over the mountain range. Although the road was improved in 1985, it often remains impassable to larger motorized vehicles, so water and other essential commodities are usually transported by horse, mule, motorbike, pickup trucks, and small cars.

The village consists of a series of farmsteads organized in a line settlement pattern, extending inland from the small lagoon that serves as the boat launching location for community fishermen. As the village grew, new homes were built along the road and now stretch from the lagoon to the foot of the mountains. By 1985, the village population had grown to approximately 855 people (Stoffle 1986,81). The majority of homes contain nuclear families, although it is common for these homes to be arranged in extended family clusters. Social networks between relatives and neighbors are horizontal and multistranded (Wolf 1966).

The people of Buen Hombre typically rely on more than one economic activity, an adaptive strategy termed "occupational multiplicity" (Comitas 1973), that is common throughout the Caribbean. Adult males engage in fishing and farming enterprises for household subsistence and cash income. Women play a significant role in agricultural production at certain critical times in the farming cycle. Occasionally, a few women accompany their spouses on fishing trips. When rainfall is adequate, women also cultivate mixed kitchen gardens, planting staple tuber and vegetable crops as well as medicinal plants and fruit trees. Women manage most aspects of domestic life. In many respects, the people of Buen Hombre use their natural resources much like the American Indian people who occupied this site before Columbus.

Terrestrial Ecozones, Resource Use and Soils

The people of Buen Hombre utilize two terrestial ecozones: (1) the hillslopes for agriculture and (2) the upland mountain forests for collecting a variety of useful wild plant resources. The numerous resources found in the hillslope and upland mountain forest ecozones are used for subsistence, medicine, shelter, and cash income. Food crops are planted on the hillslopes and in the valleys. Plants and animals are harvested for food, medicine, and construction. Timber harvested from the forested uplands is used for shelter, household fuel, and charcoal for use and sale.

The topography of agricultural lands in the village ranges from zero degree slopes at sea level to hills, valleys and mountainsides. Buen Hombre farmers describe three main soil types: black, yellow, and mixed, a combination of black and yellow soil types. All soils are considered by farmers to be very productive in years when there is adequate rainfall. Black soil is described as the most productive.

Soil samples were taken from the top thirty centimeters for the three major soil types. In most locations the topsoil layer (A horizon) is unusually thick, commonly one hundred centimeters deep. Soil physical properties were not specifically measured. Based on observations, however, Buen Hombre soils appear to have excellent physical properties. The SOLUM depth of one hundred centimeters allows ample room for root development and root exploration for nutrients. Soil structure in the topsoil appears excellent with good aggregation and a good mix of micropores and macropores that would allow for ample air exchange with the surface, good drainage, water retention, and infiltration.

Table 12.1 illustrates the results of soil testing. The pH is high in all three soils, ranging from 7.6 to 8.1. The cation exchange capacity, or the soil's ability to retain critical nutrients against leaching by water for use by plants,

TABLE 12.1. Soil Test Results, Buen Hombre, Dominican Republic. June 1990

	Black Soil	Yellow Soil	Mixed Soil
Soil pH	7.6	8.1	8
Cation Exchange Capacity (meq/100g)	39	32	41
Olsen Phosphorous (lbs/A)	27	22	18
Potassium (lbs/A)	1,486	335	579
Calcium (lbs/A)	13,642	10,863	14,147
Magnesium (lbs/A)	696	1,156	1,200
Zinc (ppm)	40	21	11.6
Manganese (ppm)	40	21	11.6
Copper (ppm)	8.4	2.1	7.4
Iron (ppm)	13	13	7
Nitrate-Nitrogen (ppm)	12.92	3.46	8.22

is well above the critical minimum of four milliequivalents (meq) per 100 grams of soil. This cation exchange capacity of between 32 and 41meq/100g is probably due to the presence of organic matter in the soil.

In the higher elevation, black soils, phosphorous, potassium, and magnesium levels are adequate for producing medium crop yields. Nitrate-nitrogen levels are higher than usual for non-fertilized soils. Zinc levels appear low, but copper levels are high and may negatively affect crop growth. The dark color of the soil is probably due to the presence of substantial amounts of organic matter.

The fertility of the yellow soils is very similar to that of the black soil, with the exception of lower levels of copper and nitrate-nitrogen. This is probably due to the lower amounts of organic matter. The fertility of the mixed soils is also similar to the black soil, except for lower levels of plant-available phosphorous. These levels are high enough, however, to permit adequate crop yields without the use of fertilizers.

It appears, then, that despite one hundred years of cultivation, soil quality has been maintained at sufficient levels. Soil fertility may be related to cropping practices of Buen Hombre farmers.

Agriculture

Agriculture on the north coast of the Dominican Republic is rainfed. Because of Buen Hombre's location in the rainshadow on the leeward side of the Cordillera Septentrional mountain range, precipitation is seasonal and unpredictable. In stark contrast to the Samana peninsula in the eastern part of the country, which receives nearly eighty inches of rain annually, the northwest coast receives a mere twenty-five inches of annual precipitation (Lang 1988,11). Brief rainy seasons occur during the summer months of August and September, and between December and January in the winter.

Fields are typically comprised of two plots, one adjacent to the homestead and another located on the forested flanks of the mountains. Dual location of fields may be related to local perceptions of crop growth and soil fertility. Root and tuber crops such as yuca are said to yield better in the black and mixed soils on the hillsides. Small game birds such as the guinea hen are hunted with rifles in fallow fields. Some farmers also retain access and use rights to plots that belong to relatives who live in interior villages. Kitchen gardens may be a separate small plot adjacent to the homestead or simply a small area in the dooryard around the house.

The local method of farming is most accurately described as slash and burn. Secondary vegetation is cleared any time from September through December. Crops are planted in November and December, and are timed accordingly prior to the advent of winter rains. Weeding occurs in intervals as necessary. Most crops are harvested in March and April. Cassava and tobacco are harvested over longer periods of time throughout the year.

Farming families in Buen Hombre cultivate yuca (cassava), maize, yams, sweet potato, several varieties of beans, squash, fruit trees such as lechosa (papaya) and lime. Tobacco is the major cash crop. Varieties of beans, pigeon pea and maize crops are planted in hillside plots. Fields adjacent to homesteads are largely reserved for the planting of tobacco crops. The yellow soils of these plots as well as of kitchen gardens are also planted with maize, beans, cassava, squashes, cotton, fruit trees, varieties of medicinal plants, herbs and spices, and other species of trees such as mesquite (locally known as cambron) which are used for shade and construction purposes. Wooden fences around field and garden boundaries support climbing vines which are used for fiber, medicinal plants, and spontaneously growing crop and non-crop plants. Living fencerows of cacti are also planted and serve as hedges around fields and gardens. Mesquite bean pods and crop residues are used as fodder for domestic chickens, pigs, goats, guinea hens, cattle, horses and mules.

The farmers of Buen Hombre practice mixed crop agriculture by intercropping. Beans and squash are interplanted with maize, tobacco and cassava. Beans and pigeon peas serve a nitrogen-fixing function for maize plants, thus replenishing soil nutrients. Fruit trees are grown in fields as well as in kitchen gardens. Edible greens, medicinal herbs and grasses which thrive in the disturbed soils between crop rows are spared and harvested from fields.

Buen Hombre farmers create, manage and maintain complex agroecosystems and field microclimates typical of rural small-scale, limited resource farmers (Altieri 1987; Gliessman 1984; Wilken 1972, 1987) by interplanting a variety of agricultural crops, selectively weeding and sparing useful plants that grow spontaneously between rows and incorporating tree crops into agricultural fields. Sufficient levels of soil fertility are maintained for longer periods of time by virtue of controlled burning of crop residues not used for animal fodder and by interplanting nitrogen-fixing bean plants with maize and tobacco. Differential heights of crop stories serve to preserve what little moisture is retained in crop plant material and soils. Multiple stories also modify shade patterns within fields. All other factors being equal, then, the farmers of Buen Hombre appear to be practitioners of sustainable agriculture.

Upland Forest Resource Collecting

The foothill woodlands are dominated by desert scrub vegetation, mainly varieties of cacti and Acacia. At higher mountain elevations, forests are comprised of pine, several types of hardwood tree species and numerous wild plants. Positive botanical identification of these species has not yet been completed. However, it is clear that these resources provide local people with many necessities of everyday life.

The people of Buen Hombre collect a wide variety of wild resources for fuel, medicine and construction. An inventory of over ninety distinct types of plants was obtained from respondents. These plants include herbs, fruits, grasses, cacti, flowers, and trees. Several of these plants are transplanted from the upland forests and slopes to kitchen gardens for easier access.

The majority of collected plants are used for medicinal purposes. Leaves, stems and roots are mixed with water and prepared as medicinal teas for treating a variety of ailments and illnesses. Trees and flowers are used primarily for shade and ornamentation. Several species of wood are used to make fish pots, traps, fences, palisades, and for the construction of houses and ramadas. Palm fronds are obtained from villages on the other side of the mountains and used as thatch for roofing. As with agricultural crops, seeds and cuttings of these useful plants are exchanged between relatives and neighbors.

The foothill woodlands and mountain forests are also utilized by the residents of Buen Hombre for collecting fuelwood. Based on observations, it appears that only deadfall timber and branches of several varieties of trees are collected as fuelwood.

Charcoal production is a supplemental economic activity in the community. Over half of the farmers interviewed (54 percent) are engaged in charcoal production, while 37 percent of fishermen interviewed make charcoal. Most charcoal is produced for cash income.

Local and External Factors Affecting Agriculture and Forest

The most significant limiting factor in agriculture for Buen Hombre is water, in the form of both precipitation and stream flow for irrigation. Like the northwestern region in general, the bimodal annual rainfall schedule is subject to considerable fluctuation within and between specific years (Georges 1990,15, 176). Needed rainfall may not occur during crucial months of the agricultural year, a condition that has been defined as "agricultural drought" (Glantz 1987,45). At the present time, the people of Buen Hombre appear to be experiencing extended drought conditions.

Subsistence goods and cash derived from agricultural produce has declined according to those interviewed. Respondents commented that 1989 was the driest year of the previous four, which were also very dry. The drought situation was confirmed dramatically during the 1990 study. Comments made by community members and government officials, as well as national newspapers, emphasized the impacts of the severe drought that has affected the entire nation. Millions of dollars in crop and livestock losses have stimulated government relief programs, including crop seedling distribution, to the hardest hit areas.

Lack of adequate rainfall has led women temporarily to abandon full-scale kitchen gardening. Small amounts of purchased water are used to pot irrigate medicinal and other plants in dooryards. The crisis in village agriculture is related not to exhausted soil fertility, but to the prolonged lack of adequate rainfall, in short, drought conditions.

It is not clear whether the drought can be characterized as "meteorological" (defined as a 25 percent decrease in long-term average rainfall) or as agricultural drought (Glantz 1987,45-46). Whatever the case, drastic conditions have begun to stimulate emigration from Buen Hombre. Compounding the effects of vagaries in climate are a number of social processes and policies that have adverse consequences for terrestrial ecozones, resources, and village economy.

Social and economic processes have also played important roles. In the mid-1980s, the Dominican government initiated a subsidized tobacco-growing program. Loans were provided to farmers to begin cultivation of tobacco as an export crop. Many, if not most, Buen Hombre farmers participated in the program. By 1987, however, the tobacco market had crashed. Since the collapse of the market, large portions of the Buen Hombre tobacco harvest remain stacked inside houses and outbuildings because it no longer receives a decent price, according to agricultural association members.

Historically, government programs have affected population-environment dynamics on the north coast. At the turn of the century, the development and expansion of the lumbering and cattle industries resulted in deforestation and land concentration dominated by large holders. The opening of the northwestern frontier region stimulated both spontaneous and directed colonization, thus increasing population and exacerbating destructive land use practices (Georges 1990). Colonists, seeking new lands to cultivate because of land concentration and population pressure in the interior, have begun to slash and burn their way up the southern slopes of the Cordillera Septentrional Range to the crest of the mountains. Deforestation has adversely affected the already arid environment's capacity to maintain and generate moisture, thus leading to desertification.

Marine Ecozones and Resource Use

Buen Hombre fishermen-farmers utilize two marine ecozones and the resources found within them. These are (1) the tidal shore ecozone and (2) the coral reef ecozone. Each of these ecozones is described below, followed by a discussion of the environmental and human factors and their impacts on each ecozone.

The tidal shore or littoral ecozone used by the people of Buen Hombre is composed of three microzones. These are (1) the beach, (2) the mangrove swamp, and (3) the lagoon.

The beach area is used as the cleaning and weighing station for fishermen returning with their catches. It is here that the various fish captured are weighed for sale and cleaned for home consumption. Intermediaries from interior towns as far away as Santiago and the capital city of Santo Domingo, as well as buyers from nearby villages, congregate at the beach on a daily basis and wait for Buen Hombre fishermen to return with their catch. While they wait, spouses of Buen Hombre fishermen prepare dishes of fish, rice, and plaintains from a stand adjacent to the weigh station for sale to waiting buyers. Intermediaries buy portions of the first class species for sale to retail dealers in urban centers. Buyers from neighboring villages purchase seafood to take back to their homes to eat.

The Buen Hombre shoreline consists of white sandy beaches interspersed with extensive mangrove swamps. This microzone constitutes the junction of sea and land. Water and heavy vegetation result in an environment rich in plant and animal life. In contrast to the arid conditions further inshore, the mangrove is characterized by high humidity.

The mangrove provides a natural nursery for numerous species of aquatic life that are harvested by Buen Hombre fishermen. Crabs, turtles, and shellfish are found in the mangroves and the shallows just offshore. These warm waters support healthy beds of seagrasses and algae which are consumed by a range of marine species.

The Buen Hombre lagoon serves primarily as the boat launch for village fishing crews. The majority of these are typical wooden yolas, the local term for small fishing vessels. Other, more modern boats of aluminum and fiberglass, powered by fifteen horsepower Johnson, Yamaha and Evinrude outboard motors, also comprise part of the local fishing fleet.

The lagoon shallows also support thick beds of seagrass which are used by marine species such as crabs, lobsters, and other shellfish as nesting and feeding grounds. During low tide, these seagrass beds are exposed just offshore in shallow waters. Frequently, fishing families walk along the shoreline in shallow waters to collect clams and other shellfish. Crabs are a highly valued resource harvested from nearshore waters.

Field observations and interview responses indicate that the residents of Buen Hombre recognize the value of the beach, lagoon, and mangrove microzones, and are conservative in their use of these microzones and resources. Mangroves are only occasionally used to collect wood poles from the dominant tree species for use as roof beams.

Over and above the wise and careful utilization of the mangrove by local people is the presence of the coral reef, which serves the function of preventing beach erosion and mangrove destruction by buffering the Buen Hombre shoreline (cf. DuBois and Towle 1985,233). Both environmental features and sustainable human practices on the local village level combine to protect the beach and mangrove microzones from large-scale degradation.

External Factors Affecting Beach, Mangrove, and Lagoon Microzones

The Buen Hombre tidal shore ecozone is beginning to undergo changes as a result of exogenous developments. In the beach and mangrove microzones, the number of national and foreign tourists has increased. A small twenty-eight room hotel has recently been constructed in the neighboring village of Punta Rucia. As a result of road improvements, the number of tourists visiting and residing in Buen Hombre has also increased. Beach front property and plots along the new road have been sold and six new single and multi-family vacation homes have been constructed. There is a direct connection between the new road and these homes because the tourists who drive for hours to spend a few days at these homes need to leave the village regardless of weather. Day tourists come more often because of the new road, but their numbers and impact are unknown.

Despite its small scale, the effects of tourism in terms of increased motorized boating for snorkeling excursions and water skiing may have detrimental effects. Nearshore waters may potentially become convenient disposal areas for non-biodegradable trash such as glass, plastic, and metal. Pollutants such as battery acid, spilled or leaking gasoline, and oil from boats, could adversely affect marine species, seagrass beds, and water conditions.

In the mangrove microzone, government interventions in the form of legislation have been initiated to protect mangroves. This legislation prohibits the use of mangroves for any unlawful purpose, including tree cutting.

Prohibition of mangrove use and tree cutting has resulted in a significant decline in wood harvesting by Buen Hombre residents. Recreational tours for tourists, however, have increasingly subjected the mangrove microzone to disturbance and pollution. As the population of small-scale fishermen increases in the surrounding area, exploitation of mangroves will likely increase, leading to degradation.

The Coral Reef Ecozone

The coral reef ecozone located off the north coast of the Dominican Republic consists of an inner reef about a quarter-mile off shore, and an outer reef located a quarter mile beyond the inner reef. For most of its length, the coral reef serves as a barrier between the deep ocean and the shore. The only major break in the reef is at the entrance to the Buen Hombre lagoon. Because it is a double reef ecozone, changes in weather, water temperature, and wave action affecting the outer reef potentially affect the inner reef. Smithsonian marine scientists have described the Buen Hombre reef as one of the best in the Caribbean in terms of both size and condition.

Fishing is one of the two major economic activities in Buen Hombre. As is the case among most small-scale coastal fishermen, the task of fishing is constrained by fluctuations in weather conditions and a general lack of mechanized technology. Buen Hombre fishermen adapt to these constraints by forming social and economic relationships that help ensure access to resources for fishing as well as subsistence.

Few males who identify themselves as farmers also fish. In contrast, virtually all males who identify themselves as fishermen also farm. Consequently, most fishermen belong to both the community-based fisherman's association and the agricultural association. The fisherman's association is composed of men who have risen through the ranks of the developmental cycle of fishing, which involves four distinct stages: (1) apprentice, (2) journeyman, (3) craftsman, and (4) beached (Stoffle 1986,95-100).

Buen Hombre fishermen use a variety of methods for catching fish. The most common method is the use of snorkel and speargun for diving in the coral reef ecozone. This method involves the ability to remain submerged for substantial periods of time in order to locate, stalk, wait for and shoot one's target. Accuracy is crucial because spears must be retrieved and refastened to the gun should a fishermen miss his target.

A 1989 inventory of fishing equipment illustrates that the thirty-four fisherman's association members employ multiple methods in fishing. Forty-one percent of association members use handlines (cordeles), which are used mainly during night fishing. Fifty percent use snorkeling gear and spearguns. Thirty-five percent of association members own and deploy nasas, or fish traps in deeper waters. Access to and use of traditional yolas and motors is controlled by 26 percent of association members, but it must be remembered that fishing crew members cooperate in boat travel to fishing locations. Twelve percent of association members use atarrayas or beach cast nets. Only two association fishermen use boat nets (trasmallos); no fishermen use beach set nets (chinchorros) as a fishing strategy. Night fishermen also use flashlights and makeshift lamps. These are submerged into the sea in order to attract fish. Social relationships, both kin and non-kin based, facilitate sharing or loaning of equipment among and between fishermen.

A wide variety of typical reef fish species are harvested by Buen Hombre fishermen. Large groupers and red snappers are first class fish high in market demand. Coral-eating parrotfish of several sizes and varieties are either sold or consumed in the household. The size and variety of parrotfish determines whether it is classified as first or second class. Delicacies such as octopus are also captured. Larger shellfish such as lambi (conch), lobster, and bulgao are captured by spearfishermen diving inside the inner reef. Third class fish species are kept primarily as subsistence fish. They bring the lowest price in the market, if and when they are sold. Occasionally, barracudas and sharks are taken. Shark is captured for sale and barracuda is generally kept for home consumption.

Fishing crews usually operate in three "shifts" because of frequent equipment failure, access to boats, or other economic commitments in the system of occupational multiplicity. The first shift is usually worked by the majority of fishermen, who begin about 8:00 A.M. return around 12:00 M. depending on weather conditions. In the early morning hours, the sea is at its calmest, allowing easier boat travel to the reefs and beyond. Returning is also easy because fishermen have the prevailing northeast wind at their backs.

The second shift begins after 12:00 M. Rowing out to the reefs can be difficult against the strong afternoon winds and rough waters. After four or five hours of fishing, the return trip home is facilitated by the same winds.

Several individuals and some crews fish at night. Their shift begins around 8:00 P.M. and lasts throughout the night. Equipped with containers of coffee and rum, a flashlight hooked up to an automobile battery, hand lines and hooks, night fishermen have the advantage of calm waters. Fish are attracted to the light and thus some of the largest catches occur at night. Night fishing is, however, the most dangerous because of the risks of running into coral heads, damaging boats and motors, and the possibility of being attacked by barracudas or sharks, should the fisherman decide to snorkel dive. The night shift is the longest because fishermen must wait until morning to bring their catch to the market, when someone is there to weigh the fish and put them on ice.

Each of the shifts, then, has advantages and disadvantages. Some fishermen will occasionally fish more than one shift, going out in the morning and then making another all-night trip (Stoffle 1986,101-102).

Buen Hombre fishermen traditionally have employed sustainable methods of fishing that appear to derive from a conservation ethic. Interviews with key experts indicate that fishermen recognize the potential adverse effects of indiscriminate fishing practices on reef fish populations. Small fish of all classes are not targeted by fishermen; only rarely are they captured in fish pots. Expert fishermen explain that small fish are avoided in order to allow them to grow to an appropriate size. Economically, small fish are not ideal for consumption or sale because of the low proportion of meat. Larger fish provide higher returns in terms of the amount of protein-rich food compared to the amount of energy expended to catch them. Avoidance of small fish and other seafood species implies that fishermen are cognizant of the effects of overfishing on population reproduction.

The enterprise of fishing entails the dual goals of providing food and income. Consequently, fishermen harvest a diversified supply of seafood. Daily individual catches usually include an array of parrotfish, grouper, snapper, crab, lobster, conch, and other reef fish. The diversity of catch clearly indicates that multiple species are deliberately and commonly sought. Buen Hombre fishermen thus employ deliberate fishing strategies for both subsistence and cash. While fishermen prefer certain species for home consumption, these species are usually part of a diversified catch. It can be argued that diversifying the catch reduces the risk of overfishing certain species.

Data suggest that these strategies can and do change, based on such factors as weather conditions and stress in other sectors of the local economy. These changes can be either short-term (day, week, month) or longer-term (seasonal). Under the current conditions of environmental (drought) and economic (crop failure) stress, Buen Hombre fishermen appear to be intensifying their fishing efforts in terms of (1) length of fishing trip, (2) more intensive exploitation of certain locations along the coral reef, and (3) a concentrated effort to capture species that are in high demand in the market economy.

One major factor affecting fishing is weather. Wind and rain play significant roles in decisions regarding whether or not one goes out to fish. If the weather is favorable, the pressure of having to fish long hours and exert great amounts of effort is reduced. On the other hand, when weather conditions are adverse, the lack of larger boats and outboard motors hinders going out to the reefs to fish. Boats and motors are too small to be safely handled in strong winds and rough waters. Consequently, fishermen may be more likely to walk along the shore to the point of the lagoon and swim out to fishing spots well inside the inner reef. To compensate for lost subsistence and income on those days when weather conditions are not favorable, fishermen may exert more effort while fishing or target specific species of seafood on those days when the weather is favorable.

Field observations reinforce this hypothesis. Following two successive days in which strong morning winds prevented crews from going out, fishermen fished much longer than on previous trips. One of the authors had participated in fishing with a crew of Buen Hombre fishermen many times during 1989 and 1990, thus providing an accurate sense of the amount of time the fishermen normally spend fishing during a morning outing. These observations correspond with observed and recorded patterns of fishing from 1985 (Stoffle 1986). In the past, first shift fishing trips normally lasted about four hours, from 8:00 A.M. to 12:00 M. On the day following adverse wind conditions, however, the first shift fished from 8:00 A.M. until about 1:30 P.M., an increase of one and a half hours over typical outings. Observations and fishermen's responses indicate that the reason for this change in fishing patterns was due to adverse wind conditions during the morning hours of the two previous days.

Other changes in fishing patterns were observed during the 1990 fieldwork. Usually, fishermen go out to a particular location along the inner reef and attempt to capture a variety of species. The following day, fishermen choose a different location along the reef. During fishing trips in 1990, participant observers noted that the crew went to a particular location during the early morning. After spending a period of time there, the crew moved to another location with the goal of capturing lobster. After spending nearly an hour searching for lobster, the crew then moved to a third location to resume spearfishing.

Fishermen were not observed going to a spot deliberately for lobster during 1989 fishing trips. Usually, lobster were taken spontaneously when encountered to supplement fish caught with spearguns. Likewise, fishermen were not observed fishing exclusively for one type of fish or seafood. During some of the 1990 fishing trips, however, fishermen stayed out longer in order to ensure that adequate amounts of specific fish were captured for sale and consumption.

A second factor that affects fishing in the coral reef ecozone is a seasonal disease known as ciguatera. The disease is apparently contracted by fish that consume algae and other ocean nutrients contaminated with a highly toxic substance. Like PCBs, the levels of toxin accumulated in fish is correlated with the size and type of fish. Buen Hombre fishermen commented that the poisoning is seasonal. The toxin first begins to appear in May, June and July. The condition begins to peak between November and December; by January, February and March, it is prevalent. Susceptible fish species are recognized by a blackening that occurs in the skin. It is not clear how many types of fish are affected by this poisoning, but the condition may affect the kinds and amounts of fish caught in a given period. These seasonal factors and their effects, as well as changes in fishing patterns and the coral reef ecozone, can be analyzed and better understood by examining fish sales records.

Analysis of Fish Sales Records

One of the primary human impacts on the coral reef ecozone is the amount and variety of seafood that is removed by fishing. The Buen Hombre fisherman's association sales records provide one direct measure of fishing impacts. Most seafood is sold to middlemen who resell the seafood to regional, national, and international retailers. The small amount of seafood that is cooked and sold on the beach comprises the only retail sales in the village. The association keeps reasonably accurate daily records of seafood that is sold to it by its members.

The records, however, do not reflect seafood (1) caught by Buen Hombre fishermen who do not belong to the fisherman's association, (2) caught by people who live in communities outside of Buen Hombre, (3) consumed by the fishermen while on their boats, and (4) taken home by fishermen for family consumption.

The analysis of seafood sales records consists of four comparisons that use data from three summer months in 1989 (June, July, and August) and three spring months in 1990 (February, March, and April). These comparisons include (1) total catch impacts, as measured by total seafood sold, (2) efficiency of fishing effort, as measured by the average amount of seafood sold per day fished, (3) targeting impacts, as measured by species sold, and (4) targeting impacts, as measured by the economic value of seafood sold.

Total Catch Impacts

The total amount of seafood caught is the most general indicator of how much fishing pressure is being exerted on the coral reef ecozone. There is not always a direct relationship between the amount of fish caught and the condition of the natural system. Some fish communities are known to be density-dependent; that is, their rates of growth and reproduction will be highest at some intermediate level of population size. Therefore, certain amounts of fishing pressure can actually strengthen certain fish populations by thinning overabundant numbers (Frost 1979). On the other hand, it is generally recognized that there is a point at which the amount of seafood caught exceeds the capacity of the coral reef system to reproduce.

Figure 12.4 presents the total weight of seafood sold, measured on the Y-1 axis, cross-tabulated by the month and season when it was sold. During the summer season of 1989, the fisherman's association sold 6,530 pounds of seafood. In June, 2,170 pounds of seafood was sold. In July, 2,214 pounds of seafood was sold; this was the highest total sold of the three summer

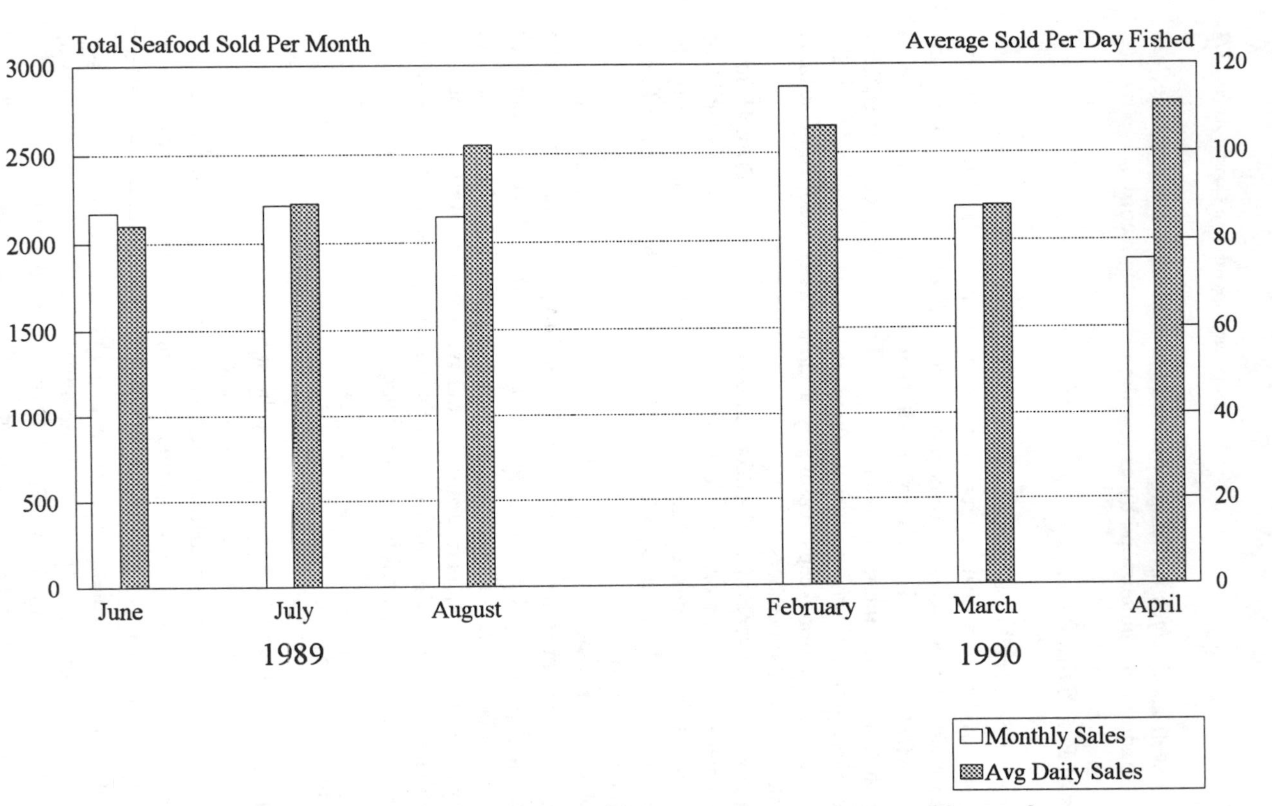

Fig. 12.4. Total seafood sold per month & season by average sold per day

months. In August, 2,146 pounds of seafood was sold; this was the lowest total sold of the three summer months. The volume of monthly seafood sales was generally consistent throughout the summer months.

During the spring season of 1990, the fisherman's association sold 6,950 pounds of seafood. In February, the association sold 2,869 pounds of seafood; this was the highest total sold of the three spring months. In March, 2,192 pounds of seafood was sold. In April, 1,889 pounds of seafood was sold; this was the lowest total sold of the three spring months. The volume of monthly seafood sales declined significantly throughout the spring months.

Comparison of the total seasonal seafood sales reveals a similarity between the two seasons. Approximately the same total amount of seafood was sold during the two seasons, with only a 6 percent (420 pounds) increase in the amount of seafood sold in the spring. Despite the relatively similar overall seasonal sales volumes, the summer monthly total sales contrasted with the spring monthly total sales. While summer monthly total sales remained virtually the same during the three-month summer season, the decline in spring monthly total sales (and by extension, total catch) throughout the three-month spring season is evident. Monthly total sales declined by 34 percent (980 pounds) from February to April.

Comparison of monthly total seafood sales makes it evident that fishermen reduce their overall fishing effort during the spring months, as reflected in the total sales of seafood. A significant factor that affects the amount of effort and time a fisherman spends fishing is theagricultural cycle. Because fishermen are also farmers, they must devote their time to farming activities during certain periods throughout the year, particularly during the harvest months of March and April. For example, tobacco, the major cash crop, is harvested in April. Fishermen-farmers must divert some of their effort from fishing in order to bring in their tobacco crop. According to association records, fishermen sold less seafood during April, implying a lower total catch and thus reduced fishing effort.

In combination with local environmental factors such as weather, agricultural commitments that form part of the system of occupational multiplicity also function to reduce fishing pressure at certain times of the year, as illustrated above. Other social factors are also involved. Demographically, the number of fishermen in the village has not varied significantly. This may be due to the shortage of fishing equipment; however, the number of fishermen in the village is socially controlled by the developmental cycle that governs access to fishing equipment and crew membership. Despite intraseasonal variation, the data suggest that total seasonal catch does not fluctuate significantly.

Both physical and socioeconomic environmental processes, then, interact to affect patterns of fishing. From the seasonal comparison of the volume of seafood caught, it appears that Buen Hombre fishermen are taking a sustainable volume of seafood.

Efficiency of Effort

The second comparison is an effort to understand whether catch efficiency varied by month or season and if this could account for the overall variances in monthly seafood sales. Catch efficiency is calculated by dividing the total amount of seafood sold in a month by the number of days that were fished during the month. Figure 12.4 presents this data with the Y-2 axis at the right of the table the average seafood sold per day fished, cross-tabulated by month and season. The figures represent the average daily seafood sales by the fishermen of Buen Hombre.

The efficiency of fish catch during the summer months gradually increased. In June, the fishermen sold an average of eighty-four pounds of seafood per day of fishing; in July they sold an average of eighty-nine pounds, and in August they sold an average of 102 pounds. Between June and August the fishermen of Buen Hombre increased the amount of fish they sold by an average of 21 percent (eighteen pounds) per day fished.

During the spring months, the efficiency of the fish catch fluctuated. The fishermen sold an average of 106 pounds of seafood per day of fishing in February; an average of eighty-eight in March; an average of 111 in April. Between February and April, then, fishing efficiency varied by an average of 26 percent (twenty-three pounds) per day fished.

When the fishing efficiency findings are compared with the total seafood sales findings it is clear that they are not correlated. For example, the February seafood sales were the highest during the spring sample months, yet did not yield the highest fishing efficiency rate. Conversely, the April seafood sales were the lowest during the spring sample months, yet yielded the highest fishing efficiency rate.

The data raise the question: "Why is efficiency not related to the amount of catch as measured by seafood sold to the association?" One possible explanation is that when the fishermen are able to fish every day, there is not as much pressure on them to expend a great deal of effort. When multiple economic commitments and variations in weather conditions converge during certain times of the year, fishermen-farmers are not able to engage in fishing as much as during other months. Because agricultural commitments reduce the amount of time spent in fishing, fishermen must make sure that they take full advantage of the days when they are able to fish. These data demonstrate that fishermen are able to catch more seafood when they feel the need to do so.

Targeting Impacts By Species

One means of understanding the motivations of fishermen to increase efficiency is to examine which seafood species they target. Except for sharks, the seafood resources of the coral reef system are generally available on a daily basis. Therefore, variations in the types of fish caught and sold, therefore primarily should be a function of fishermen decisions.

Figure 12.5 presents an analysis of high, medium, and low classes of seafood, cross tabulated by the weight sold and the month and season when it was sold. This comparison shows an increase in sales of higher classes of fish and more expensive types of seafood. For example, fishermen sold 893 pounds of first class fish in the summer and 1,194 pounds of first class fish in the spring. This represents an increase of 34 percent (301 pounds). There also was an increase in the amount of lobster sold, from 1,294 pounds in the summer to 1,559 pounds in spring; an increase of 20 percent (265 pounds). Sales of octopus, another high class seafood product, increased from 671 pounds in the summer to 866 pounds in the spring; an increase of 29 percent (195 pounds). Higher class seafoods (first and second class) were the only classes to increase in sales from the summer to the spring. Overall, fishermen sold 2,858 pounds of high class seafood in the summer and sold 3,619 pounds of high class seafood in the spring; an increase of 27 percent (761 pounds). Sales of small lobster, barracuda and second class fish remained the virtually the same between the spring and summer seasons

All other types of lower class seafood decreased in sales. Third class fish sales declined dramatically from 419 pounds in the summer to 221 pounds in the spring; a decrease of 47 percent (198 pounds; see figure 12.5). In addition, conch sales declined from 70 pounds in the summer to 35 pounds in the spring; a decrease of 50 percent (35 pounds). Crab sales declined from 132 pounds in the summer to 94 pounds in the spring; a decrease of 29 percent (38 pounds). Overall, fishermen sold 621 pounds in the summer, 350 pounds in the spring; a decrease of 44 percent (271 pounds). These data demonstrate that, despite the fact that fishermen were selling approximately the same amount of seafood during the summer as in the spring, they targeted different classes of seafood in the two seasons. The preference for higher classes of seafood in the spring appears to be associated with monthly variations in fishing efficiency.

High fishing efficiency should be associated with capturing more valuable species of seafood. This question was explored by comparing the lowest (June) and highest (April) fishing efficiency months according to the percentage of total sales that were high, medium, and low value. Sales in the low efficiency month of June were comprised of 52 percent high value, 36 percent medium value, and 12 percent low value species sold. Sales in the high efficiency month of April were comprised of 56 percent high value, 41

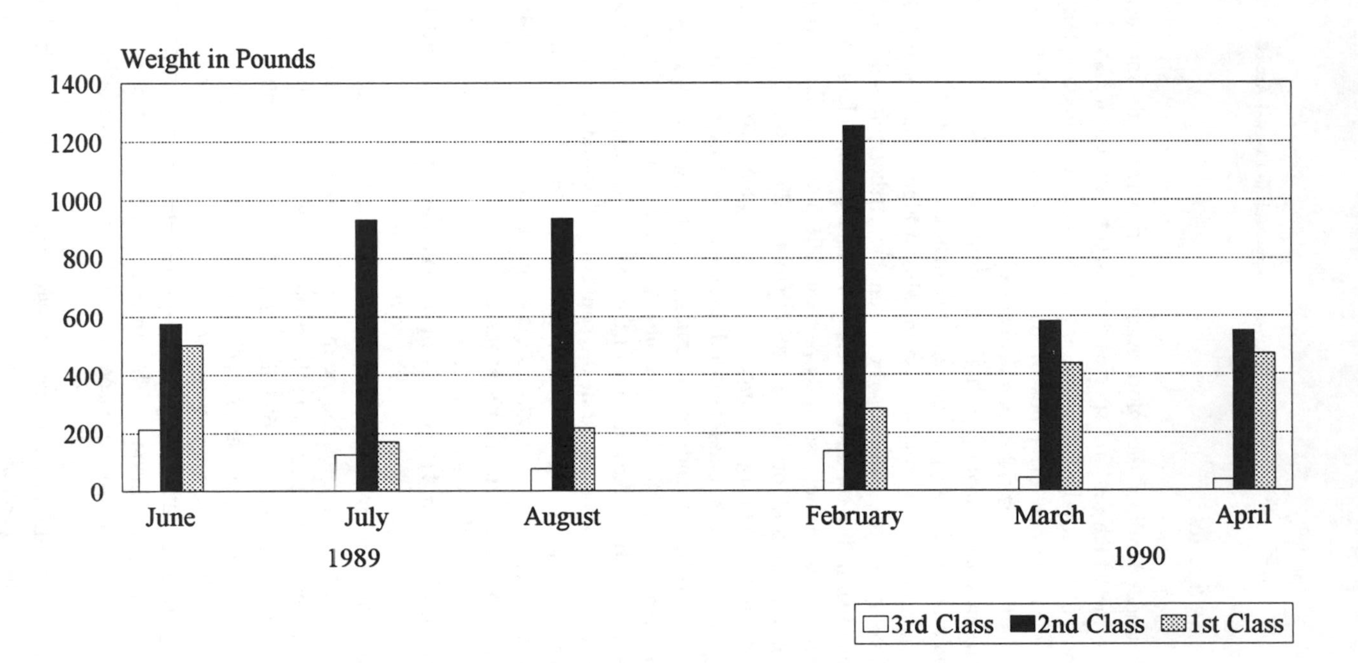

Fig. 12.5. Classes of seafood sales by season
Note: Excludes shark and barracuda

percent medium value, and 3 percent low value species sold. These figures suggest that higher fishing efficiency is associated with targeting higher value seafood.

Targeting Impacts By Economic Value

The need to generate sufficient income potentially motivates fishermen to target different seafood species at different times. This analysis explores the hypothesis that the fishermen change their fishing pattern from catching a diversity of the species that are available to one of targeting specific types of seafood in order to generate additional profit. The efficiency of effort analysis indicates that fishermen do target higher value seafood at certain times, which results in a decrease in the sales of lower class seafood. The analysis below calculates the economic implications of these higher value catches.

During the summer months, the fishermen sold a catch valued at RD $53,578. During the spring, they sold a catch valued at RD $61,068. Although the total weight in pounds of seafood sold in the spring was only 6 percent (420 pounds) more than the total weight sold in the summer, the spring catch was worth 14 percent (RD $7,490) more than the summer's catch.

It is possible that increases in total catch weight and income were caused by increased fishing effort during the spring. The association records show, however, that during the summer months of 1989, fishermen fished seventy-one out of ninety-two possible fishing days; that is, they fished 77 percent of the possible fishing days. In the spring months of 1990, fishermen fished sixty-nine days out of eighty-nine possible fishing days, 78 percent of the possible fishing days.

Conclusions

In summary, fishermen sold roughly the same amount of seafood overall but made more money during the spring season. This was accomplished by targeting high value species and improving efficiency of fishing effort. These data show that the Buen Hombre fishermen can and do engage in fishing practices that increase the amount of high value seafood removed from the coral reef ecozone when it becomes necessary to do so. The question remains as to why they do not do so all the time. The answer provided by the fishermen is that they realize that fishing for only high value species will eventually destroy the coral reef.

Despite a general desire to be economically better off, the fishermen of Buen Hombre have weighed this desire for short-term economic benefit against an even stronger concern for the long-term sustainability of the ecozone. Fishermen resolve conflicting desires for economic betterment and conservation by fishing for a mix of species and at lower daily levels whenever they can in order to protect the coral reef for themselves and future generations. When they do engage in nonsustainable fishing practices, they do so only on a short-term basis in order to compensate for (1) fishing income lost due to the effects of adverse weather conditions on the ability to fish, (2) competing labor demands, such as peak seasonal commitments to agricultural activities as part of the system of occupational multiplicity, (3) crop losses due to drought, and (4) to purchase basic essentials such as water and other staple foods.

The local adaptive strategy of engaging in multiple occupations (fishing and farming), based on mixed production of diverse commodities, including seafood and crops, serves to reduce the risk of economic failure. Perhaps an under-recognized adaptive function of occupational multiplicity is that such a system potentially serves to reduce the risk of environmental degradation in terms of overuse of terrestrial and marine ecozone components of coastal ecosystems.

Local resource use and management practices of Buen Hombre fishermen-farmers are currently being threatened by the destructive practices of outsiders. Like most small-scale fishermen (Cordell 1988a, 1988b, 1988c), the people of Buen Hombre perceive the coastal waters as part of their community territory. Interior village and port city commercial fishermen compete for access to reef and sea resources with fishermen of Buen Hombre. In addition, foreign fishing fleets from Puerto Rico have exploited Dominican waters. In the words of a Smithsonian marine scientist who has observed the practices, large-scale competitors are "reef rapers" (Walter Adey, personal communication 1985) because they use destructive and perhaps illegal net fishing techniques. The Caribbean Fishery Management Council has adopted a fishery management plan for reef fish fisheries of Puerto Rico and the United States Virgin Islands that includes regulations on net size. While the Dominican government also has national regulations on fishery practices, the north coast is relatively isolated. Moreover, manpower for enforcement is generally lacking.

The burgeoning tourism industry also affects the coral reef ecozone. Even in small-scale resorts near Buen Hombre, there already appears to have been an increase in coral harvesting, collected by tourists as souvenirs. As the industry continues to grow and expand beyond the boundaries of port towns, increasing numbers of tourists will intensify their search for "wilderness" areas, thus subjecting the Buen Hombre coral reef microzone to extreme levels of disruption. Together, tourists, commercial fishing fleets and growing numbers of small-scale fishermen have the capability to destroy one of the largest living reef zones in the world.

How can increased pressure on the coral reef ecozone be reduced or eliminated? One approach to ameliorating current conditions of environmental destruction has been termed "sustainable development" (World Bank 1989;

World Commission on Environment and Development 1987a, 1987b) because of the intimate interdependency between the local people and the ecosystems that they inhabit. This approach seeks to conserve natural environments and, at the same time, sustain the economy and society of human populations who reside within the ecosystems and must interact with its resources on a daily basis to survive. The latter goal is achieved by combining local systems of traditional knowledge with economic production plans that involve natural resource utilization, in order to provide food, shelter and income for rural populations. The former is achieved by combining scientific knowledge with traditional local conservation techniques. Co-management (Pinkerton 1989) of natural resources by sharing authority with local people is one potential strategy for achieving sustainable development.

Two sustainable development alternatives that appear to reduce pressure on coral reef zones are (1) mariculture, or nearshore cultivation of marine species such as crab or shrimp and (2) the introduction of deep water fish pot technology. Buen Hombre was a site for a pilot demonstration of crab mariculture technology, developed by the Smithsonian Institution's Marine Systems Laboratory, in 1985 (Rubino and Stoffle 1989, 1991; Stoffle 1986; Stoffle, Rubino and Rasch 1988). For a variety of reasons beyond the scope of this paper, the project was terminated in 1987 (Stoffle, Halmo, and Stoffle 1990a, 1990b). Social assessment of the feasibility of the mariculture technology predicted that the intervention would fit with the local system. An ex-post social assessment found that it had reduced fishing pressure on the coral reef microzone. A deep water fish pot project was initiated in Antigua on the heels of rejection of mariculture by local fishermen. This technology also has reduced pressure on the nearshore marine fishery and coastal zone.

These sustainable development interventions using appropriate technologies are more effective in protecting degradation of coastal ecozones than regulating fishing technology and practices. Development of commercial capture fisheries and increasingly large-scale tourism, as is currently the trend in coastal development, can only lead to ecosystem destruction. As one development practitioner has put it:

> the policy implications are clear: capitalize a more rational kind of development which favors not the outsider but the indigenous population, which favors not the destructive cultural tradition . . . but the more ecologically viable and economically productive . . . systems of local people already skilled in the management of . . . resources. (Partridge 1984,78).

Participation and understanding of local resource use and management strategies is key to implementing programs that serve both development and conservation goals. McCay has noted:

> there is very little appreciation of the sociocultural realities of the systems being managed and even less appreciation of the potentials of existing informal or indigenous forms of resource management . . . Perhaps these, as well as indigenous technologies, might be built upon, resulting in resource management, as well as development, "from below" (McCay 1980,11).

Development planners, policy makers, and project managers in government, national and international agencies must come to grips with recognizing that local communities have the capacity to manage their own resources and development in a sustainable way.

Chapter XIII

Population-Environment Dynamics in a Constrained Ecosystem in Northern Benin

Valentin Agbo
Nestor Sokpon
John Hough
Patrick C. West

Introduction

Population-environment relations are not uniform everywhere. Theories of how population-environment interactions vary under different circumstances are therefore needed. This chapter explores a middle range theory (Merton, 1968) of differences in population-environment relations under conditions of "constrained ecosystems." Using a rather extreme case from a village in Northern Benin (Tannougou), this theory is elaborated and illustrated. It is beyond the scope of this chapter to present comparative evidence of variation within different types of constrained ecosystems, or between highly constrained ecosystems and those that are less constrained.

All ecosystems are constrained to some degree - that's why we have the dismal science of economics. Here we shall deal with concepts and theoretical propositions of more highly constrained ecosystems. Within the general category of highly constrained ecosystems we can distinguish between at least four types that are often found together: geographically constrained (abutted against a mountain range or ocean); ecologically constrained (e.g., poor soils), institutionally constrained (e.g., state policies that force higher man-land ratios; class relations that marginalize peasants etc.); and economically constrained (e.g., lack of diversification and external trade - such as isolated subsistence economics). The basic theoretical proposition we would posit is that the greater number of types of ecosystem constraints involved and the greater degree of constraint of these components, the more tightly linked population-environment relations will be; i.e., the more intensely population pressures will negatively effect the environment and the more intensely the environmental degradation will redound upon the human population. The concept of "tight linkage" is borrowed from Perrow (1984).

The concept and theory of "constrained ecosystems" is not static but dynamic. It is often rapid change in one or more of the above dimensions that becomes the most constraining aspect that increases the tightness of linkage between population and environment and causes dysfunctional disequilibriums. In our case, institutional constraints (both population distribution policy and national park displacement policy) changed rapidly, causing a rapid increase in population and a rapid decrease in land-base within an already tightly constrained ecosystem in its geographic and ecological dimensions. Under rapidly changing conditions that rapidly increase constraints to the ecosystem, human adaptive patterns that were formerly adaptive become rapidly maladaptive, leading to a rapid degradation of the ecosystem (e.g., declining soil fertility) and thus to increasing ecological constraints (adapted from Burch 1971). For populations without the ability to add other productive resources, this may cause them to increase fertility rather than decrease fertility as an adaptive response to increase labor inputs (adapted from Cottrell 1955). This of course adds further to the downward spiral of circular tragedy in population-environment relations, and secondary increases in ecological degradation and ecological constraints. Again, we cannot present comparative evidence to fully test these theoretical propositions and hypotheses. Rather we present a case study of an extreme case that illustrates this model in an almost worst case scenario.

This study was conducted in the commune of Tannougou (a group of nine villages) in northwest Benin. The main focus of the paper is the relationship between population and environment in Tannougou, the largest village and the administrative center of the commune (fig. 13.1). The Tannougou area is under severe ecological, geographic and administrative strains. Its basic population dynamics are affected both by natural growth and administratively mandated in-migration within this ecosystem. The land area is bounded on the southeast by the Atacora Mountains, and on the northeast by the Penjari National Park and Hunting Preserve. In the mid-eighties, due to a change in park policy, the local people were confined to a three kilometer strip of land between the park and the road that parallels the Atacora Mountains thus creating a severely constrained ecosystem--i.e., a sharp increase in institutional constraints (fig. 13.2).

The official figure for the population of the commune (combined villages) was 3,601 in 1988. According to the survey in 1989, the population of the village of Tannougou was 497 and population density 38 per square kilometer (CARDER-Acatora, *Survey of Households*, 1989). In addition to a natural increase in population at a rate of 2.5 percent (Hough 1989, 1976) official figures show a huge jump in forced in-migration to the commune between 1976 and 1977 (fig. 13.3). However, there is strong reason to believe that some or much of this official increase was artificially inflated because communes were expected to be a certain size despite land area and

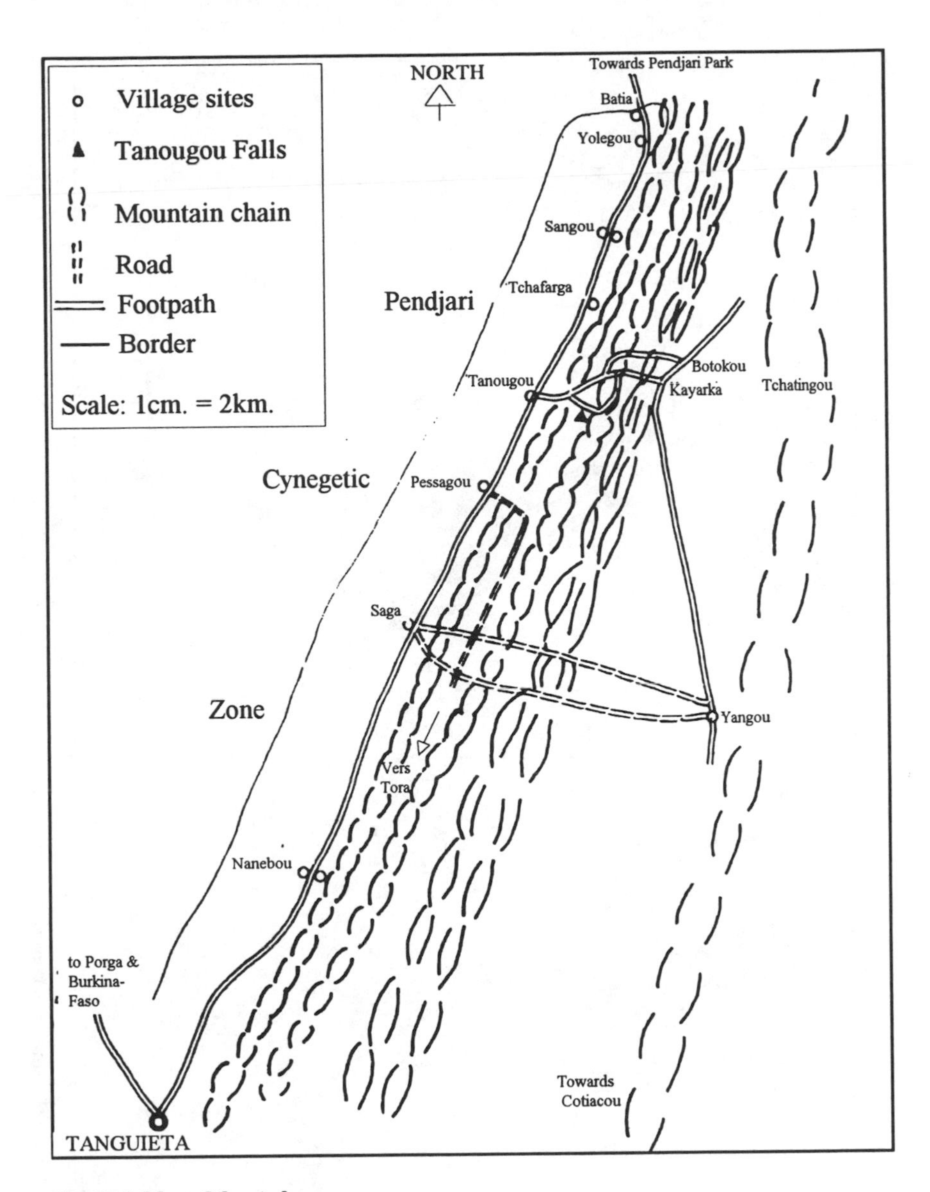

Fig. 13.1. Map of the study area
Source: Agbo and Sokpon, undated

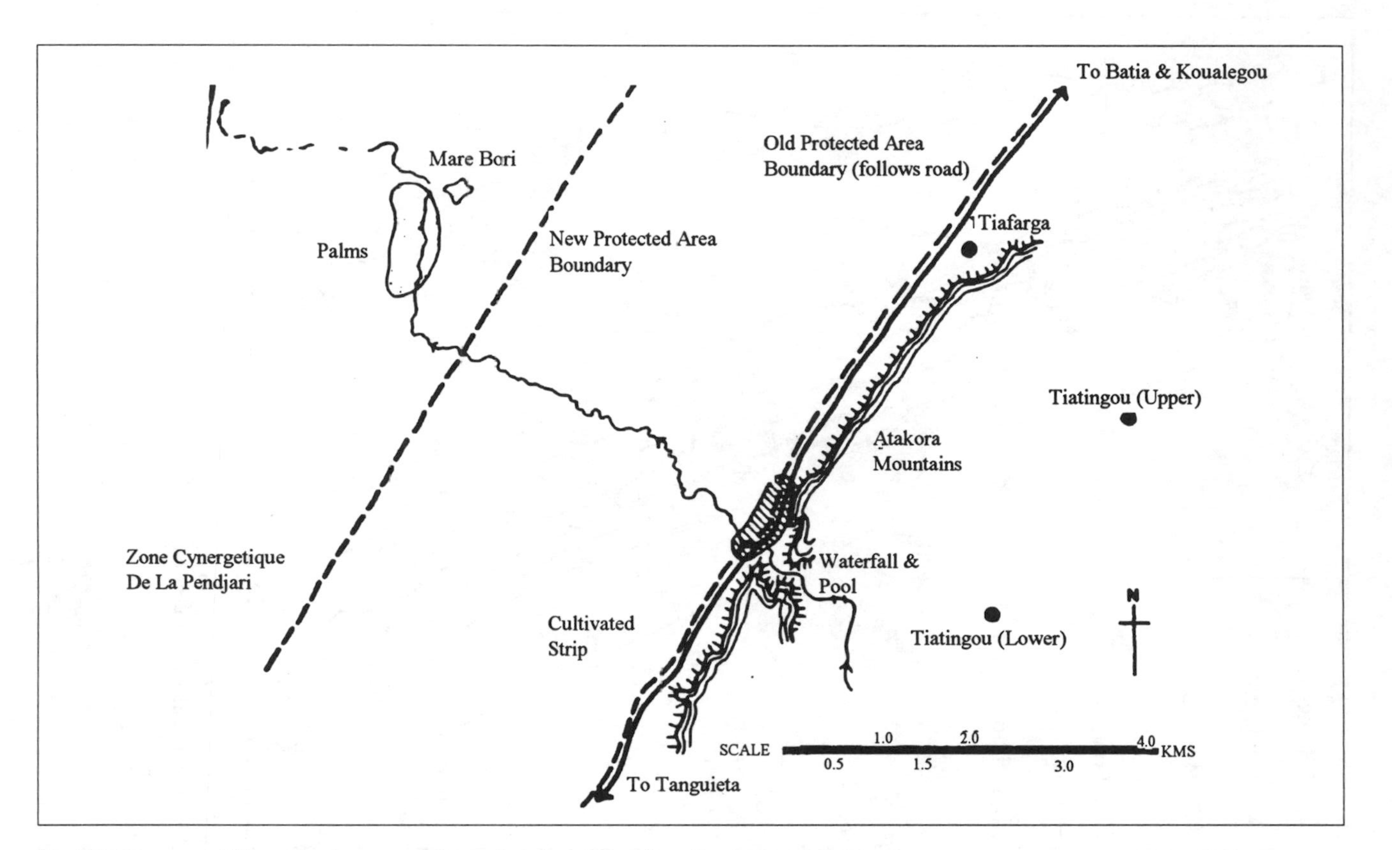

Fig. 13.2. Tanougou Village showing mountains and protected area boundary

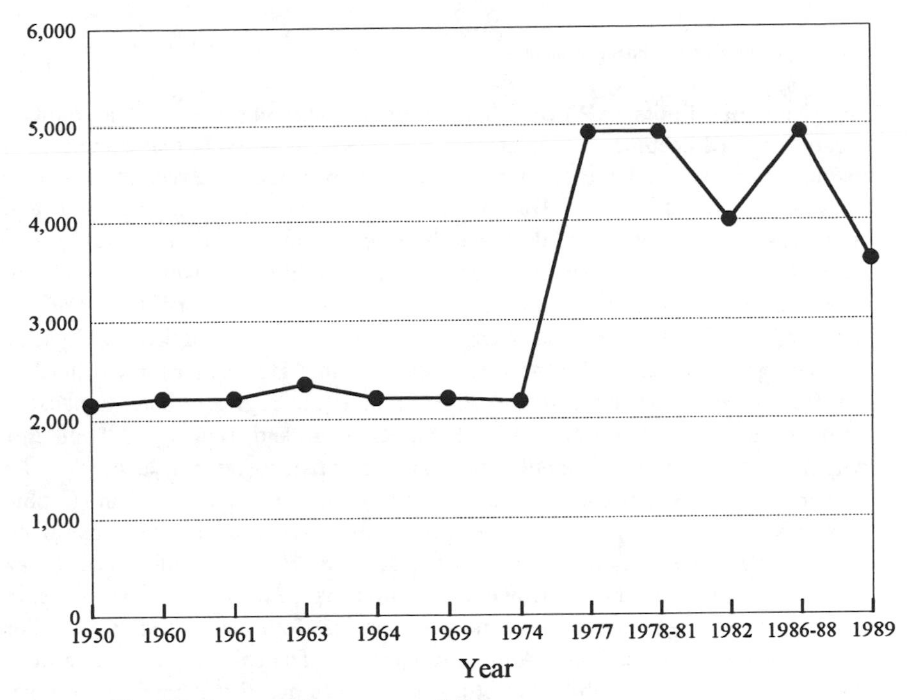

Fig. 13.3. Tanougou Commune population
Source: Government/Provincial Administrative Office, Tanguieta

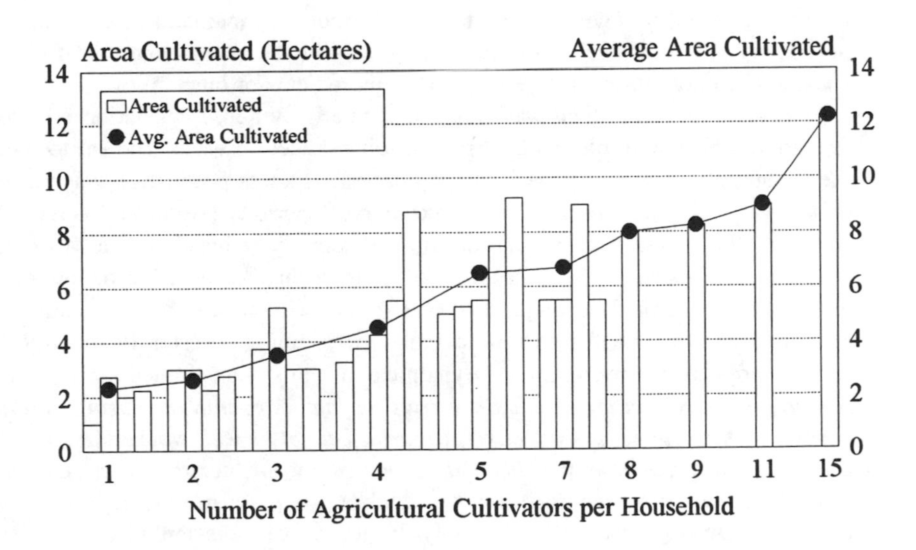

Fig. 13.4. Agricultural cultivation

environmental limits. When administrators could not meet actual quotas, creative record-keeping was practiced. However, it is likely that some forced in-migration occurred and further increased the population (a second change in institutional constraints). Due to the severe pressures on the land and resources in this ecosystem, there has been some out-migration pressure. For example, some young men migrated over the mountains towards the Borgou province of northeastern Benin and to Nigeria in search of better arable farmland. These migration dynamics were further complicated by the increasing desertification in the north, the nomadic Houssa pastoralist herders sought to settle permanently in the Tannougou region, thereby causing conflicts among ethnic groups and higher man-land ratios. All of this combines to create severe population strains on a narrow strip of land.

Perhaps the culmination of administrative, geographic, and demographic problems would not have been so potentially disastrous if the constrained ecosystem at stake (this three kilometer wide strip) were made of rich soils in a region of high rain fall. However, rainfall is meager in northern Benin (averaging only 942.63 millimeters over the last fourteen years), which lies just to the south of the Sahel (Agbo and Nestor). The soils are also poor since they are overused. Within this study area there are three types of tropical soils: (1) non-solidified ferruginous soils in the colluviums of quartzites and micashists; (2) solidified ferruginous soils on quartzschists situated on the long slopes with low declivity; and (3) ferruginous hydromorphous soils on quartzschists. They are harsh, barren tropical soils that have always required shifting agriculture with long fallow periods to maintain their meager productivity. Traditional agricultural practices in the region involved four to six years of cultivation followed by ten or more years of bush fallow.

In earlier times the land was richer and relatively dense vegetation covered the mountains and arable land. Agricultural production was sufficient to meet the consumption needs of the local population. Excess production was stored collectively in granaries until the next harvest periods (Agbo and Sokpon). But with the increase in population and the sharply reduced land area of the Tannougou ecosystem, the duration of fallows has been reduced and crop yields have declined (Hough 1989, 1981). Given the basic parameters described above, one foresees a formula for looming disaster in the tightly linked population-environment dynamics of this constrained ecosystem. *Already decreases in crop yields are causing famine conditions during certain periods in the year when collective granaries cannot be well stocked* (Agbo and Sokpon). *Malnutrition has been documented by the health center and has caused several diseases such as kwashiorkor and marasmus, especially in children.* During times of famine only 26 percent of the sample has sufficient grain reserves to last until harvest time. In times of famine, when granaries are empty for the rest of the local population, the people are often reduced to surviving on leaves of trees and other plants that have some minimal nuturitional value. The most common leaves are *Sesamum indicum*, *Seasamum alatum*, *Hibiscus sbdarifa*, and young leaves of *Adansonia digitata*, *Allium cepa*, *Corchorus triens*, and *Corchorus olitorus*. Until a better balance is established in the population-environmental relations, careful attention should be given to the preservation of these species in the ecosystem.

Given these ominous implications, further details of the human population and its interaction with farming systems and the natural environment need to be examined. Data, including observations in the field and transect measurements of selected environmental parameters, were collected from a sample survey of the village Tannougou (a 56.6 percent sample). Results indicate that in addition to soil depletion, other needed resources are being increasingly degraded. Rising population coupled with this constrained ecosystem is both the chief cause of this degradation, as well as the ultimate recipient of the tragic consequences. How can this circular tragedy be reversed within such narrow confines? Can an equilibrium be created between a population and its environment under these severe conditions of a constrained ecosystem?

The People of Tannougou

Originally populated by the Wama Tribe, the Tannougou area is now primarily populated by the Gourmantche. The Wama Tribe still occupies the village of Pessagou within the Tannougou commune (figure 13.1). The village of Tannougou is predominantly Gourmantche, with a few Wama residents. Having been isolated from the outside world until very recently, both tribes hold similar traditional religious beliefs. There is potential for both traditional resistance to environmental management changes and for environmental adaptation arising from respect for the environment embodied within traditional religious myths. The age structure of the population is very young, with 63 percent of household heads between the ages of twenty-four and fifty. Thus, potential for further increase in birth rates exists depending on a variety of conditions.

Agricultural Farming Systems

As noted above, traditional systems of shifting agriculture have been confined to a small, narrow strip of land with reduced fallow periods and declining agricultural yields. Rising population within this constrained ecosystem has placed further strains on the agricultural system.

The primary crops cultivated include Igname (genus dioscoracea) or sorghum which come at the head of the rotations. Various rotations include the following:

(1) Sorghum + maize + peanuts (15 percent of the sample)
(2) Sorghum + beans (98 percent of the sample)
(3) Igname + maize + cowpeas (20 percent of the sample)
(4) Igname + maize + rice [in lowland areas] (54 percent of the sample)

Notes: The total percentages exceed 100 percent because households pursue different rotation patterns on different plots of land.
The peanut crop noted above is a high yielding groundnut, grown with the aid of fertilizers. It is the principal cash crop of the village. Years of low rain have reduced cash income (Hough 1989, 1977-1978).

The total area cultivated by the peasants within this sample in 1990 was 155.25 hectares. Among the thirty households in the sample this is an average of only 5.2 hectares per household. The larger the number of workers per household, the larger the area cultivated (fig. 13.4). Thus, both total population and population per household further strains the limited agricultural resources of the ecosystem and in turn reduce fallow time and soil fertility. As noted above, cultivation of plots had been four to six years followed by ten or more years of bush fallow. Given the population pressures, the duration of land cultivation for the sample was reduced to three to four years and fallow time was reduced to five to six years. These shortened agricultural cycles correspond to phases II and III of itinerant agriculture according to Greenland (1974). Further declines in soil fertility and rotation times portend disaster. These trends are further complicated by traditional cultural and religious restrictions on the use of some plots of land. Population increase coupled with this small land area are the direct cause of this environmental degradation, and looming human tragedy. Chemical fertilizers could ameliorate the situation, but the peasant farmers of Tannougou are too poor to afford these agricultural inputs and outside assistance is needed to implement this strategy to reverse this negative cycle of population-environment dynamics. Changing park policies to provide the local people access to more land from the park could also help to reverse this circular tragedy but park policy is not likely to change in the near future. Again, support from the outside conservation community would be needed to affect these changes.

In some situations, declining food productivity leads to a decline in the birth rate, alleviating some of the population pressure on the environment. However, in other agricultural systems, such environmental decline will cause increased population growth because farmers need more children, especially males, to eke a meager subsistence from land with decreasing productivity. Though there is insufficient data to determine whether either of these trends may actually be occurring, the high fertility rate of 2.5 suggests the latter cycle of environment-population dynamics. However, one can ask which is more or less likely to occur in more constrained ecosystems. Our theory above would suggest the latter and a resulting downward cycle that is likely to beget yet more tragedy.

Shifting Agricultural and Wood Resources

A key component of the environmental ecosystem are the agricultural systems, their associated soils and rainfall, and also their interactions with other aspects of the agricultural and natural ecosystems. A chief factor is the forest and tree systems, both those that occur and recur naturally as well as planted agroforestry systems. To analyze this dimension a tree vegetation transect was run from the mountain system to the three kilometer boundary of the park.

The starting point for this analysis is the amount of bush fallow forest systems that are cleared for use in the shifting agricultural systems in the flatlands. With with the decrease in fallow periods, the size and biomass of scrub forest in the fallow areas have been reduced. Any soil nutrification functions and potential wood products to meet domestic needs are being substantially reduced.

Within the sample, the number of hectares of bush fallow that has been cleared has remained relatively stable over the four years prior to the field work. In 1986 an average of 1.4 hectares was cleared; in 1987 an average of 1.7 hectares; in 1988, 1.5 hectares; and in 1989 a slight dip to 1.25 hectares. Lack of fertile land in the three kilometer strip of land has caused some incursion of shifting cultivation into the gallery forest that lays at the base and slopes of the mountain chain. This has led, along with fuelwood cutting in the mountains, to deforestation, erosion, and some increase in flooding that sometimes damages crops on the flatlands. This chain of environmental impacts can again be traced back to the population pressures which in turn can be tied to rapid changes in institutional constraints.

For this sample, on the recently cleared land an average of thirty-seven trees per hectare were preserved and protected while 426 were cut down. The forest species protected by the population during land clearing are for the most part the karite (*Vitel laria paradoxa*), the nere (*Parkia clappertoniana*), the tamarind (*Tamarindus indica*), the vene (*Pterocarpus erinaceus*), the lingu (*Afzelia africana*), and the kapok tree (*Bombax costatum*). The nere and the karite are especially protected by the population, because their fruit play an important part in the diet of the people of Tannougou.

It has been the traditional practice to burn the fallow land as an easy way to clear it for crop production. However, this practice further damages soil fertility and burns unprotected trees. This in turn places further stress on other forest resources that are being increasingly depleted for firewood, construction and other domestic uses. Again, formerly adaptive patterns become muladaptive under changed population-environment relations in an increasingly constrained ecosystem.

One possible environmental management adaptation to this condition would be to first deliberately plant fallow land with fast growing leguminous agroforestry species that naturally increase nitrogen in the soil, thus helping to reverse the trend towards soil fertility decline. When it comes time to clear the fallow land perhaps the practice of burning could be changed to one of harvesting the trees for domestic use and reducing the negative impacts of burning on soil fertility and pressures on deforestation elsewhere.

In the meantime, deforestation is occurring from a variety of areas within the local area. Ninety percent of the families in the sample harvest wood poles for construction of tatas (homes) and storage areas from the fallowlands and the mountains, while 57 percent harvest them in the fields. In this sample, 83 percent of the families gather fuelwood in the fallowland, 73 percent in the fields, and 90 percent in the mountains (fig. 13.5). All wood species found in the study area are used as fuelwood with the exception of Afzelia. The fallowlands are the preferred sites for the collection of thatch used to cover the tatas and storage areas. Eighty-seven percent of the sample collect thatch on fallowlands, which is primarily from Andropogons.

Fuelwood cutting is a major strain on the ecosystem and is directly tied to population levels and population growth. On a per meal basis an average of 37 kilograms of wood is used for the preparation of a local drink, "tchoukoutou", made from fermented sorghum. Sixteen kilograms of hardwood is used to prepare an average meal. On a per household basis, the larger the household, the greater the amount of fuelwood utilized (fig. 13.6). Overall, the greater the population, and the smaller the land area open to fuelwood gathering (prohibited in the park), the greater the deforestation in the environment. If deforestation becomes extreme, it will, in turn, negatively affect the standard of living for the local people.

While one usually thinks of fuelwood as the main drain on wood resources in peasant village communities, poles for housing construction in Gourmantche culture place equal if not greater pressure on forest resources. Housing for a household is a cluster of huts called tatas--a group of round or rectangular huts forming a circle with a millstone shed and a community hall in the center. The walls are made of banco, built around a framework of wooden posts, poles, bamboo and, sorghum stems. On the average one post is used for the roof and sixty-nine poles for each structure. Four tata structures in a compound would quadruple the number of posts and poles needed. The larger the size of the household, the larger the number of units in the compound and the greater number of poles that are required *overall*. But on a per capita basis, smaller tatas cause greater deforestation. For instance a large tata unit having

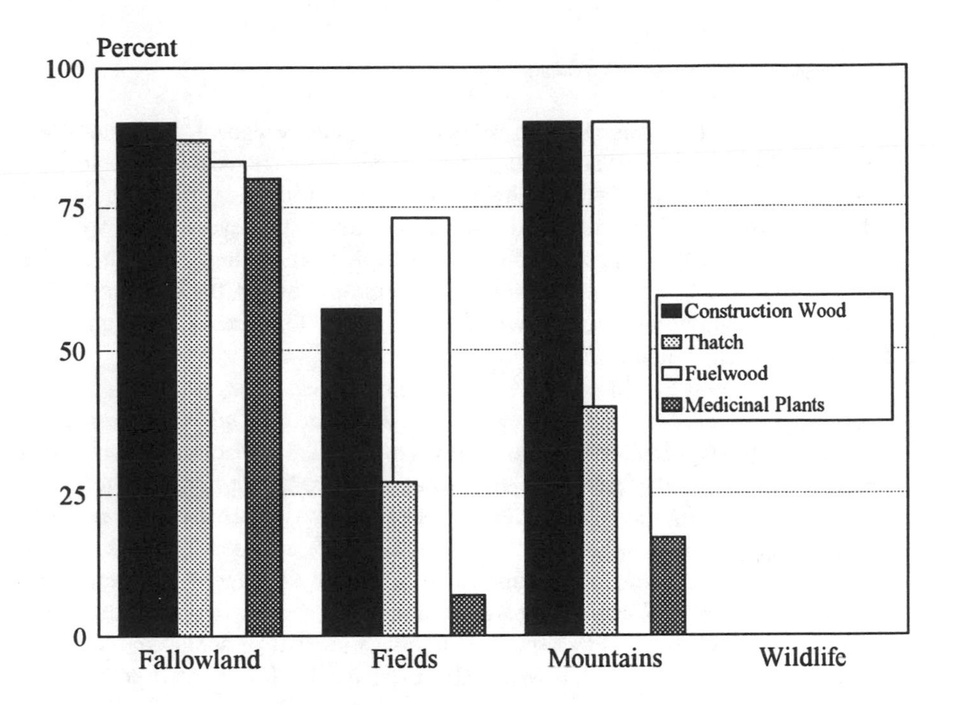

Fig. 13.5. Cultivation of various land areas satisfies needs for wood, thatch, and medicinal plants.
Note: The total in each category surpasses 100 percent because many villagers use more than one area at any given time to meet their needs for wood and other plant products.
Source: Data collected during field investigation.

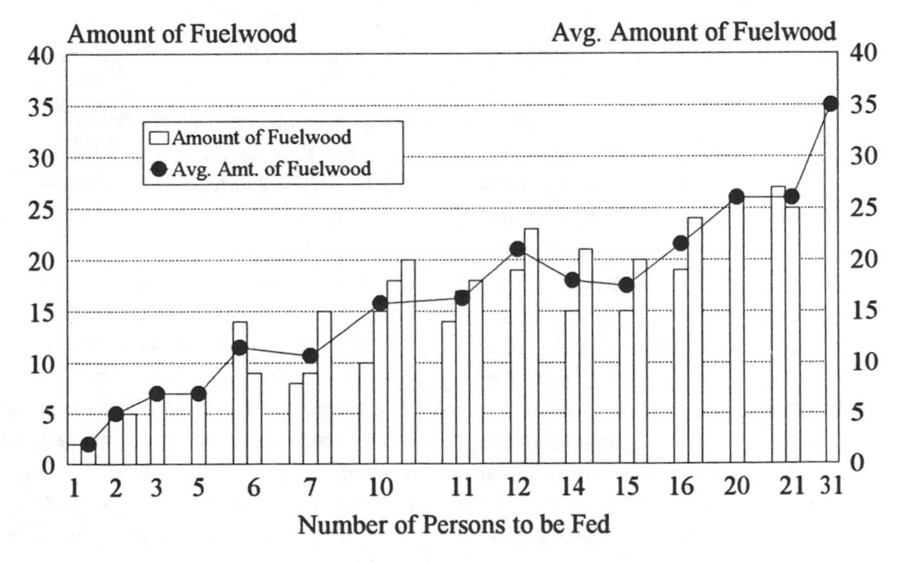

Fig. 13.6. Ration of fuelwood

seventeen household members uses 197 poles, an average of 11.5 poles per person. In contrast, a tata unit having nine members uses 179 poles, an average of 19 poles per person. According to tradition and due to limitations of the durability of the design, tatas require new poles every three years in order to be rebuilt. The greater the population of the village, of course, the greater the number of tatas requiring rebuilding every three years. The resulting demand pressures on wood resources in this area are a significant strain on the ecosystem.

Certain species are in high demand for these uses. *Anogeissus leiocarpus, Pterocarpus erinaceus, Vitellaria paradoxa* and *Prosopis africana* are desired for center posts because they are very sturdy hard wood, can bear much weight, and are also resistant to termites. *Burkea africana, Pteleopsis suberosa, Terminalia avicinoides, Anogeissus leiocarpus, Hexalobus monopetalus, Bambusa vulgaris*, and related species are used for poles. There are a few houses made of corrugated iron built by villagers returning from abroad, but the tata housing pattern is deeply rooted in the culture and functional to its cultural identity. It is likely to persist for some time to come. Large populations in relation to highly restricted land areas will continue to place strong deforestation pressures on the ecosystem.

There is some indication that the villagers realize the degree of deforestation and its potentially negative consequences. They have been engaged in reforestation and further support and encouragement needs to be given to these efforts. Reforestation efforts by species type for the sample is shown in table 13.1 for the period 1987-1990.

Protein Cycles in a Constrained Ecosystem Under Population Pressure

The Tannougou villagers are not total vegetarians, and eat much wild game and tend large flocks of domestic animals. Yet there are a number of curious facts in their current dietary and animal husbandry patterns. Their current diet consists mainly of carbohydrates from farinaceous sorghum, corn or millet paste, and igname (dioscoracea) in a variety of forms. Protein are primarily from vegetable production, mainly beans and voandzeia; rarely are they of animal origin. Part of the explanation for this lies in cultural and ritualistic reasons. Animals are rarely slain and eaten except for ritual ceremonies and for illness (including illness from hunger and malnutrition). This second reason implies the storage of reserve protein in animal protein for desperate times, an adaptive survival strategy used by many peasant and tribal peoples. Yet there is a surprisingly small number of domestic animals in a typical village in Benin (table 13.2). This may be due in part to do the fact that baboons raid the farmyards and kill domestic animals. The villagers, however, are officially prohibited from killing the baboons. Given the small land area and the relatively large population, the villagers have chosen vegetable

TABLE 13.1 Current State of Reforestation in the Village of Tannougou

SPECIES	1987	1988	1989	1990
Khaya senegalaensis (Calicedrat)	13	5	-	
Parkia clappertoniana (Nere)	35	57	12	30
Citrus sinensis (sweet orange trees)	3	0	0	0
Mangifera indica (mangoes)	7	8	7	20
Anacardium occidentale (Anacardiaceae)	20	0	0	0
Psidium goyava (myrtaceae)	7	0	0	0
Eucalyptus camadulensis (Eucalyptus)	30	9	6	0
Acacia auriculiformis (Acacia)	8	9	0	0

Source: Data from field research

production over animal production as the former yields greater nutrition per land area. This may be an indicator of the degree of strain that the agricultural production system is under. However, this leaves the villagers with less protein reserves to help them through lean times and may contribute to the periodic famine and malnutrition noted above. One possible adaptive strategy is to work with the villagers to plant agroforestry species that have fodder potential. This would lessen the conflict between land use for vegetable production and animal production since there would be less need for grazing land and raising grains for animals. The National University of Benin is conducting research on fodder agroforestry. Perhaps its results could be applied in the north with dryland, drought resistant species.

Fishing

Fishing from the Mare Lake in the park was once a major source of supplemental protein (Hough 1989). However fishing is now confined mainly to the stream coming from the Tannougou waterfall. The most common species caught are *clarias* and the *tilapias* and various dams, nets, and traps are used. Certain forms of damming with clay affect the ability of villagers to get clean potable water. If there were more sources of water there would be

TABLE 13.2. Livestock and poultry in the village of Tanougou, May 1990

HOUSEHOLDS	BOVINES	SHEEP	GOATS	PIGS	POULTRY*
1					8
2	3	7	1	1	7
3					7
4		6	4	2	9
5		1	3		5
6	5	9	17	3	15
7		2			
8			3	2	6
9		1	15		7
10		6	1	6	12
11		11	9		7
12	6		8	4	20
13				1	1
14		1	1		17
15			2		3
16		1	4		2
17		2			3
18			9		
19			4		10
20		8	7	6	2
21			3	1	11
22		2	1	2	12
23	3	2	1		22
24			1		6
25			6	6	8
26	6	4	8	5	18
27	2	14	3		16
28			1	2	5
29	2	3	11	6	2
30			2		1
Σx	27	80	125	47	242
x	4	5	5	3	9
Σx^2	123	628	1109	213	3046
δ n-1	1, 8	4, 0	4, 5	2	6, 0
N	7	17	25	14	28

*Chickens and guinea-hens

potential for tillapia fish farming but this has lower potential in this arid climate.

Poaching

The other source of potential meat protein is poaching wild game from the National Park. Villagers poach for economics reasons--given their great poverty--and out of resentment of the National Parks Service which forced them into this precarious state. There is some subsistence consumption of poached bush meat, but it is mostly small animals such as hares, cane, giant rats, ground squirrels, and various birds (Hough 1989, 1979). Once, hunting was almost exclusively a source of subsistence protein for cultures in this region (Hough 1989). Today, most poached animals are, instead, sold clandestinely. It is unclear why poaching is now more of a commercial activity by local people. It could well be that the poached meat is worth more to the local economy as a cash supplement to subsistence. The animals most frequently killed are buffaloes (*Syncerus caffer*), the bubales (*Acephalus buselaphus majoro*), the buffon cobs (*Adenota Kob*), the fassa cobs (*Kobus de fassa*), and the antelope-horses (*Hippotragus equinus*). Monkeys and wild boars are also killed as a common currency. Clearly, this illegal hunting is in some way an adaptive strategy to "diversify the economy" in a constrained ecosystem. The park service might consider taking a more active role in ecodevelopment activities that would help the local people. Alternatively, perhaps, it should legalize hunting for local people and more rigorously prohibit hunting by high status hunting classes and administrators who enjoy special privileges.

Under Utilized Strategies to Diversity the Economy

Poaching is but one attempt to diversity the economy and to alleviate the severe strains of high population levels in a tightly constrained ecosystem. Ecotourism is a legitimate strategy that could be pursued given the natural attractions (e.g., national parks, wildlife, and a high, beautiful waterfall at Tannougou). Any such strategy, however, must be designed to include the participation of the local people, to primarily benefit local people, and to minimize the social impacts of western tourists on isolated native peoples. Currently tourists do pass through Tannougou but local villagers do not benefit economically (Hough 1989, 1980). Thus, theres a potential for diversifying the economy, providing cash income, and perhaps reversing the environmental degradation in soil fertility and deforestation.

Health and Sanitary Conditions

Finally, it is important--for humanitarian reasons and as a prerequisite for effective population control--to assess the health of the population in relation to the health of the ecosystem. In addition to the debilitating effects of famine and malnutrition noted above, there are a variety of diseases that affect the population including meningitis, conjunctivitis, and others. Of all these however, it may be simple diarrhea, dysentery, and related abdominal ailments that are the most ominous. These tend to occur most frequently during the rainy season when the stream, the sole source of potable water, is polluted. Diarrhea is known to be one of the most simple, yet most deadly causes of infant mortality in developing countries. Though there are no exact figures on infant mortality or trends over time for the village, those which have been collected are ominously suggestive.

The people of Tannougou practice a combination of modern and traditional medicine. While traditional medicines are often scorned by modern medicine, some may have more healing powers than acknowledged and villagers prefer them to modern pharmaceutical pills. The protection of the ecosystems in which these medicinal plants grow (table 13.3) is thus of high priority and the erosion of soil fertility and forest ecosystems may endanger some of these plants. Interventions must therefore respect and work with these important aspects of the people's culture.

Conclusions

From Malthus to Hardin we have been conditioned to think that wherever population is outstripping resources there is a need for population control, usually by whatever draconian means are deemed necessary. However, it is important to separate the empirical causal relations between population and environmental degradation from single solution quick fix population control measures. The demographer Judith Blake said, "lift pressures to reproduce; don't impose pressures not to reproduce." If infant mortality rates can be lowered by stopping the causes of diarrhea, people may not feel the need for many children in order to ensure the survival of a few. With more land (from the park service) the villagers' farming would not cause the current degree of degradation, longer fallow periods could be reestablished, and more fuelwood preserved. If fodder agroforestry could be reintroduced perhaps animal production could increase without sacrificing vegetable production. Allowing the villagers to shoot (and eat) marauding baboons would increase the number of domestic animals. Granaries would fill up and famine would decline. This might well pave the way for the introduction of voluntary population control since parents would not feel the need for as many children to tend the fields, guard against baboons, or to ensure the survival of a few to provide care for

TABLE 13.3. Vegetable species used for medicinal purposes

LOCAL NAMES	SCIENTIFIC NAMES	ELEMENTS	ILLNESSES
Ikokopieni Busebiu	Acacia sieberiana Anogeissus leio	Stem leaves & bark	Child growth Jaundice
Mangui saabu	Mangifera indica	Leaves &	Jaundice
Busabu paradoxa	Vitellaria Fruit	Bark	Diarrhea
Bupugubu	Tamarindus indica	Bark & fruit	Diarrhea
Kunamu saakpetiegu	Annona senegalensis	Leaves	Dysentry
"	"	Fruit	Aid to birth
Budubu	Parkia clappertoniana	Leaves	Skin desease
"	"	Boiled fruit skins	Diarrhea
Bunagnibu	Nauclea latifolia	Roots	Painful menstruation
Bupulubu	Blighia sapida	"	"
Kunamu saakpetiegu	Annona senegalensis	"	"
Kupopolu	Seduridaca longepeduncu lata	"	"
Bugapu	Diospyros mespiliformis	Leaves	Dysentry
Kugbelu	Pteleopsis suberosa	Bark	Dysentry
Buwaribu	Lannea kustingii	Bark	Colic
Liceduli	Terminalia avicinoides	Leaves	Eye diseases
Kulokogu	Swartzia madagascarensis	Bark	Dysentry
"	"	"	Snake bite
Ligogobli	Acacia gourmanensis	Boiled bark & karite butter	Cough
Licecieli	Pseudocedrela kortschii	Bark	Skin diseases
Bunakpagabu	Detarium microcarpum	Leaves	Children's tonic
Npopolu	Securidaca longepedunculata	"	"
Busabu	Vitellaria paradoxa	"	"
Bumiribu	Xomenia americana	Fruit	Abcess
Bufobu	Bombax costatum	Bark & bone	Kidney ailment
Bupugubu	Tamarindus indica	Bark	Toothache
Kunabanigu	Bauhinia thonningii	Leaves	Dysentry
Buwanbu	Vitex doniana	Leaves & bark	Diarrhea & dysentry
Lidjbalu	Mayetenus senegalensis	Leaves	Toothing
Amantui tunan	Cajanus cajan	Crushed leaves & pimento	Measles

parents in their old age. This seems to be the more responsible, enlightened, and sociologically sound strategy for bringing population-environment dynamics back into a better and more healthy equilibrium. This is especially true in constrained ecosystems under severe population stress.

The above case illustrates an extreme of the theoretical model presented in the introduction. Rapidly changing institutional constraints (forced in-migration policy, and restrictive park policies increased population and greatly reduced the land base squeezed up against the geographic constraint of the Attacora Mountains. The remaining lands were ecologically fragile and insufficient to support the population. An already ecologically constrained ecosystem became further constrained. Rapid ecological degradation increased and further constrained ecosystem. Birth rates did not decrease and remained high. Malnutrition and famine were the culmination of the "tragedy of the constrained ecosystem." Under such conditions there was extremely tight linkage between population and environment. This extreme case should be contrasted with others in comparative analysis to determine generalization to other similar cases (with different particulars), and with cases with less severe constraints on the four dimensions to test for comparative differences and variation hypothesized in the theoretical model.

While such comparative theory testing is badly needed, the scientists involved in this study also feel a responsibility to help the immediate situation of the people we have been studying. We cannot in good conscience use starving people as subjects and then walk away from them. Currently, a team is in the field working on solutions. Perhaps elaborate theory building will help these and other villagers some day and it is our sincere hope that it will; but the children of this village are hungry today and their need is now. When asked how best to save the world, a wise old sage said, "one village at a time."

Section IV

Emergent Ideas: Theory and Method

Two new ideas have emerged from the overall project and from the task of putting this volume together. One is a new theoretical framework, the other a methodological innovation.

Transitions: A Theoretical Framework

In the process of putting together the papers from the symposium for this volume, we were struck by the fact that the population-environment *dynamic* we were observing was usually cast as a profound *transition*. Bill Drake's appreciation of the beauty of the logistical curve that marks the demographic transition, indicates the possibilities of representing this in mathematical language. He suggested we work together to develop the idea of the transitions, which were suggested as well in agriculture, urbanization and even in what we have come to call the toxification of the environment. After a few drafts, it became evident that this was largely Drake's project and he eventually went on to develop this theoretical framework and to complete the paper.

We believe that while the other papers in the preceding section tell a good story of what is emerging in our thinking on population and environment dynamics, this theoretical framework may well provide a tool for much productive work in the future.

The transitions we see here are those from one state of relative equilibrium to another. As Drake notes, this overemphasizes the stability of the population-environment condition at both ends of the transition, but there is utility in this oversimplification. In any specific transition itself there is a substantial change on that dimension of behavior. Generational differences are remarkable, for example, in the period of fertility decline, or in the rapid movement from rural to urban areas. The differences between the lives of parents and those of their children are often very dramatically during a transition. It is this speed of social change and the intergenerational discontinuity that mark the *transition* as important in itself.

But there are two other ideas that emerge once one focuses attention on the transition. One is the combination of transitions that are found together, especially in the modern process of population-environment changes. This leads Drake to characterize the modern processes as a *family of transitions*.

Demographic, agricultural, urban and toxicity transitions occur together and it is the combined, familial character that marks our era as both distinctive and troubling. Drake also notes, however, that the way the members of that family fit together, coming either sequentially or concurrently and lasting for longer or shorter periods, is not by any means mechanically fixed. Thus the number, sequencing and duration of the transitions represent important dimensions of variance that are subject to empirical observation and scientific analysis.

Second, he proposes that during the transition itself, the society is in some ways more vulnerable. The social structure is weakened, permitting new rules or new norms to be generated and to become institutionalized, or *infused with value.*[1] The outcome can be a healthy process, or it can be very destructive. Thus the outcome of social change triggered in this period of vulnerability is also open to observation and scientific analysis. Both ideas lead to important arenas of research for the future, which we attempt to identify in the final chapter.

Local Monitoring of the Population Environment Dynamics

Our overall population-environment dynamics project began with a general idea that has become more fully developed both in the field and as a result of the symposium. The initial idea concerned monitoring population-environment dynamics especially at the local level. More specifically, it involved new ways to display the areal distribution of population and environment conditions. This is the subject of the paper by Zinn, Brechin and Ness on *Population-Environment Monitoring Systems (PEMS).*

It is not difficult to agree with Harold Jacobson[2] that the emergence of remote sensing represents a technological innovation that may have as profound an impact on the social sciences as did the emergence of the large scale sample survey technology half a century ago. This has given us a new way to perceive the population and environment connection by looking down on it from above. Here one can see directly the spread of the urban scene, the loss of a forest, or the encroachment of a desert. But computer enhancement also permits us to see the areal distribution of human behaviors and even of conditions not fully visible on the ground.

Since this perception of areal distributions does not require special training in statistics, it is more directly available to the lay, less educated, or even uneducated population. That is, it should be available to local communities in the less developed countries where so much development work is currently aimed. But the development of remote sensing technology today involves the massive costs of rockets, satellites and large computers. Thus, the view from

1. The term of Phillip Selznick. See his *TVA and the Grassroots*, (1949) for an early and very influential exposition of this central idea of what is meant by institutionalization.
2. Personal communication in PEDP seminar.

above that may be rapidly changing our view of the world, is not immediately available to small groups of villagers or local level development agents in the less developed countries of the world.

The technology described in the paper by Zinn, Brechin and Ness is specifically tailored to be a low-tech, inexpensive, and highly portable computerized mapping system. It will permit any local group to enter any specific data on land use, individual health, crop cover, or population growth into a computer program and then to display these conditions on a local map of their own area. This permits the integration of data from multiple monitoring processes in one simple map, hence the term PEMS.

Developing the computer program for this process was the simple part. The more difficult task is to organize ground level use of the technology. At present there are projects in Indonesia, Mexico and Zimbabwe in which the PEMS team is working with national, provincial, district, and even small village populations to identify important data sources and to show how these can be displayed in map form in a very inexpensive and portable system.

Thus these two papers present ideas that have emerged from the process of the overall project and from the more specific task of putting together the symposium papers for publication. They represent an *emergent*, something we consider new and innovative. We also expect that both the theoretical framework and the new monitoring technology will continue to be developed. They are central elements of the next steps in our continuing effort to grapple with both the theoretical and practical issues of population environment interactions.

Chapter XIV

Towards Building a Theory of Population-Environment Dynamics: A Family of Transitions

William D. Drake

Examining Complexity Versus Manageability

There are many ways to think about the population-environment dynamic, each with its own strengths and weaknesses. Any view of this dynamic must be an abstraction because the full intricacies of the real world are too difficult to model. Yet, in the process of abstraction, much of what is important can be lost. In climate prediction, for example, models now available are both inadequate in detail and already have exceeded the limits of computer capacity and data accessibility. In deforestation, the models presently estimating loss of forest cover are admittedly crude and unable to accurately predict change. They also lack longitudinal data necessary for calibration. In general, though significant modelling advances have been made recently, unmanageable complexity abounds in virtually all sectors. *Therefore, one of our intellectual tasks is to search for a balance between the complexity required to portray intricate relationships adequately and the simplicity required for analytic manageability.*

Another feature which increases this complexity is the realization that now, more than ever, *explicit* recognition must be given to the interconnectedness among sectors and across scales. For example, not too long ago, local air pollution was caused primarily by local polluters. Now it is often determined by emissions hundreds of thousands of miles away. A few decades ago, most farmers of the world relied mainly on locally available inputs. With advances in mechanization and new high-yield grains, inputs come from other sectors of society often far away. As political conditions vary in the Middle East, the price of energy changes, altering the most significant inputs now used in agriculture: petroleum and fertilizer. Thus, the rural farmer operating in the agricultural sector is now closely linked with the global energy sector and both are dominated by conditions which may be continents away. Implicit recognition of this interconnectedness is no longer sufficient. Models that span sectors are required or faulty results will be obtained. And with this understanding comes the need for a common calculus to help relate entirely different sectors.

The problems for an investigator in population-environment dynamics are further complicated by the desire to provide analysis useful to policymakers. This focus on decision-makers often imposes additional constraints. Most decisions must be made in less time and with less resources than needed for adequate analysis--even with the existing application models. More vexing still is the need to keep the analysis simple enough for the decision-maker to utilize.

This chapter offers a somewhat different formulation of the population-environment dynamic. The formulation attempts to focus upon critical time periods in the evolution of societies using a common framework which can be applied across many sectors. It is hoped that by this focusing, the dilemma of complexity versus manageability discussed above is accommodated and, perhaps, constructive insights are advanced.

The Dynamics of Transitions

One way of viewing the complex dynamics of population and the environment is to visualize these relationships as a **family** of transitions. That is, not only is there a demographic and epidemiological transition, but also a deforestation, toxicity, agricultural, energy, urbanization, technological and educational transition as well as many others. Transition implies change and, of course, change is ever present. But the definition of transition here is meant to describe a specific period of time which spans the shift from slow to rapid change in the sector and then usually a return again to relative stability. While return to stability can result in a positive outcome, it also can describe a very undesirable condition. For example, under adverse conditions, the deforestation transition can result in a very stable desert. This return to relative stability does not necessarily mean that the determinants of the new equilibrium are the same as before the transition. Often, in fact, there is an entirely new set of interactions in operation. Further, stability is relative to the prior period of rapid change and does not necessarily imply permanence.

Transitions are part of a **family** in several ways. First, they exist in many different sectors of civilization and, in each case, have some common properties. These common traits are especially evident in modern society, where the unprecedented population growth rates of the past two centuries have been associated with massive social and ecological change. Second, transitions frequently affect society in a similar way. For each transition there is a critical period when society is especially vulnerable to damage. During that period, rates of change are high, societal adaptive capacity is limited, in part due to this rapid change, and there is a greater likelihood that key relationships in the dynamic will become severely imbalanced. The trajectory that society takes through a transition varies, depending upon many factors operating at local, national and international levels. Third, transitions interact with each other. Often, a society experiences several transitions simultaneously, creating additional problems because they can amplify each other. The timing of transitions, especially as they relate to one another, also is critical. And because timing affects societal vulnerability and is frequently influenced by public policy, special attention will be given here to procession. Finally, transitions not only occur in many different sectors but also at different scales, both temporal and spatial.

While transitions have similarities, they also have differences. In fact, it is argued here that each local setting, while embedded in a larger environment, may have its own singular context that can result in a unique passage through a transition. Focusing upon why these differences exist could provide insights useful elsewhere, especially if the voyage through the transition resulted in a favorable outcome. Sometimes, the dynamic of the transition seems irreversible, particularly if a critical threshold has been reached. However, these transitions are also subject to influence by public and private programs. As a society emerges from a transition, it can encounter conditions ranging from substantial improvement to catastrophic harm. Whether a society passes through the transition into a satisfactory state is determined by many factors, largely defined, and often remediable by human intervention. It should be the purpose of public policy to formulate, develop and implement these positive interventions.

If the population-environment dynamic can be described as such a family of transitions occurring in different sectors, then investigating the interconnectedness between them may be facilitated. Consequently, the concept of a theory of transitions is developed here in the hope that it will be useful in studying these dynamics. First, the breadth of this phenomenon will be demonstrated by describing its characteristics in nine sample sectors. Next, the general features of these transitions will be developed to see if complexity might be reduced by drawing upon their common attributes, thereby facilitating comparisons among social systems in different time periods and at different scales. Finally, by employing these comparisons, insights might be gained on how to prioritize social interventions.

Examples of Transitions

This section presents a brief description of nine transitions, selected to span a wide range of examples so that general characteristics are illustrated. Seven transitions represent different sectors of society and two are examples that depict means for implementing change. Many other examples could be cited and one of the tasks of theory building in this area is to construct a nomenclature system for classifying transitions.

The Demographic Transition

A good place to begin is with a review of the ideas behind the widely observed demographic transition. At the onset of this transition, births and deaths are both high and in relative equilibrium with one another. Historically, births have exceeded deaths by small amounts so total population has risen only very gradually. Occasionally, famine or an epidemic caused a downturn in total population but, in general, changes in rates were low. During the transition, however, death rates drop dramatically, usually due to a change in the health condition of the population. This change in health is caused by many, often interrelating factors. After some time lag, the birth rate begins to drop and generally declines until it is again in approximate balance with the death rate. In Ness' chapter, "The Long View: Population-Environment Dynamics in Historical Perspective," documents how different societal conditions affect the trajectory of these two indicators. It is important to note that the trajectory followed by the birth and death rates determines the extent and timing of the population increase. If the two rates track one another, as they did in the past, population growth is significant but manageable. On the other hand, if they are broadly divergent as they have been recently in the Third World, there is a population explosion and society experiences all the stress and human misery created by this condition.

The graphs in figures 14.1 through 14.3 portray the demographic transition under several different conditions. Figure 14.1 represents the transition which occurred in North America during 1950-1990 and is projected to occur through the year 2025. Each figure is scaled equally so that direct comparisons can be made among regions. For North America, the vulnerable period of transition was bridged as a result of the conditions described earlier in Ness'chapter. Total population growth was modest as shown in figure 14.4 and, as a consequence, the stage was set for dramatic increases in the standard of living. Figure 14.2 shows the demographic transition in Southeast Asia during the same period. In this case, total population increased substantially as shown in figure 14.4, but there appears to be an equilibrium approaching which has already allowed for substantial increases in the quality of life and shows great promise for the future. Finally, figure 14.3 portrays the demographic

Demographic Transition

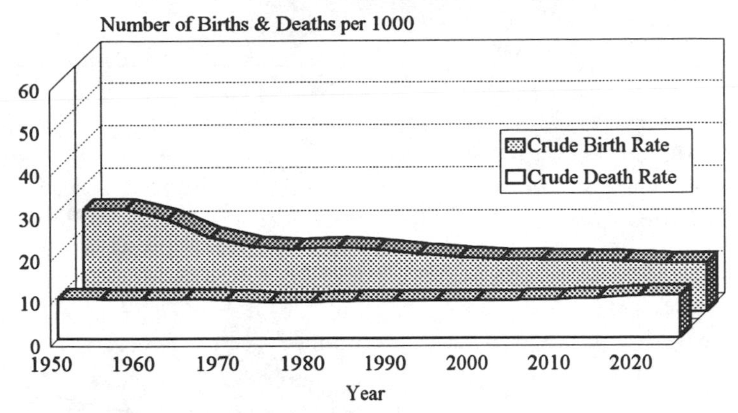

Fig. 14.1. North America

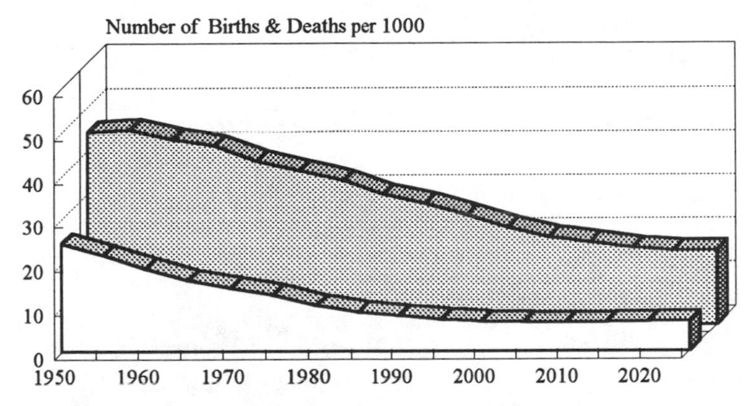

Fig. 14.2. Southeast Asia

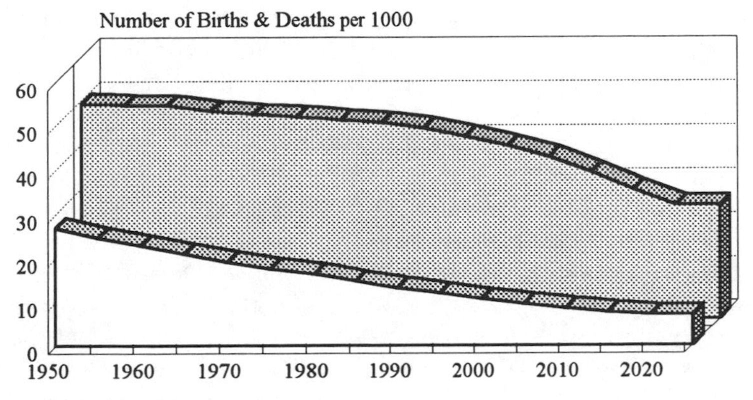

Fig. 14.3. Africa

Demographic Transition (cont'd)

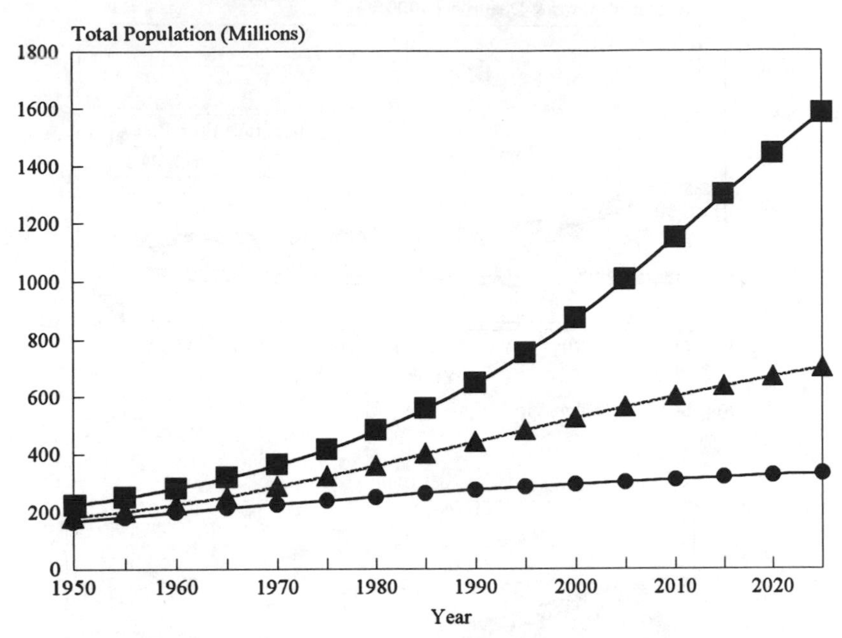

Fig. 14.4. Total population: North America, Southeast Asia & Africa

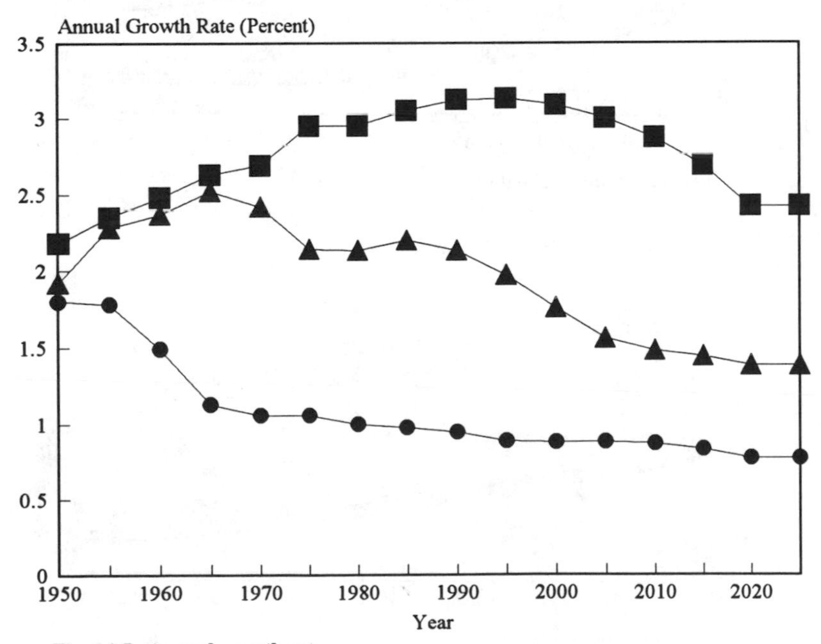

Fig. 14.5. Annual growth rate

transition for Africa during the same period. Here, society is faced with an extremely serious situation. The death rate has dropped somewhat but births remain high resulting in a population explosion of very substantial proportion. Figure 14.5 presents the annual growth rates for all three regions.

There are many determinants of the trajectory followed by the demographic transition. In summary, during the early stages of the transition, changes in population are determined primarily by reductions in the death rate whereas in the latter stages the dominant change variable is reduction in the birth rate. While birth rates eventually decrease to match reductions in death rates, the response time can be greatly influenced by public policy. In North America, where death rate reductions were more gradual, birth rates decreased without much influence by public policy. In Asia, which experienced sharp reductions in death rates due to the application of Western medical technology, public programs have been needed in order to try to bring birth rates down in a corresponding way. Unfortunately, Africa has benefited primarily from Western help in death rate reduction with little policy impact as yet on fertility reduction. The linkages between this transition and others will be discussed later in this chapter.

The Epidemiological Transition

The term epidemiological transition was coined to describe the changing source of mortality and morbidity from infectious diseases occurring primarily in the younger age groups to degenerative diseases in older age groups. As with the demographic transition, there is considerable volatility during the transition. At the onset, infectious diseases usually begin their decline as a result of rising standards of living and extensions of health care and sanitation by the national or local government. In the late twentieth century, this has been in concert with international and voluntary organizations, especially in the developing world. Single vector programs such as immunization programs are often implemented first because they are capable of ready extension and do not require as heavy a commitment to education and other sustained infrastructure, especially in rural areas. These single vector programs are then followed by broader-based health care which demand heavier investment in human and physical infrastructure. Programs that focus on gastroenteritis and growth faltering, for example, require a broad approach including simultaneous education of the entire family especially the mother, provision of oral rehydration therapy, improved water supply, and a referral system for dealing with cases that do not respond to family level treatment.

Depending upon the effectiveness of implementation of these multi-sectoral programs, death rates often decline dramatically, especially in younger age groups, and the stage is set for the first part of the demographic transition described earlier. As a consequence of this decline in death rates, the *relative* importance of degenerative and other diseases such as cancer and heart disease occurring in older age groups gradually increases. Because of this shift in importance, societies that have moved through this transition tend to focus new activities on degenerative diseases while still maintaining an infectious disease program. While the cause of death changes during the transition, a successful move through this period results in significantly lower overall death rates in all but the oldest age categories.

But an entirely successful move through this transition does not always happen. For example, today, Mexico City is experiencing the worst of both worlds due to an overwhelmed health care delivery system and heavy loads of toxins in the air and water supply. There, the primary cause of death is a mixture of infectious and degenerative diseases with the overall death rate for all age categories far above what would have been apparent in a successful epidemiological transition. It is important to note that there are many other places in the world where severe problems are found during the epidemiological transition. Eastern Europe, for example, is facing similar problems due to lack of infrastructure and a recent history of industrialization without attending to the toxic emissions associated with industrialization. Africa, while not generally experiencing heavy toxicity, continues to have high death rates in younger ages because of a shortage of infrastructure and underdeveloped food supply.

Figures 14.6 through 14.8 portray the epidemiological transition under several conditions. Figure 14.6 describes thematically, the composite death rate (z), as the sum of infectious diseases (x), and degenerative diseases (y), in a successful epidemiological transition. The overall death rate is dropping as the cause of death shifts from infectious to degenerative diseases. Figure 14.7 is a similar representation for a flawed transition such as Eastern Europe or Mexico City. The expected shift in cause is delayed and in severe situations, the overall death rate increases. Figure 14.8 shows a different representation of the epidemiological transition for the United States over the period 1900 to 1970.[1] The proportional mortality ratio for children under five dropped dramatically, while the same ratio increased in the over fifty age category.

The epidemiological transition, like virtually all other transitions, has a vast assortment of underlying determinants, many of which are amenable to modification from public policy. Perhaps a more careful examination of the wide variation of experiences, both favorable and unfavorable, could lead to a better understanding of how to overcome flawed transitions.

---

1. Derived from "A Century of Epidemiologic Transition in the United States", by Abdel R. Omran, in *Preventative Medicine*, Vol. 6, No. 1, March 1977, pp. 37-38.

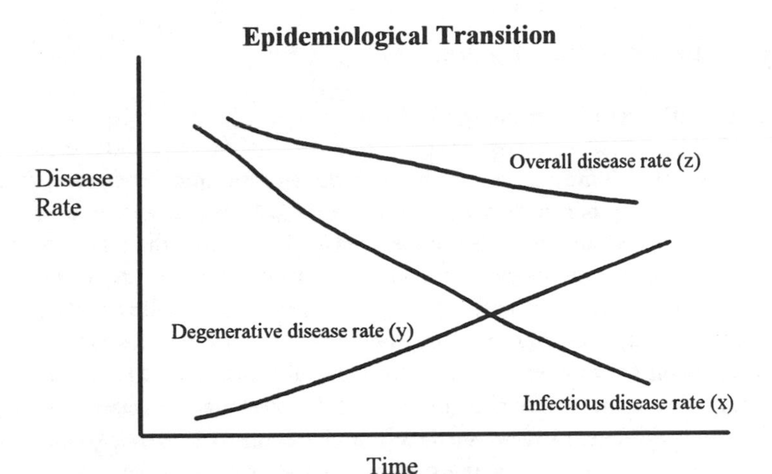

Fig. 14.6. Successful epidemiological transition

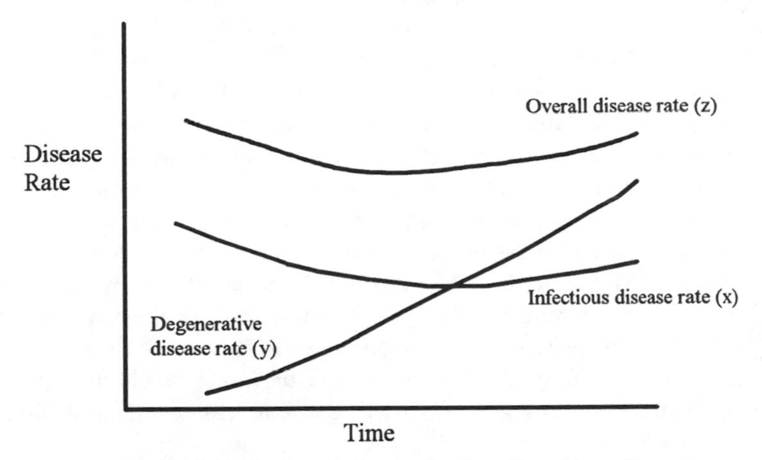

Fig. 14.7. Epidemiological transition for a flawed transition

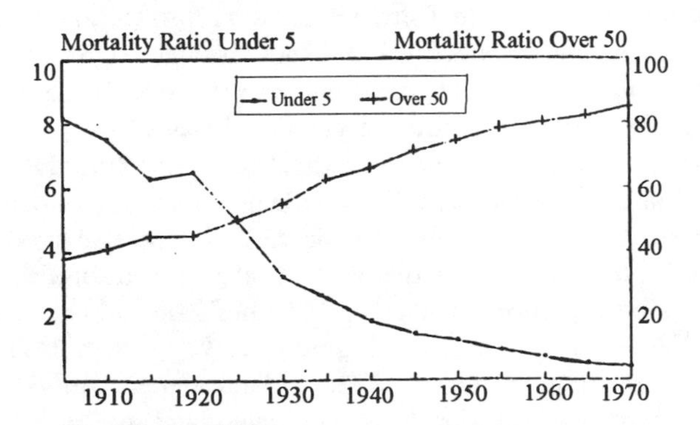

Fig. 14.8. United States
Source: Omran, Abdel (1977) p.37.

The Agricultural Transition

For several hundred years, worldwide agricultural production has been rising in relative harmony with population. Overall, increases in production have kept up with and at times, have even outpaced growth in population. The two factors that have been responsible for these increases are 1) extension of land under cultivation and 2) improvement in land productivity. The agricultural transition described here begins as the principal source of increase in production shifts from extension of land to improvement in productivity per hectare. It continues through the period of dramatic increase in productivity due to new grain varieties and heavy application of fertilizers and pesticides and stabilizes when diminishing returns in yield per hectare are experienced.

Ruttan's chapter points out that prior to 1900 almost all increases in food production was obtained by bringing new land into production. In contrast, by the end of the first decade of the next century, almost all increases must come from higher yields per hectare. During the 1960s and 1970s, the Green Revolution heralded remarkable increases in grain yields, especially in some parts of Asia and Latin America. Now, however, this pattern is changing so that the increase in yield per hectare is beginning to lessen. Between 1950 and the peak year of 1981, the area planted with grain increased by twenty four percent worldwide. However, since 1981, there has been a worldwide *decline* in planted acreage of eight percent. Furthermore, most of the world's increase in cultivated land over these last forty years occurred in only three countries: China, the Soviet Union and the United States. And recently, all three of these countries have reduced acreage planted in grain because of their belief that the expansion was ecologically unwise.[2] On the other hand, during this same period, total output of grains increased, thus confirming an increasing intensity of yield per hectare.

As other evidence of these patterns, Ruttan presents a worldwide consensus among experts indicating that future gains in production will be through improvements in land productivity and not through expansion of cultivable land. Perhaps more important, this same expert consensus believes there will be a diminishing marginal increase in yield per hectare.

The shift from expansion to intensification of land productivity becomes even more distinct at the local and individual farmer level. Landholdings are often fixed so that there is only one source of increase in production: yields must take over the increase. Otherwise, rural to urban migration must occur with a likely contribution to another transition, the urbanization transition, which will be discussed shortly. Figures 14.9 through 14.11 show the agricultural transition in terms of total land under cultivation. There is considerable evidence that worldwide cultivated land has peaked at about 700

2. *The Global Ecology Handbook 1990*, p. 70.

Agricultural Transition

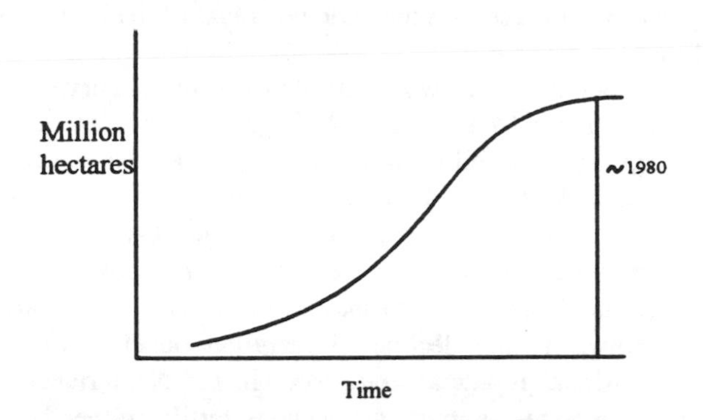

Fig. 14.9. Total land under cultivation--worldwide (thematic)
Source: USDA (for 1950-87); Brown, Lester. The Changing World Food Prospect: the Nineties and Beyond, Worldwatch Paper 85. Washington, D.C.:Worldwatch Institute, Oct. 1988, p.18 (for 1988). Global Ecology Handbook, p.71.

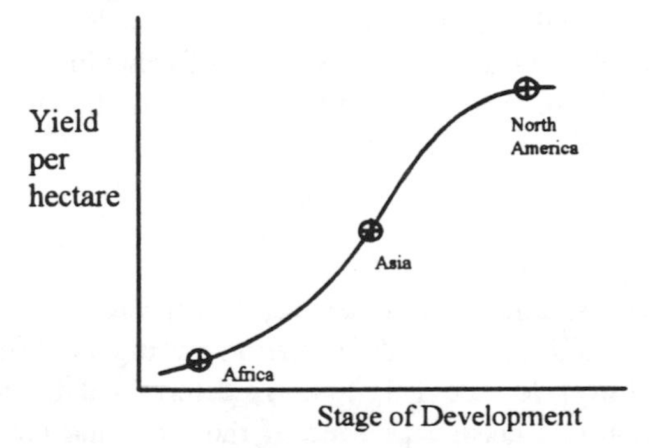

Fig. 14.10. Yield per hectare (thematic)

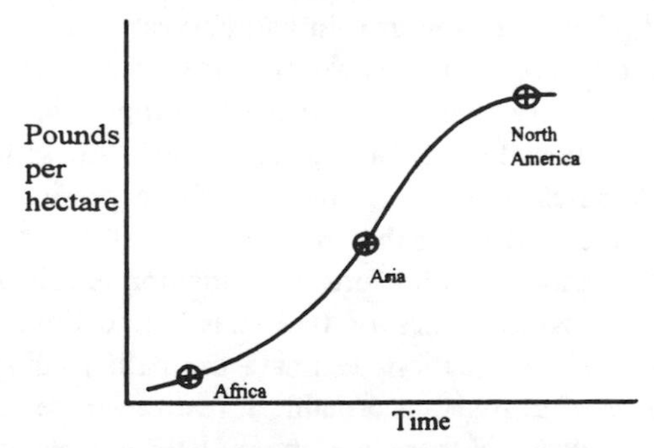

Fig. 14.11. Pesticide use in agriculture (thematic)
Source: USDA (for 1950-87); Brown, Lester. The Changing World Food Prospect: the Nineties and Beyond, Worldwatch Paper 85. Washington, D.C.:Worldwatch Institute, Oct. 1988, p.18 (for 1988). Global Ecology Handbook, p.71.

million hectares. Figure 14.10 presents a thematic portrayal of the agricultural transition in terms of yield per hectare.

The positioning of a region of the world on this transition curve varies considerably as shown in figures 14.12 to 14.14. Figures 14.12, 14.13, and 14.14 present the cereal grain yield and fertilizer use per hectare between 1950 and 1985 in North America, Asia and Africa respectively. At present, North America has yields per hectare and fertilizer usage many times that of Asia and Africa. Clearly, Africa is in the initial stages of the transition, Asia is somewhere in the middle and North America may be approaching the point of diminishing returns. However, the decline of marginal benefits does not necessarily mean that additional inputs are reduced. In the Netherlands, for example, there is nearly ten times as much nitrogenous fertilizers per hectare applied to arable land as in the United States.[3] It is interesting to note, however, that over the last decade, the Netherlands application rates have dropped significantly. Diminishing returns are also being experienced in other factors of agricultural production. Figure 14.11 is a thematic portrayal of pesticide usage in agriculture. Again, in the United States, pesticide usage per acre has already peaked which may be a positive precursor of behavior in other parts of the world and may bode well for another transition, the toxicity transition.[4]

The Forestry Transition

At the onset of the forestry transition a large percentage of a region is under forest cover. During the transition, rapid deforestation occurs and finally forest cover stabilizes at a lower level. This level is determined by many factors including the local region's needs, the state of the local and national economies, climate and soil characteristics. In most settings, this transition ends in a steady state equilibrium balancing growth and harvest. That relative equilibrium can occur at extreme conditions, however. In Haiti, for example, there is a virtually treeless stability, while in much of Europe, the United States and Japan there is stability with extensive forest cover. As with most transitions, how society handles the vulnerable transition period often profoundly determines the quality of life for the region.

Grainger provides a useful model of the forestry transition in chapter 5. His model describes the **net** characteristics of the transition calibrated for tropical forests. In this model the equations estimate the path leading to a steady-state under the premise of a given acceptable forest cover per capita. However, as Grainger acknowledges, there are more determinants of this trajectory than the model can presently accommodate. The **net** characteristics

3. *The State of the Environment*, Organization for Economic Co-operation and Development, Paris, 1991, p. 180.
4. U. S. Environmental Protection Agency, 1988, 80, *Global Ecology Handbook* 1990.

Agricultural Transition

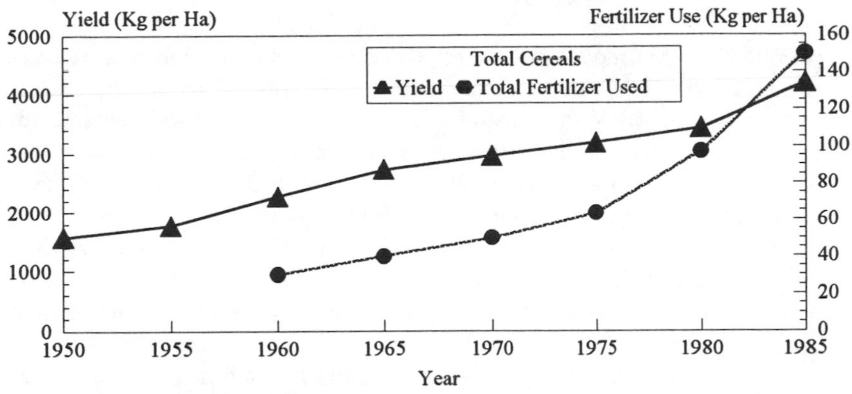

Fig. 14.12. North America

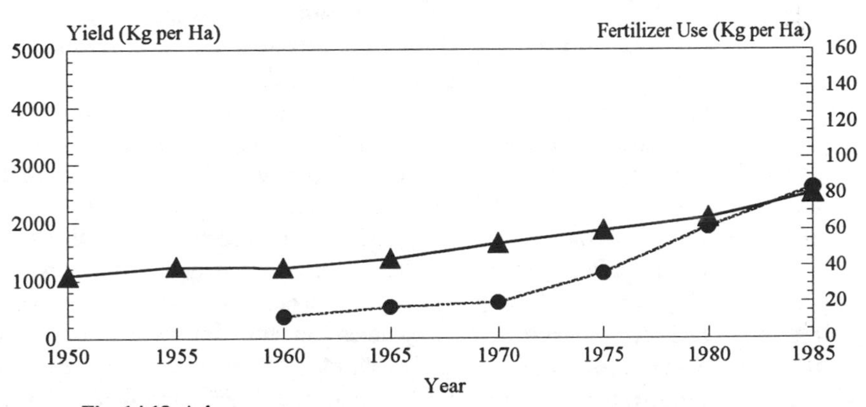

Fig. 14.13. Asia

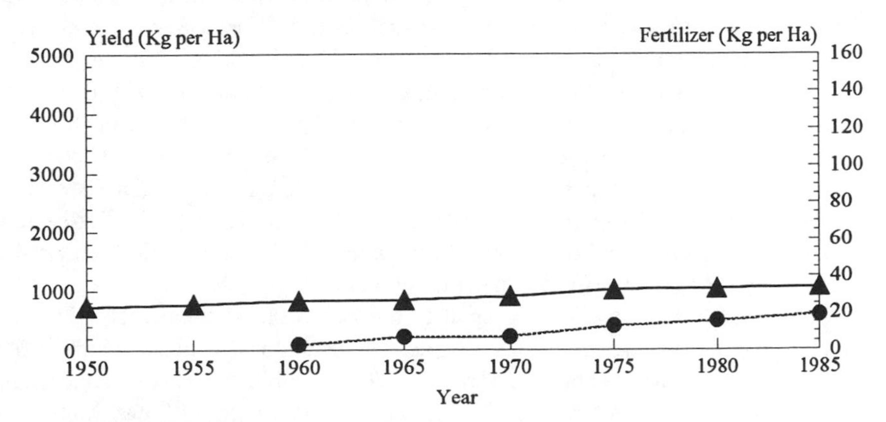

Fig. 14.14. Africa

portrayed by his model are, in reality, a composite of both deforestation and regeneration. The rate of deforestation is determined by factors such as the technology employed, demand for forest products in general, market desirability of the species available for harvest, local demand for agricultural land, accessibility to transportation systems and fuel wood needs of the region among others. The rate of forest regeneration is affected by many of the above variables and, in addition, soil type, nutrient and water availability, temperature and forestry practices used during the original harvest, to name a few. Although some of these variables are fixed by local conditions, others are amenable to influence by public policy.

The forestry transition, therefore, assumes many forms depending on local conditions. Figures 14.15 and 14.16 describe two different historical trajectories. The composite, forest growth, is the sum of deforestation and reforestation. Figure 14.15 shows the nature of the transition for a region with fertile soil, available water, good forestry practices and limited demand for fuel wood, such as exists in the midwestern United States today.

An example of favorable regeneration can be found in the state of Michigan and offers a useful historical perspective when compared to the present day developing world. In 1873, the city of Chicago was destroyed by a great fire which created intense demand for forest products to rebuild the city. Western Michigan was then covered with hard and softwood species especially suitable for construction and the distance to the center of the burned out city was only fifty miles by water. New rail and ship technology had been developed recently which greatly increased efficiencies. In short, many factors converged to generate conditions of rapid deforestation. By the turn of the century, almost all forest cover in lower Michigan had been harvested. Then, as undergrowth sprung up, fires broke out and with no fire fighting infrastructure available, most of Western Michigan remained on fire continuously for over a decade. Partly as a result of these fires, the U.S. Forest Service was created which extinguished the burns and commenced reforestation. Now eighty years later, the annual growth in board feet of timber in Michigan exceeds the level of growth prior to the fire of 1873. An equilibrium, which shows promise of yielding forest products on a sustained basis, has been achieved. The state has played an important part in this story by providing technical assistance and other incentives through its Department of Natural Resources. Furthermore, soil and climatic conditions have been favorable to reforestation and luckily, heavy erosion did not follow the original harvest.

Figure 14.16 provides a thematic representation of the forestry transition under more difficult conditions. If there is heavy local demand for fuel wood, steep slopes, infertile soil and limited water availability, the forestry transition can take a very different form. Figures 14.17 through 14.25 show the wide variation in **net** deforestation as well as, the sum of both deforestation and

Forestry Transition

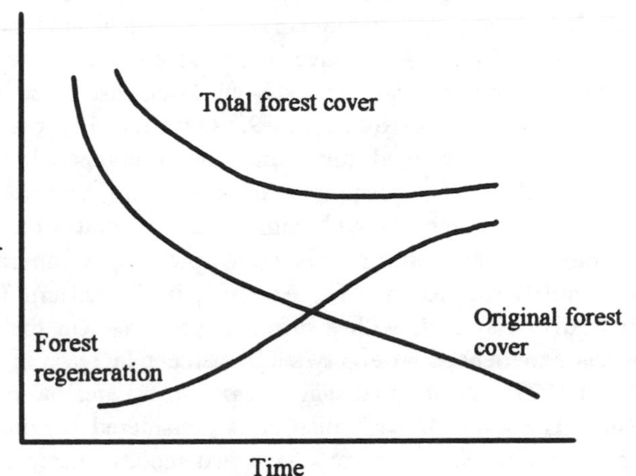

Fig. 14.15. Forestry transition under conditions of of fertile soil, available water, and limited demand

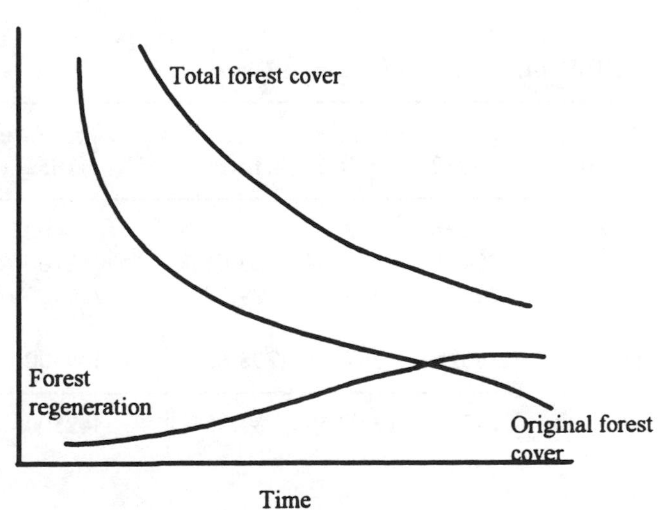

Fig. 14.16. Forestry transition under more difficult conditions of infertile soil, heavy local demand, and limited water

regeneration, among different regions in the world.[5] North America and Europe have remained stable since 1850 whereas Tropical Africa, Latin America and South Asia have experienced major declines. Especially interesting is the comparison between Southeast Asia and South Asia as portrayed in figures 14.18 and 14.19. Over this 130 year period, forestation in South Asia has declined thirty-nine percent compared to only seven percent in Southeast Asia. The precipitous drop in South Asia has occurred in the last forty years and coincides with high rates of population increase. As with many transition phenomena, when the spatial or temporal scale is reduced, more volatility is apparent. For example, in the Eastern Terai of Nepal there are four districts, each with a different pattern. On the whole, the Eastern Terai has experienced an eighty-seven percent increase in population between 1952 and 1972. During that same period, forest area has dropped by fifty-four percent. However, if each district is considered separately, there is much wider variation. Some districts have had modest increases in population and similar small changes in forest cover. Others have had population increases even greater than the average with corresponding steeper deforestation trajectories as evident in table 14.1.[6] Historically, the forestry transition has been largely the mirror image of the agricultural transition as land was cleared to extend cultivation. In the future, however, with increases in agriculture coming mainly from intensification, this association will be less apparent.

TABLE 14.1 Eastern Terai of Nepal

District	Population 1952	Population 1972	Increase	Forest Area (acres) 1952	Forest Area (acres) 1972	Decrease
Siraha	176915	302307	70.88%	30400	19700	35.20%
Sapatari	254915	312564	22.61%	32200	29900	7.14%
Morang/ Sungari	228952	524991	129.30%	230400	119900	47.96%
Japata	80252	247698	208.65%	149500	35500	76.25%
Total	741034	1387560	87.25%	442500	205000	53.67%

5. *World Resources 1987*, p. 272.
6. *Nepal: A State of Poverty*, by David Seddon, Vicas Publishing House, Delhi, 1987, p. 100.

Forestry Transition

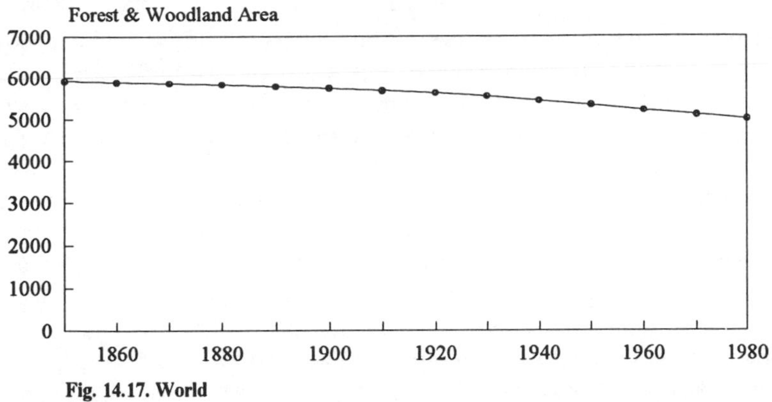

Fig. 14.17. World

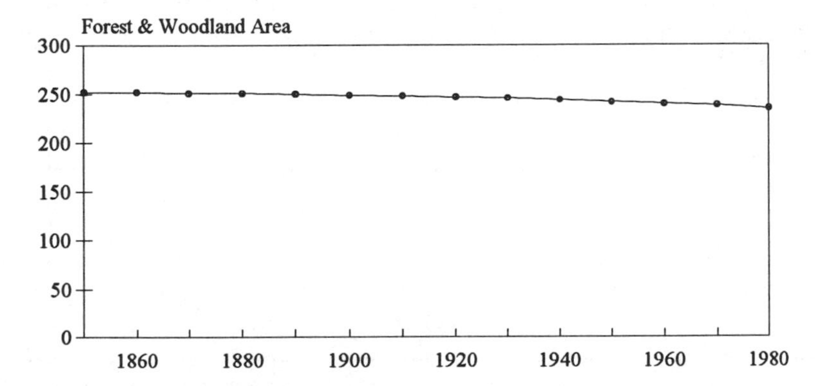

Fig. 14.18. Southeast Asia

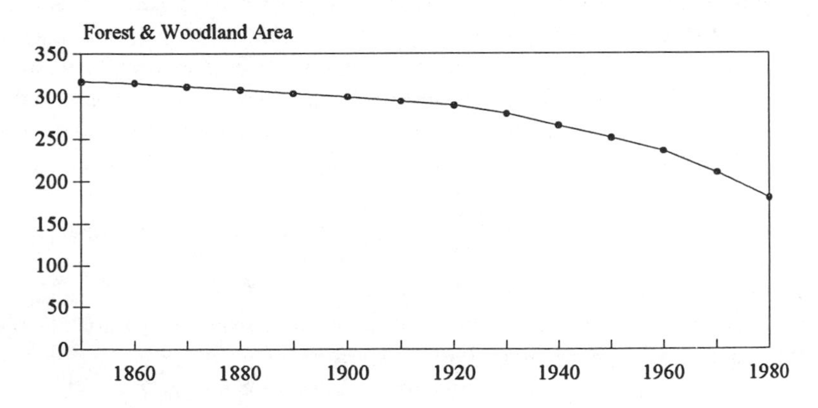

Fig. 14.19. South Asia

Forestry Transition

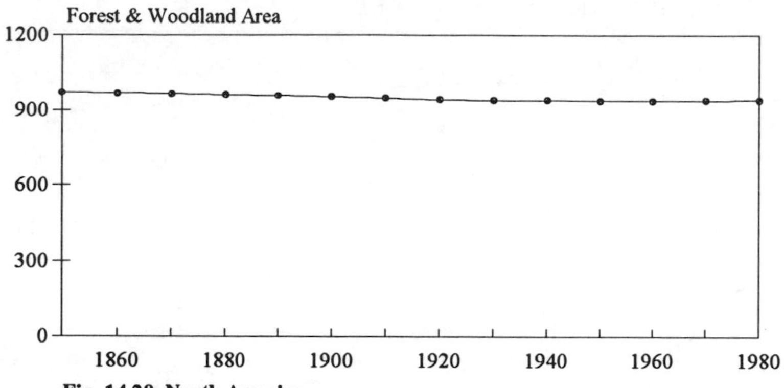

Fig. 14.20. North America

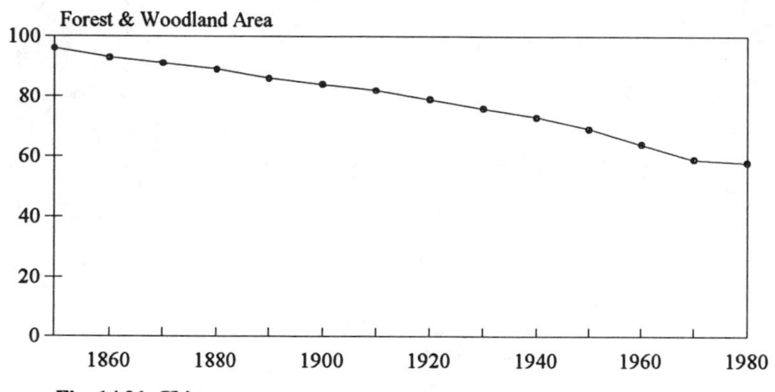

Fig. 14.21. China

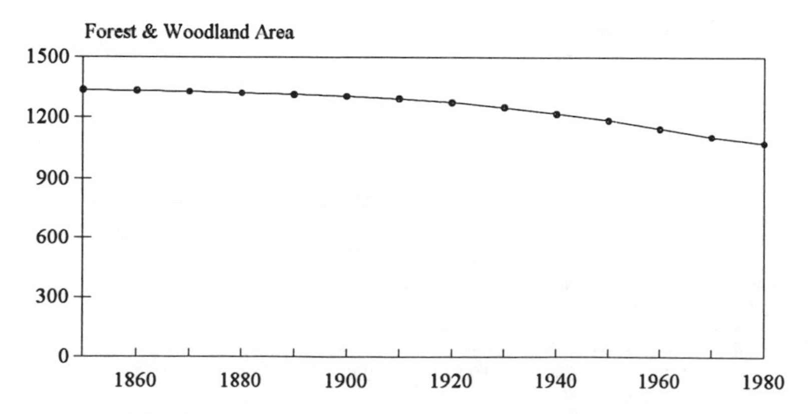

Fig. 14.22. Tropical Africa

Forestry Transition

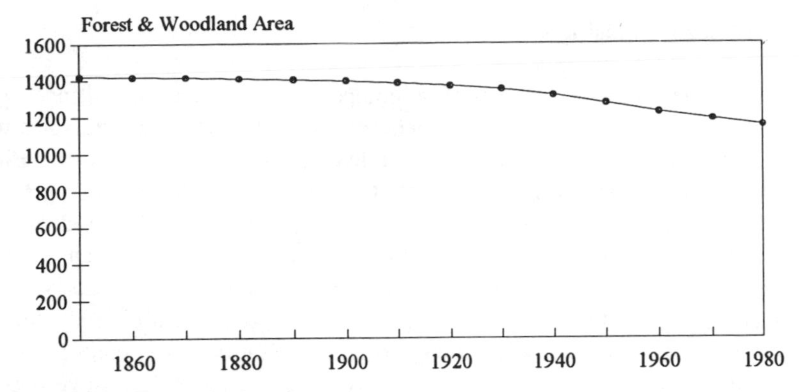

Fig. 14.23. Latin America

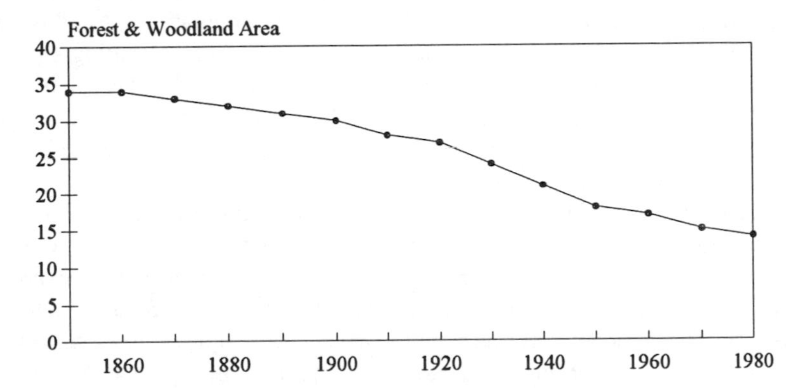

Fig. 14.24. North Africa & Middle East

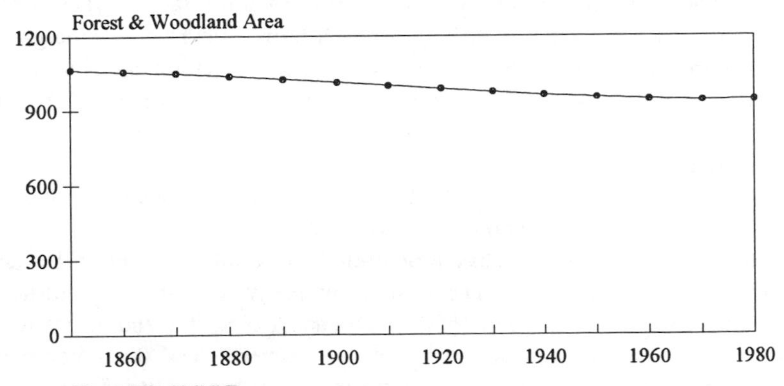

Fig. 14.25. U.S.S.R.

Note: All figures in million hectares
Source: World REsources 1987, p. 272

The Toxicity Transition

The toxicity transition is itself a composite of several transitions: global atmospheric, local air pollution, surface water, ground water and solid waste, to name a few. Again, there are at least two sets of factors operating in tandem. The transition begins with low levels of industrial or agricultural production and correspondingly low levels of toxicity. As production and population increase, toxic by-products increase to levels which eventually become unacceptable to the general public. This, in turn, causes a public demand for pollution abatement. After an environmentally costly time lag, remedial steps are taken which helps to bring pollution under control. As with other transitions, it is possible to conceive of the corrective change occurring only after remediation is no longer helpful. In this instance, the new stability could be achieved by the demise of the polluting population. At a less extreme condition, the handling of nuclear wastes now may be presenting society with its first major problem that is technically unsolvable. The toxicity transition is strongly related to at least five other transitions: the demographic, epidemiological, agricultural and, yet to be discussed, the urbanization and technological transitions.

Like other transitions, society is most vulnerable to the toxicity transition during the rapid growth period of industrialization and its associated pollution. Some societies weather this transition better than others, but all experience it in some way. Although technical solutions are often hard to achieve, the length of time delay between public demand for action and societal response is related more to institutional factors. In Western Europe and North America there are now well developed mechanisms for providing this feedback. Because of widespread pressures brought by environmental constituencies there, considerable progress has been made in remediation. The democratic political processes are helpful, but perhaps more important are the grass-roots environmental constituencies which operate both within and outside formal governmental channels. By contrast, regions which have ineffective or non-existent feedback mechanisms suffer greater harm during the toxicity transition.

Another factor which affects the vulnerability of a society is the level of development and availability of infrastructure to respond to toxic waste problems. If the region has low capital accumulation and/or has suffered exploitation, it has less capacity to respond even if there is political will. Although it can be argued that an absence of capital and infrastructure is largely the result of past decision priorities, nevertheless, the deficit exists and efforts must be made now to deal with the problems. A brief review of some of the problems facing regions with limited infrastructure and minimal feedback may be helpful.

Today, Eastern Europe and the U.S.S.R. are experiencing frightful levels of pollution generated as by-products of heavy industry expansion since World War II. Emissions per unit of production of sulfur dioxide (SO_2) in Eastern Europe are approximately sixteen times that of their neighbors to the west: Sweden, France and the former West Germany. In the U.S.S.R., SO_2 emissions per unit of production are ten times the Western average. One lignite-burning electric generating plant in a small town of former East Germany emits 460 thousand tons of SO_2 annually, more than Denmark and Norway combined.[7] Nitrogen Oxides (NO_x) emissions are also higher but the disparity is not as great. Half the cities of Poland, including Warsaw, discharge untreated sewage directly into the nation's waterways. Many large Soviet cities have no sewage treatment at all and, as a nation, only thirty percent of effluent is properly treated. In Hungary, 700 cities rely totally on bottled water or have it piped in from neighboring cities because their sources are severely contaminated with pesticides and nitrates.[8]

The stark contrast among different regions does not stop with environmental concerns alone. Heavy loads of toxins are taking their toll on public health as well. In Czechoslovakia, life expectancy has dropped as much as five years in heavily polluted areas compared with relatively clean regions. In Russia, overall sickness rates of children in areas of high pesticide use are five times those of children in relatively clean areas.[9] In Poland, illness such as tuberculosis, pneumonia, bronchitis, and leukemia are more common in polluted regions. It is believed these trends explain why life expectancy for Polish men aged 40-60 has fallen to 1952 levels.[10] Close linkage between increases in child mortality and pollution does not prove causality. If not conclusive, this is strong evidence at the least.

In China a particularly difficult toxicity condition is confronting policymakers. For over a thousand years, fish farms have successfully utilized untreated sewage from urban areas as both a source of water and nutrients. With industrialization, however, heavy metals and other toxins have entered the effluent stream and are going directly to the farms. Fish consume the toxin-laced nutrients and amplify them biologically before becoming a principle protein source for humans. The problem is now well understood in China but with almost no sewage treatment infrastructure, the solution appears daunting. While the overall health of the population has improved considerably over the last few decades, recent incidents have shown the

7. "Restoring the East European and Soviet Environments", by Hilary F. French, in *State of the World 1991*, p. 95.
8. Ibid, p. 97.
9. USSR State Committee for the Protection of Nature, *Report on the State of the Environment*, translated for the U. S. Environmental Protection Agency; Yablokov, "State of Nature."
10. Kabala, S. J., "Environmental Deterioration in Poland and the Impact on Public Health: Difficulties in Assessment," Center for Hazardous Materials Research, University of Pittsburgh, May 1990.

potential impact of these toxins on the population. In the summer of 1986, an outbreak in Shanghai of twenty thousand cases of Hepatitis A was caused by consumption of shellfish contaminated by raw sewage. Another study showed that the incidence of liver cancer has dramatically increased in the past three decades with the suspected cause being the heavy use of chlorine in purifying water. Only recently has it become clear that when chlorine is applied to ground water containing suspended humic material a suspected carcinogen is formed.[11]

The figures 14.26 through 14.28 provide a thematic portrayal of the toxicity transition under different conditions. They show the transition plotted in terms of toxic emissions in relation to industrial production. As production increases, so do emissions; however, as public concern turns to action, toxin levels begin to drop. Figure 14.26 depicts conditions of minimal public mobilization and remediation. Figures 14.27 and 14.28 show the toxicity transition under conditions of limited remediation and more forceful action. The shaded area represents the total toxic material emitted with which society must eventually deal in some manner.

Another way to portray the toxicity transition is in terms of the percentage of the environment that is degraded. Figure 14.29 shows a thematic representation of the environmental degradation over time which is related to industrial production and population. The vertical axis can represent any one of the individual subcategories of the toxicity transition. For example, if the vertical axis is the percentage of surface water polluted by a given set of toxins, Western Europe and North America may have passed through the vulnerable period of rapid degradation and may be gradually moving downward again. For other water borne toxins, however, that same region is probably in the midst of the critical period. Surely, Eastern Europe and Russia are in a very different state. Other elements of the toxicity transition such as air pollution, ground water, and solid waste can be portrayed in a similar vein. There is, of course, an upper limit of one hundred percent for any given type of degradation. In terms of human habitation, however, this upper limit of tolerance, beyond which life is impossible is certainly below that amount, at least for some pollutants. Absent public policy, there is no reason to be sure that society will not inadvertently exceed one or more of these human habitation boundaries.

Figure 14.30 describes the toxicity transition in terms of emissions per unit of production. Some countries of the world have passed through the vulnerable period for certain toxins. Japan, Western Europe and the United States have made considerable progress in a few areas. However, strong

11. South China University Memorandum 1990 translated to W. D. Drake during visit to China in October 1990.

Toxicity Transition

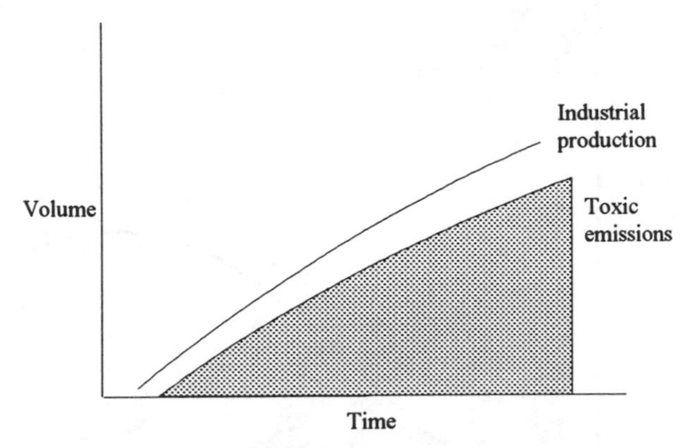

Fig. 14.26. Toxicity transition with minimal feedback and remediation

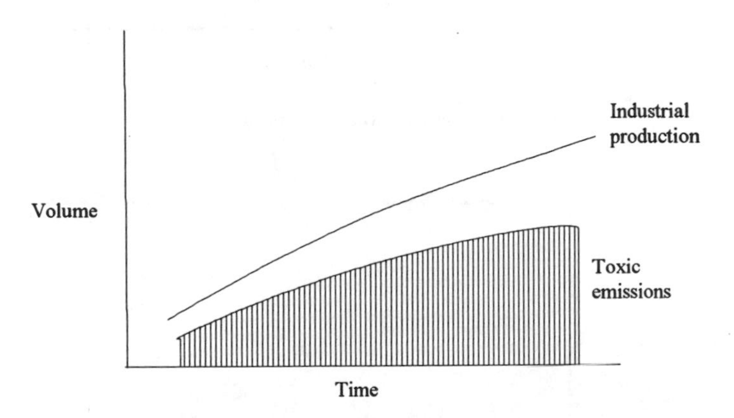

Fig. 14.27. Toxicity transition with limited remediation

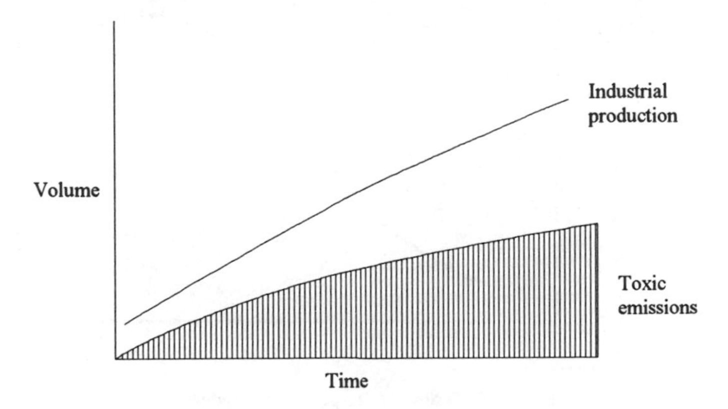

Fig. 14.28. Toxicity transition with forceful remediation

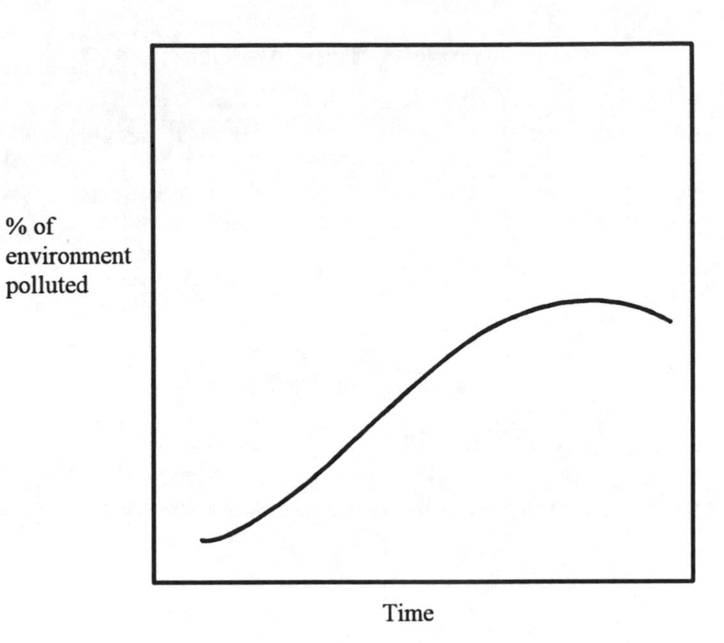

Fig. 14.29. Percent of environment polluted by toxins

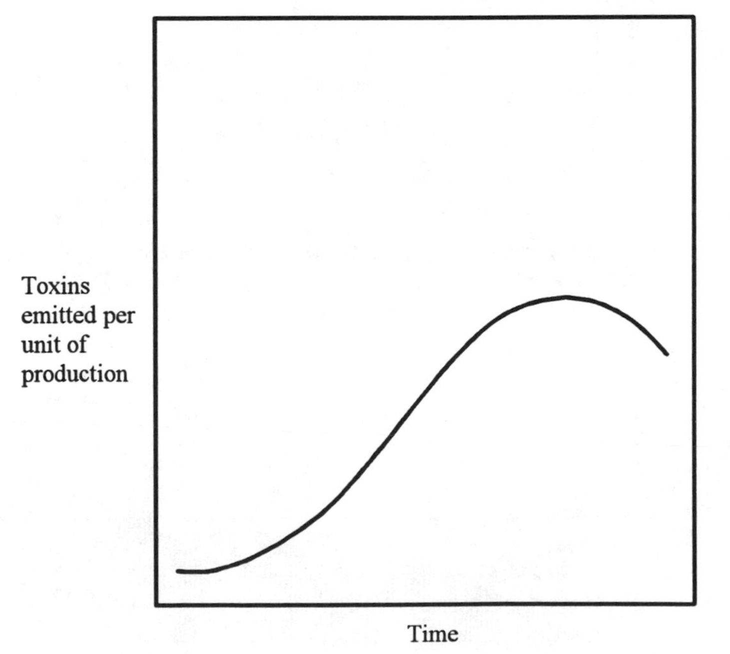

Fig. 14.30. Toxins emitted per unit of production

public policy seems to be absolutely necessary for progress to be made. A tragic example of how the absence of strong public policy leads to degradation even when there is clear evidence of harm created by avoidance, is emerging next to the southern border of the United States. Over the past decade almost two thousand factories - most of them owned by U. S. corporations - have been built along the Mexican side of the border. Recently, a series of spot samples showed that seventy-five percent of the sites were discharging toxic chemicals directly into public waterways. In one case, a plant owned by General Motors is discharging xylene, a very toxic solvent in an amount which results in a level of pollution 6300 times as high as the standard for U. S. drinking water.[12] These corporations know better because they are based in the U. S. industrial environment. Yet, the absence of strict public policy apparently resulted in their reversion to practices extremely harmful to the environment.

The Urbanization Transition

The urbanization transition is a visible and dramatic member of the family of transitions. All of the world's industrialized societies already have become urbanized and now all of the Third World is urbanizing as well. The transition is driven by the dual forces of rural to urban migration and central city population growth. The early stages of the transition are characterized by rapid growth of urban population. In later stages, growth declines and may reverse. Caution must be exercised about drawing conclusions about city population change. In some instances, trends may be an artifact of organizing data along political boundaries. That is, urban growth continues in suburbs or other urban fringe areas not enumerated as the central city. It is also possible that the city may appear to grow more rapidly merely through extensions of its boundaries. There is a large body of literature on this transition dynamic published primarily in the field of urban planning documenting the determinants and consequences of the phenomena but for our purpose, it is sufficient to briefly summarize some of the key factors affecting the urbanization transition.

Rural to urban migration is a product of many forces. The first set of forces are "pull" conditions such as availability of jobs in the urban area and the perception of better educational and health opportunities, especially for children. Family ties to those who have migrated earlier often facilitate a move. In some studies it has been shown that the establishment of an "outpost" family member in the urban area is critical to successful moves and is one reason why there is an acceleration in the migration rate during the middle of the transition. Another set of forces are "push" conditions. These

---

12. "Love Canals in the Making", *Time Magazine*, May 20, 1991, p. 51.

may include harsh drought or other environmental difficulties, lack of rural infrastructure needed for implementing improvements and insufficient landholdings to support an expanding family. The other force driving the urbanization transition is natural population growth in the region itself. This part of the dynamic is driven by the demographic transition and therefore is most heavily influenced by education, affluence and availability of contraceptives.

While all of these forces are present in any urbanization transition, their relative importance changes dramatically from region to region. Today, in Europe and North America, almost all growth is the result of population increases and migration *among* urban regions.[13] In the developing world there is more of a mix between all determinants mentioned earlier with Africa and India having the heaviest percentage of "push" forces operating. Figures 14.31 through 14.33 present the urbanization transition for several pairs of countries over the years 1950 to 1990 with projections to the year 2020. In these figures, the urbanization transition has already occurred in the United States and Sweden whereas Mexico, India and Benin are in different stages of the transition.

In terms of the population-environment dynamic, the urbanization transition often acts as an amplifier in so far as it interacts with other transitions. For example, Mexico City faces greater peril because it is in the midst of the demographic, toxicity and urbanization transitions simultaneously. Furthermore, these transitions are all changing at a very rapid pace which raises the level of societal vulnerability. In the absence of careful public policy, vulnerability can lead to serious harm. Alternatively, forceful and effective public action can result in significant improvement.

The Fossil Fuel Transition

The fossil fuel transition is a special case of the energy transition. Historically, many energy transitions have occurred in different regions and time periods. Ness' chapter, "The Long View: Population-Environment Dynamics in Historical Perspective," describes the transformations in the sixteenth century brought about by sail and later, steam power. Today, we are in the most universal and perhaps critical energy transition: fossil fuels. Studying this transition is especially instructive because the record on different societies' passages through the vulnerable period is varied and appears to be heavily influenced by public policy.

---

13. Presently, in the United States, the farm population represents approximately two and one half percent of total population.

Urbanization Transition

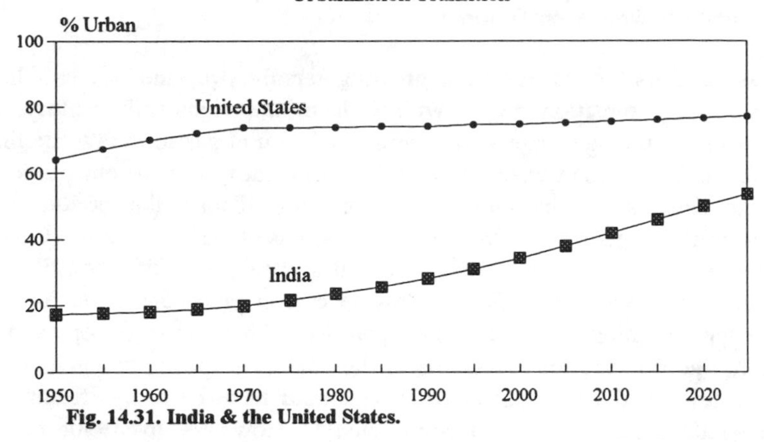

Fig. 14.31. India & the United States.

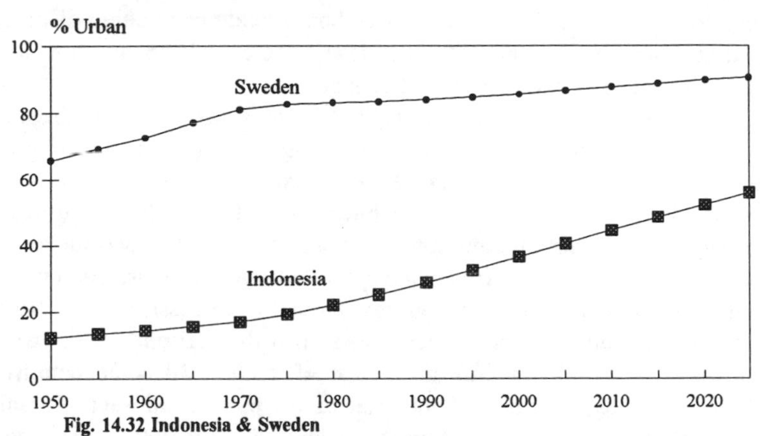

Fig. 14.32 Indonesia & Sweden

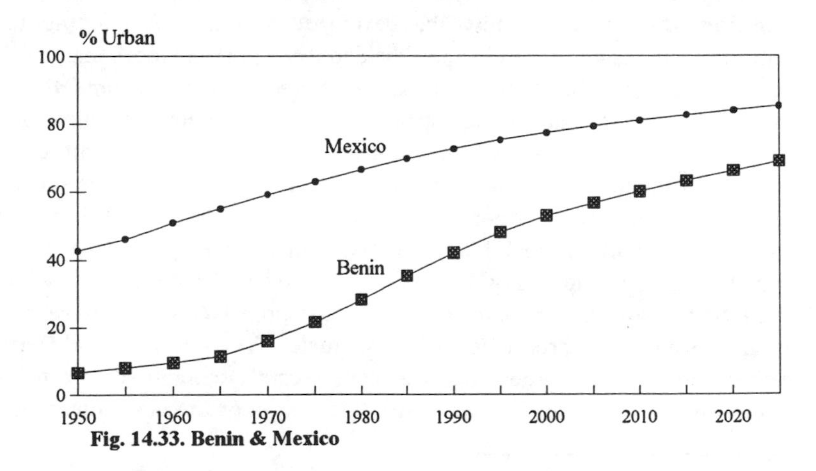

Fig. 14.33. Benin & Mexico

Source: Prospects of World Urbanization 1988

Use of fossil fuels has been growing rapidly over the last two hundred years. The transition began with coal in the eighteenth century, added petroleum in the late nineteenth century and natural gas in the twentieth. The starting point of this transition is difficult to identify but for our purposes the precise date is unimportant. It can be argued that the period of high vulnerability began in the 1940s when emissions of carbon dioxide (CO_2) were at levels well below tolerable limits in relation to their contribution to greenhouse gases. In 1950, worldwide carbon emissions from fossil fuels were approximately 1.5 billion tons per year with Western Europe and North America accounting for two-thirds of the total. According to some estimates, if this level had been sustained, there would have been significant but not deleterious impact on global conditions.[14] However, by 1990, nearly six billion tons of carbon in the form of carbon dioxide were being disgorged into the atmosphere each year. At this level, there is little question about the harmful long-term effects on global change.

Projections of fossil fuel usage over the next several decades vary considerably. However, there is wide agreement that consumption will continue to rise and that the mix between coal, petroleum and natural gas will change. Overall, coal usage will decline in favor of petroleum and natural gas, primarily because carbon emissions are lower for the latter two fuels. In 1989, natural gas accounted for 16.9 percent of total carbon emissions but provided over twenty-four percent of the energy derived from fossil fuels. Virtually all projections on future usage expect total British thermal unit (BTU) production to increase but shift towards the "cleaner" fuels. In some countries with plentiful supplies of coal and minimal or no petroleum, such as China and India, coal production will continue to expand, but this will be the exception.

An important issue is how the developed world is responding to this situation. The answer depends upon the country. Perhaps more than any other issue, the response portrays a society's current values. Some developed countries have made substantial strides in reducing their consumption while others have continued as before. Energy consumption per unit of Gross National Product (GNP) now varies considerably among the developed countries. Japan, the former West Germany and the United Kingdom all consume about half as much energy as the United States per unit of economic output.[15] Japan spends only five percent of its GNP on energy while the United States spends ten percent. Some developed countries have recently chosen to limit their production of fossil fuels. The former West Germany, together with fourteen other countries have recently committed to a reduction of their carbon emissions by twenty-five percent over the next fifteen years.[16]

---

14. *The Global Ecology Handbook*, Beacon Press, Boston, 1990, p. 200.
15. Ibid, p. 202.
16. Flavin C. and Lenssen, N. "Designing a Sustainable Energy System", *State of the World 1991*, Worldwatch Institute, Washington D. C. 1991, p. 21.

Such a commitment to reduce emissions at the national level represents significant political will. Figure 14.34 portrays the cost of gasoline to the consumer in selected countries. Not surprising is the high correlation between the cost of this fossil fuel and overall energy efficiency.[17] But the United States, the largest consumer of fossil fuels, is not among them. Clearly, there are institutional problems which have caused this country to become stuck in the fossil fuel transition to the detriment of the nation and to the world at large.

The Technological Transition

The technological transition, like the toxicity transition, can be subdivided into a large number of individual technologies. Almost every technological innovation that has diffused throughout society and between societies has had a major impact on the population-environment dynamic. A full treatment of this subject is beyond the scope of this chapter but it may be useful to indicate the relationship of technology transitions to the others discussed above.

Technological innovation frequently builds upon itself. For example, in the communication sector, the telegraph was, to some extent, stimulated by the need to improve upon semaphore technology. The telephone was an extension of telegraph, which then expanded to broader band width communication such as video. At some point, in the 1960s, the benefits from another technology, computers, began to play a major role in communication technology. Each innovation results in higher efficiency and generally, reduced cost per unit of service which then fosters greater diffusion throughout society.

Sometimes, innovations are required to continue the transformation provided by the underlying technology. For example, it has been estimated that it would take every person in the world to staff the telephone switchboards in the United States alone if the communication industry was still using the technologies of the 1930s.[18] It is only through electronic switching systems and other computer-based technologies that demand for that service can be accomodated.

Certainly, improved technologies has both helped and, at times, harmed the environment. Better communication fosters quicker responses to environmental and population problems. But it also promotes the diffusion of harmful elements as well. Advertising of environmentally harmful products is facilitated by good communication as easily as is information about how to recycle consumer products. When the diffusion into society of all technologies is taken as a whole, it constitutes what many call modernization. Improved

---

17. Data taken from International Energy Agency, *Energy Prices and Taxes*, First Quarter 1989, Paris, OECD. Also in *Global Ecology Handbook 1990*, p. 207.

18. Bell Laboratories communication.

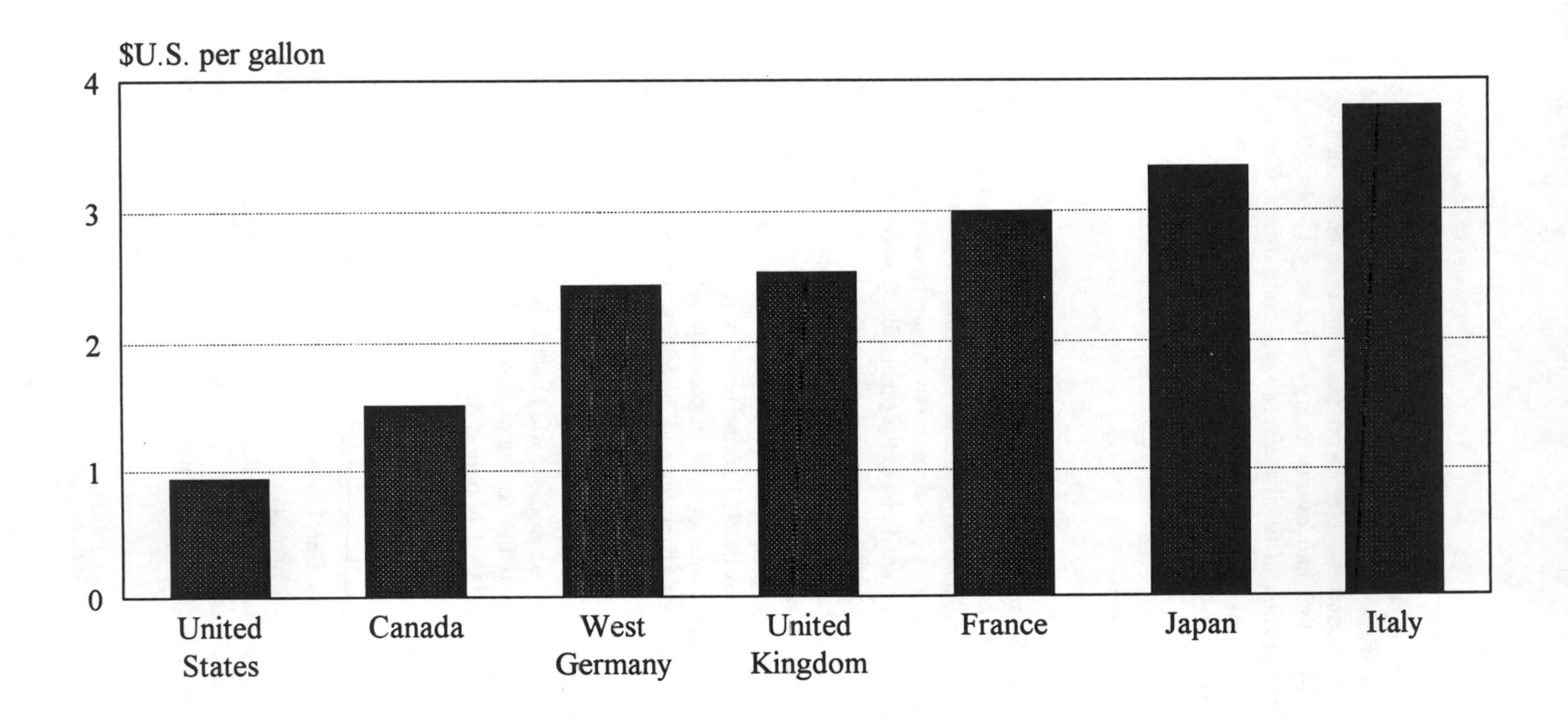

Fig. 14.34. Gasoline price levels with taxes in selected countries, 1989
Source: Global Ecology Handbook p. 207; International Energy Agency, Energy Prices and Taxes, First Quarter 1989 (Paris: OECD, 1989).

efficiencies, resulting in lower cost per unit of output, often are related to actual and perceived improvement in the quality of life.

Transportation provides a useful example of a technology which has had a profound impact on the population-environment dynamic. Figures 14.35 and 14.36 describe thematically, the transportation transition in several ways. Figure 14.35 shows the diffusion of each successive transportation innovation. The progression begins with foot travel, passes through stages including horse, sail, rail, auto/bus, and finally ends with air. As each new technology is introduced, it supplements prior modes and replaces prior technologies. As each new technology is introduced, it becomes the primary mode of transportation, although never completely eliminating previous modes. Figure 14.36 portrays thematically the average per capita distance travelled per unit of time. The transition reaches a plateau which is limited, at the least, by available time and cost. Average speed is presumed to follow a similar general trajectory. However, any given society probably has a unique shape determined by factors such as level of industrialization, resource base and cultural patterns.

While figures 14.35 and 14.36 describe transportation, other technologies such as communication or computation capacity could be substituted and the same thematic curves drawn. At any time, there is a mix of various technologies **within a sector** which is drawn upon by society as it develops. This mix changes over time and determines the relative effectiveness of carrying out a given task. In turn, this effectiveness affects the manner in which the passage through the previously discussed transitions occur. In general, as technologies become more potent, the rate of change becomes higher and the passage through the transition faster. Paradoxically, technological innovation increases the rate of change, thereby increasing societal vulnerability. But at the same time, it represents a predominant expectation for resolving the problems created by change.

The Educational Transition

Perhaps the most important transition of all is the educational transition. The relationship between educational attainment of a populace and its level of development is well known. All modern societies strive towards this objective. However, the means of reaching the goal are as varied as the cultural and economic settings of the world.[19] The educational transition

---

19. There are many historical examples of society's attempt to control lower classes by impeding access to education. A notable, but by no means only, example is the United States treatment of the Black population in the eighteenth and nineteenth centuries.

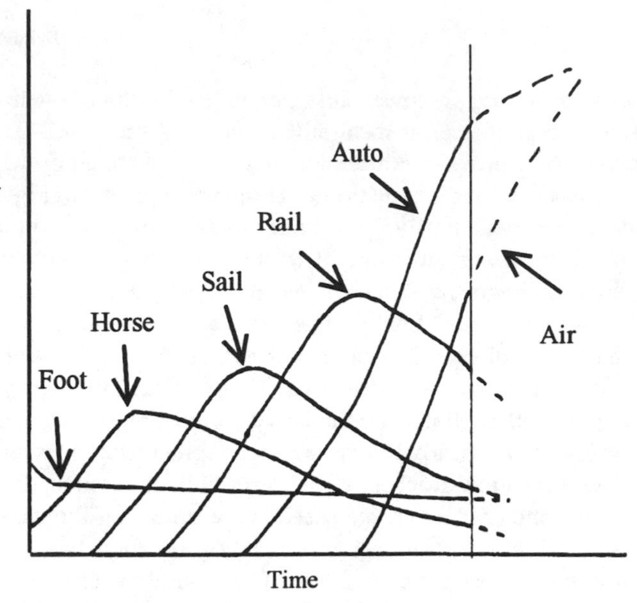

Fig. 14.35. Transportation usage by mode

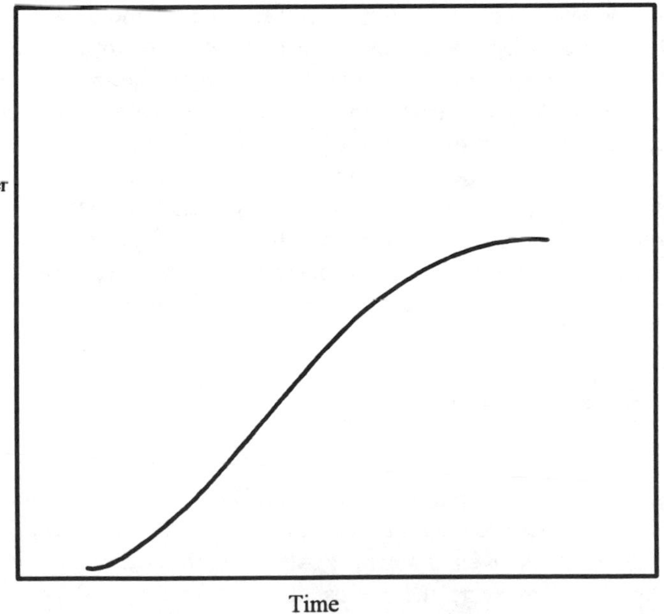

Fig. 14.36. Distance traveled per year

discussed here is not intended to embrace the entire history of a society, but rather, to focus upon the period of rapid diffusion of universal elementary education throughout themajority of the populace. There are many different ways of defining and measuring diffusion of education. They range from the percentage of the population attaining a certain grade level, to types of educational skills, to male-female ratios or other demographic and geographic dispersal indicators. Each country has a different pattern and, especially in modern times, has been heavily influenced by public policy and the resultant commitment of public resources.

Education, particularly in rural areas, has a major impact upon the effectiveness of virtually all other assistance interventions, regardless of sector. Countless studies reviewing the success or failure of these programs conclude that achievement is based largely upon concomitant improvements in education. For example, an innovative program intended to improve child nutrition and promote family planning in Indonesia, concluded that educational development occurring at the same time as the program, accounted for a preponderance of the positive change observed.[20] The same dominant effect is apparent in all present-day developing countries. North America and Europe went through the transition in the late 1800s. Asia is presently in transition and most of rural Africa has yet to experience it.

Towards the Development of a Transition Theory

Other Transitions

There are many other transitions in the population-environment dynamic which have not yet been articulated. Some have physical characterization that represent the population-environment dynamic and others reflect the **means** of change in society. In the physical category, the toxicity transition is actually a group of several transitions: global atmospheric, local air pollution, surface water, ground water and solid waste. In a similar vein, the fossil fuel transition is a subfamily of the energy transition.[21] On the other hand, transitions describing **means** of change include educational and technological. As discussed earlier, the technological transition is, in fact, a large collection of different technologies, each with their own transition.[22]

20. "KB-Gizi; An Indonesian Integrated Family Planning, Nutrition & Health Program: The Evaluation of the First Five Years of Program Implementation in East Java and Bali", BKKBN & Community Systems Foundation, October 1986.

21. Rich Stoffle has suggested that the coastal marine boundary can be thought of as a distinct region which experiences a characteristic transition.

22. The shape of a specific technology transition is also driven by a complex of interrelated variables. Many of the variables that determine the rate of change are driven by the determinants

There is yet another class of **means** transitions which call for articulation, namely, that of human institutions. An example of an institutional transition is the bureaucratic transition. It can be argued that almost every large institution has experienced it at some point in evolution. The transition occurs when a confluence of events and/or pressures is exerted upon the institution and result in a *perceived* need for greater central control. Layers of organizational structure and staff are then imposed upon the system, which connect more tightly the center to its periphery, thereby permitting greater control by the center. The transition to bureaucracy is both powerful and dangerous. Powerful because through tight control, the periphery can be made to do what the center wishes. Dangerous because the center may be wrong and so the wrong thing gets implemented well. And dangerous also because the bureaucratic structure takes on a life of its own. No longer is it interested only in accomplishing the specific goals for which it was created, but also tries to maximize its own well being. In the process of this self-maximization, a large number of low productivity personnel are added to the organization, thereby reducing its overall effectiveness. Other institutional transitions worthy of investigation include a financial/capital, international trade, judicial/equity and perhaps consumption transitions.

Determinants of Transition Trajectory

A full treatment of the causes of variation in the trajectory of a transition is beyond the scope of this chapter. However, it is possible to characterize the dimensions discussed in the preceding pages in general terms. There are three dimensions of any transition trajectory: rate of change, magnitude of change and period of change, all in relation to other transitions occurring at the same time. Each of these dimensions is, in turn, influenced by the full range of contextual variables such as cultural/religious beliefs, available technology, lifestyle, consumption patterns, level of economic development, natural resource base, educational attainment of the populace and political institutions. An important agenda for future work is to clarify the linkage between these determinants and a transition's trajectory.

---

of the diffusion of technology which is well described by Everett Rogers in *Diffusion of Technology*, 1969.

General Characteristics of Transitions

The previous section described nine examples of transitions in the population-environment dynamic. In this section, general characteristics of the family of all transitions will be developed.

Each class of transition, whether it be demographic, toxicity, forestry, agricultural, urbanization, fossil fuel, epidemiological, technological or educational has a similar pattern. It is this perception that has caused us to posit the existence of a **family** of transitions possessing some common attributes useful in analysis.

Similarity of Trajectory Across Sectors

The first common attribute of all transitions is their trajectory. They all begin in relative stability, then move to the volatile transition period where change is rapid, and finally return again to comparative balance. Analytically, these are clearly nonlinear relationships but ones with properties that lend themselves to well-understood mathematical functions. Specific functions which apply to these conditions will be suggested in the Appendix to this chapter.

Applicability of Transitions Across Scales

The second attribute has to do with scale. One of the most interesting and at times vexing aspects of studying the population-environment dynamic is that many phenomenon manifest themselves, sometimes differently, at various levels of the geographic and temporal scale. For example, data depict one demographic transition for an entire continent, a different one for a country within that continent and still other transitions at the regional level. Local conditions may delay or advance the onset or completion of the transition in relation to the larger body. Thus, moving through the demographic transition, for example, can take more or less time depending on the geographic scale.

This same variation seems to exist in all other population-environment transitions which have been investigated. Although, national or regional-level determinants often set the stage for the local dynamic, it is the local conditions which determine the timing, magnitude and specific trajectory of the transition in that setting. For example, in Mexico City the breakdown of local health care provision and heavy local and regional industrial pollution is interacting with regional climatic conditions and water supplies to impede successful passage through the toxicity transition. In a similar way, the toxicity transition of present-day Eastern Europe has regional dynamics overlaying local ones. In fact, one can select a familiar locale and visualize several toxicity transitions occurring there at the same time, all with both local and non-local elements. In any setting, air or water quality is almost always the consequence of both local and remote pollution sources. The same is true for agricultural, forestry and other sectors.

One can think of our world, seemingly so chaotic, but instead consisting of a multitude of well defined transitions in many sectors, each with its own local characteristic. Different transitions begin at different times and places, often overlapping and either reinforcing or dampening one another. As passage through the transition occurs, a new dynamic is established both within and between the sectors. Sometimes this new dynamic yields remarkable societal gains. Unfortunately, at other times, sectors may interact in a harmful manner resulting in a broadened and extended period of social vulnerability.

Timing of Transitions

The third attribute of a transition is the timing of its inception. This may be one of the most important dimensions of a transition because timing affects the interaction among different transitions. Further, timing is susceptible to influence by public policy. For example, the differences in the trajectory of the demographic transitions in North America and Africa has been determined, to some degree, by the application of medical assistance to Africa which steepened the drop in death rate. With little assistance in birth control, however, the society has maintained a traditional and even increasing birth rate. In a similar vein, government intervention which opens up a previously inaccessible area to settlement will affect the deforestation, toxicity, agricultural and probably the epidemiological transitions as well. How governments implement the order of permission to settle and the availability of infrastructure assistance will affect the quality of life of the inhabitants of the new settlement in a profound way.

Societal Vulnerability

The fourth attribute is the special vulnerability society seems to bear during transitions. A primary cause of this vulnerability is the speed of change during the high velocity portion of the transition. Adaptive capacity is impeded because there is little time for systems to adjust and often there are limited feedback mechanisms operating which otherwise could help this process. Moving through a transition often requires revising these old feedback mechanisms to reflect the new dynamic.

Another cause of vulnerability is the likelihood that key relationships become out of balance during the transition. In the demographic transition, births become out of balance with deaths. In the forestry transition, harvests often are out of symmetry with regeneration. In the urbanization transition, immigration is out of balance with development of urban infrastructure. In the toxicity transition, industrialization is often out of balance with effluent remediation. More important still, different sectors become out of balance with each other and restrict positive change. For example, if a toxicity transition results in such a polluted environment as to impede the forestry transition from regenerating, a double disbenefit occurs.

There is an interesting interplay between temporal and spatial scales, and system adaptivity which, in turn, leads to yet another source of vulnerability. Different ecological and social systems operate at widely different scales.[23] For this reason, one of the most important questions facing an observer of the population-environment dynamic is to decide which scale and consequently which variables are most appropriate for analyzing any given problem.

A provocative example of how temporal scale and variable definition relates to vulnerability is found with farmers in South Asia. In remote villages of Bangladesh, rice farmers plant at least twice a year. Each season has its own crop and for the monsoon season, especially in deltaic regions, the variety of choice is called Aus. This rice, a deep water strain with the unusual attribute of attaining heights in excess of twenty-five feet, can better deal with the fluctuating water level of the monsoons. As the water deepens, the rice stalk grows sufficiently rapidly to keep the plant above water. Even in areas where the flood peak is twenty feet above ground level, it is possible to grow a crop successfully. By harvest time, the floods have receded to the point where farmers can gather their crop. For this crop, the farmer does not care about typical agricultural variables such as total rainfall or timing of the onset of the wet season; rather, he is only concerned with the *increase* in water level each day. If the water level increases more than five centimeters per day, the rice plant cannot grow fast enough to keep ahead of it and will drown. If the water level increases at a slow enough rate, it does not matter if there is two or twenty feet of water. In this case, the Bangladeshi farmer is vulnerable to difficulty only if the rate of change becomes too great. This is perhaps a microcosm of the problems experienced by society in general. Too often, the focus of our attention is on the level of a variable rather than its rate of change.

Another contribution to societal vulnerability during a transition is the amplifying effects created by transitions occurring simultaneously in several sectors. For example, in Mexico City, the epidemiological transition is

23. For a fascinating discussion of the wide range of temporal and spatial scales operating in the population environment dynamic, see "Scale Relationships in the Interactions of Climate, Ecosystems, and Societies" by William C. Clark, *Forecasting in the Social and Natural Sciences*, pp. 337-378, International Institute for Applied Systems Analysis, 1987. In this work, Clark points out that when investigating population-environment dynamic issues, data spans nine orders of magnitude on both temporal and spatial dimensions. Faulty problem formulation and interpretation arises when scales are mixed inappropriately.

occurring at the same time as urbanization and toxicity transitions, leading to a very difficult passage through the epidemiological transition. Rapid rates of change in several sectors can more easily overpower the available infrastructure leading to the next source of vulnerability during transitions: capital availability.

Capital or investment capacity and the choice of targets for investment can either amplify or dampen societal vulnerability during a transition. If financial resources necessary for remediation are available, the effects of rapid change may be somewhat mitigated. Africa, which is struggling with a difficult demographic transition, has almost no capital available and will therefore undergo great hardship. Similarly, the former Soviet Union and Eastern Europe are struggling to find financial resources to deal with their flawed toxicity transitions. Even though now there may be the political will to deal with such issues, further damage will occur before solutions can be implemented. By contrast, the former East Germany may be able to avoid additional harm and rebound from similar defects much more quickly because of its ready access to capital.

Another dimension of transitions which affects societal vulnerability is the degree of interconnectedness between different transitions or between different locations experiencing the same transition. How closely is the local village connected to the regional and national economy? How much does what happens in one location determine what happens in another? How much does the demographic transition affect the epidemiological transition? There is no question that interconnectedness is increasing worldwide. Under some circumstances, linkage creates dependencies which, in turn, increase vulnerability. However, it can also work in the opposite direction: these same links to a larger domain can act as a safety net. If there are connections, resources can be brought to the stressed area more easily to mitigate the local adversity.

The final and perhaps most important dimension of transitions affecting vulnerability is feedback. As a society passes through transitions, its ability to respond to adverse conditions is largely based upon information available to it and the ability to act on that information. When the rates of change increase, the demands for feedback on system status goes up as well. Rapid change calls for rapid feedback. Fortunately, the information explosion and new information processing technologies, such as low cost microcomputers have come into being and are gaining popularity. Investigating transitions especially near their inflection points may help indicate when and at what level feedback must be provided to decision makers. We have dwelt upon feedback in many ways during the course of this inquiry and more will be said about the topic of feedback in chapter 14 and in the conclusions of chapter 17.

Societal Opportunities During Transitions

The last attribute of transitions which, because it is a large topic deserving special attention, only will be mentioned here; is opportunities which can occur during a transition. In this chapter, emphasis has been placed upon the societal vulnerability during transitions. But there is another side to this condition. There is some evidence that occasionally, society responds to difficulties encountered during transitions in ways that would be impossible to undertake at other times. For example, sometimes the *perception* of crisis generated, in part by the rapidity of change, results in political and social action very helpful to the environment. Recent adjustments in Eastern Europe and Russia are good examples of political action which could be beneficial to the population-environment dynamic. At other times, it seems that the very technologies which have been used by society to increase rates of change, thereby causing vulnerability, also can be used to provide a remedy. A future research agenda should, undoubtedly, include investigation of how transition periods can be better utilized by societies to promote constructive change, based largely upon exemplars already experienced.

Analytic Properties of Transitions

Many characteristics of transitions are common across all sectors and geographic scales. In an earlier section, the trajectory of most transitions was described as beginning slowly, picking up momentum during the most volatile period of change, and then slowing again to a new stable state. To an analyst, this characteristic implies that there are probably feedback loops operating among the variables under investigation. Positive feedback means that the relationship picks up speed as a result of the level of the variable. For example, if population growth is simply a function of total population, then population increases would accelerate as the total population increases reflecting positive feedback. Negative feedback means that the relationship slows down in relation to the distance it is from a given level of attainment. For example, if population growth was simply a function of the difference between a chosen limit and the current level, growth would diminish as the limit was approached.

When a relationship begins slowly, picks up speed in the middle stages and then decelerates later, it is possible that feedback is still operating but under a different and perhaps more complex form. There is a point of inflection where feedback processes reverse themselves and rates of increase begin to diminish. Such an "s-shaped" relationship is often found in the population environment dynamic. There are several well understood mathematical functions which may be useful for portraying these relationships. They have existed for several centuries in the hard sciences and have been in operational use by some of the social sciences for well over a hundred years. In the study of the population-environment dynamic they may be helpful in portraying transitions in a way that facilitates comparison and thereby increase our understanding of common characteristics as well as deviations from typical patterns. Of special interest are techniques and functions that reduce complexity and at the same time provide a reasonably accurate portrayal of reality. The appendix to this chapter introduces this topic and indicates which function would be most helpful under differing circumstances.

Policy Implications of a Transition Theory

But what are the policy implications of viewing the population-environment dynamic as a family of transitions? First, consider transitions *within* a given sector and at a given geographic scale. We know there are transitions which some societies have already experienced while others have yet to endure. If the nature of these experiences can be captured in general form and as exemplars useful elsewhere, it is more likely that knowledge can be transferred to other settings where a transition is first starting. Of course, each civilization or local culture has its own unique characteristics but any one emerging transition may be comparable to some of those which have occurred before because conditions are similar.

Second, there may be useful comparisons *across* different scales. It has already been surmised that a national-level transition, perhaps now in process, is actually comprised of a myriad of local transitions also in process or which have recently occurred. But there may be other locales in the region for which the transition has yet to happen. If similar patterns emerge because of similar local conditions, a useful prediction could be made about the nature of the passage through the transitions yet to appear.

The next potential use of transition theory is to facilitate analysis across sectors. There is, of course, no good reason to expect the trajectory of, say, a forestry or agricultural transition to mimic an epidemiological transition. However, for any society at a given time, there may be similarities in the **rates** of change across sectors. Developed economies when conditions are favorable, have slower rates of change in their agricultural sector than developing economies. Predominantly rural based cultures, may be expected to have urbanization transitions which are steeper than non-rural cultures. In short, it is worth testing to see if patterns can be empirically determined which would be helpful in predicting the shape of future transitions, given a level of development and stated degree of intervention.

Fifth, transition theory may help to deal with the related dynamic of cycles. Recently, much attention has been given to investigating various cycles evident in the global system. Most notable is the carbon cycle which traces the path of the carbon molecule through its various transformations in the global system. But there are many others occurring simultaneously including the nitrogen, sulfur, hydrogen and fresh water cycles. Transition theory, by providing functions for some of these relationships could help in the modelling of global cycles.

Finally and perhaps most importantly, transition theory may permit more informed public and private intervention. At one level we find ourselves believing that the trajectory of a transition is somehow fixed by an immutable law of nature. But at another level we know that this is not the case. Public and private policy can make a difference as we have seen from some of the cases discused in this book. Rates of change can be influenced by policy redirection and consequent resource allocation. To the extent that we can link historical rate differentials with historical policy implementation, a more informed determination can be made about which intervention mix works best in dealing with problems facing society today.

Linkage Between Transition Theory and Risk Analysis

There is yet another policy implication of transition theory. The subject deserves separate attention because it is linked to a related but distinct policy tool, risk analysis. A brief review of risk analysis may be helpful before showing its connection to transition theory.

Risk analysis is a technique based upon the notion that each problem can be structured in terms of the risk it poses to either the environment or the people. First, this methodology estimates the form, dimensions and characteristics of the risk using the techniques of risk assessment. Then, risk management or remediation is applied to reduce the risk. Environmental risk assessment, together with its analytic methodologies, permits discussion of disparate human and environmental conditions with a common language. For example, child health risk resulting from exposure to certain atmospheric toxins can be compared to the risk of exposure to infectious diseases derived from impotable water. In turn, these two can be compared to the risk of long term damage emanating from poorly educated mothers or lack of transportation to health centers. In a similar vein, the different risks of species extinction predicted by alternative forestry practices in the tropics can be assessed. And the risk to biodiversity created by broader use of pesticides can be compared with alternative cropping practices designed to raise yields without pesticides. In short, the decision maker and policy analyst can range across a wide spectrum of alternative scenarios when considering solutions to a given problem. The common metric is expressed in terms of probability and the goal is to select the combination of strategies that provides the largest payoff for a given level of resources. Different risk reduction options can be evaluated on a common basis and their relative efficiency compared.

Considerable progress has been made in determining the risk of many environmental conditions but much remains to be done. In the toxicity sector where the approach was first widely applied, the predominant technique for determining human risk is the dose-response methodology. First, large doses of a suspected harmful product are given to laboratory animals over a short period of time. Then, an extrapolation is made to estimate the effects on the animal of a low dose over a long period of time. Finally, these findings are related to humans. While this technique is used throughout the world, there are some Herculean assumptions that must be made in translating high doses applied to rats over short times to low doses applied to humans over long times.[24] Nevertheless, public policy is being determined, in large part, based upon these studies. This implies there is a large potential gain to be had by improving this methodology.

Because of time delays and the irreversibility of some risks, it may be very important to prioritize them based upon the dimensions of time and space. We know that for some environmental problems the temporal dimension can be very long. Pollutants can exist in the environment indefinitely. Depleted species of wildlife may never recover from loss of habitat and progress to extinction resulting from human activity begun centuries before. In spatial terms some environmental problems like elevated radon are local, while others such as stratospheric ozone depletion, have the potential for affecting everyone in the world. Assessment of risk therefore needs to take into account both the degree of reversibility and the time frame of impact. Toxins which have long lives and which are amplified up the food chain pose a threat to future generations and should be weighted accordingly. An example pointing out both the tenuous basis of risk assessment and effects of time delays may be helpful.

Over the last few decades miniscule amounts of Polychlorinated Biphenyls (PCB) have entered the Great Lakes. Through biological amplification they have now gained a significant presence in sports fish such as lake trout. In the early 1980s the state published a sports fish eating advisory based upon PCB risk determined by the dose response-methodology. These advisories were extremely conservative because they contained ample safety factors to compensate for uncertainties in the methodology. However, in 1986 a study was conducted on pregnant women in Western Michigan which showed that women eating approximately twelve fish meals during their *entire* pregnancy had babies with significantly lower birth weights. A follow-up study

24. There is an alternative approach to the dose-response methodology which involves direct determination of risk based on epidemiological studies. This approach, which also has considerable difficulties in implementation, is only now becoming more acceptable.

published in 1990 shows these same children at three years of age had significantly reduced attention spans, a precursor to future learning disabilities. None of these frightening results was predicted by the *conservative* application of the dose-response methodology.[25]

In spite of the need for further development, risk analysis is being applied in a broad range of population-environment dynamic sectors. They include stratospheric ozone depletion, greenhouse gases (carbon dioxide, methane, etc.), species extinction, loss of biological diversity, herbicides/pesticides, acid deposition, oil spills, groundwater pollution, drinking water pollution (heavy metals, carcinogens etc.), and ambient air pollution such as carbon monoxide, benzene, lead, arsenic, acid aerosols, and carcinogenic hydrocarbons. This work will surely continue to expand and become ever more important to policymakers.

The question of how risk analysis relates to transition theory rests with how general are the resultant descriptions of transitions. If a transition can be described in relation to various risk levels, then policymakers can discuss intervention options in terms of changes in the transition trajectory and at the same time have an estimate of risk reduction.

Next Steps in Developing a Theory of Transitions

The motivation for formulating a theory of transitions has been, in part, to try to strike a balance between portraying the complexities of the world while maintaining analytic manageability. The true test of this concept, therefore, rests with whether or not these relationships operating at the population-environment interface are clarified by the theory. The approach has seemed to be useful as this initial inquiry has moved forward. However, much needs to be done before there is a better understanding of its benefit. There are several avenues of exploration in further developing a theory of transitions. They will be mentioned briefly here and then expanded upon in the concluding chapter.

Developing a taxonomy of transitions is an important next step. Nine example transitions were presented in this chapter. However, these are by no means exhaustive and need to be more fully described. In addition, an orderly nomenclature useful for classification should be developed. Second, empirical evidence for each type of transition should be obtained at all levels of spatial and temporal scale. A careful exploration of relationships at different scales may yield generalizations useful for predictive purposes. For example, initial investigation by the author suggests that it may be useful to formulate a

25. "Effects of In Utero Exposure to Polychlorinated Biphenyls and Related Contaminants on Cognitive Functioning in Young Children" by Jacobson, Jacobson and Humphrey, *The Journal of Pediatrics*, January 1990.

theorem relating rates of transitional change as spatial scales vary. A third avenue of inquiry is to obtain empirical evidence on specific geographic regions which are experiencing multiple transitions simultaneously. If the theory has validity, it is to be expected that such regions would exhibit special vulnerability and therefore be subject to heightened likelihood of difficulty.

The word "theory" has been used here to describe a broad systematic strategy for linking comparable observations in order to seek useful similarities. To move in the direction of a true theoretical construct requires more careful investigation of the extent of underlying generalizations.

Perhaps the most important and potentially beneficial next step is to study the causes behind transitions and their associated societal vulnerability. A better understanding of these causes could help foster useful public policy suggestions which, after all, is the fundamental goal of our entire study.

APPENDIX

The previous chapter mentioned the application of analytic techniques in determining the character of transition trajectories. In this appendix, a review of possible functions useful for this purpose is provided. Then, the implications of using mathematical techniques to describe transitions is introduced.

Exponential Functions

The most common function used to portray feedback is the exponential function. It is useful in describing conditions having purely positive or negative feedback. Generally they assume the form:

$$Y_t = ae^{bt}$$

where $a = Y_0$ is the Y_t intercept and b can assume either a positive or negative value. If the exponent is positive then there is positive feedback and the value of Y_t climbs to infinity. If the coefficient is negative, feedback is negative and the value of Y_t slows to an infinitesimally small increase. Figures 14.37 through 14.41 show the character of all the unbounded functions important to transition theory and figure 14.37 provides the shape for both a positive and negative exponential function.

Exponential Function With a Limit L

A variant of the exponential function is one which has a limit embodied in it. In this case, either a positive or negative exponential form moves toward a limit L defined in the following manner.

Let $Y_t = q - (q - a)e^{bt}$

where $a = Y_0$ is again the Y_t intercept and the function converges to the limiting value $L = q$ as shown in figure 14.38. This function is especially useful when there is a known upper or lower limit of convergence.

Logistical Function

The logistical function is useful in portraying "s-shaped" relationships as we often find in the population-environment dynamic. The equation for this form is:

$$Y_t = q/(1 + ae^{bt})$$

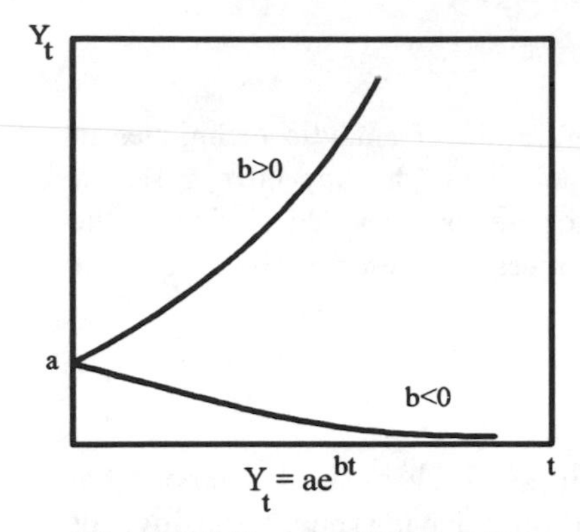

Fig. 14.37. Exponential function

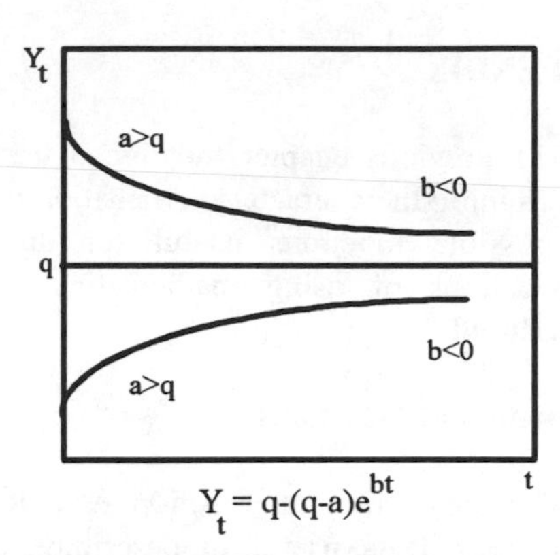

Fig. 14.38. Exponential function with a limit to L

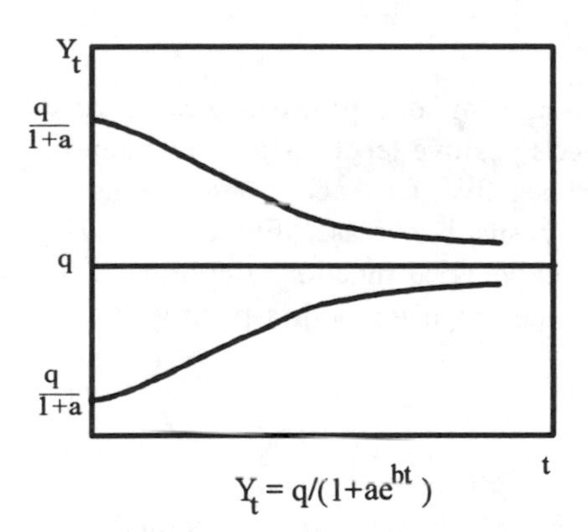

Fig. 14.39. Logistical function

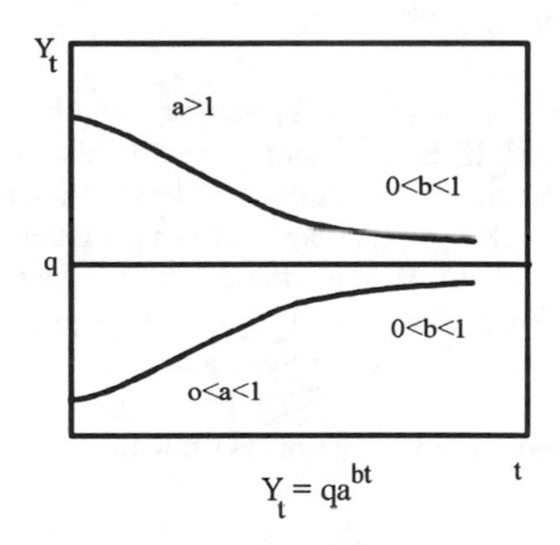

Fig. 14.40. Gompertz function

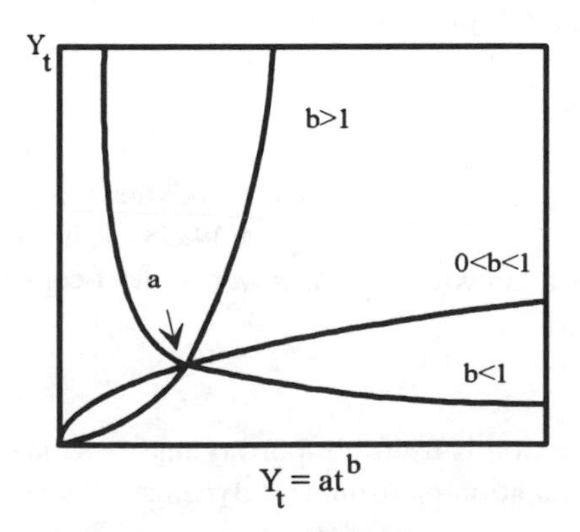

Fig. 14.41. Power function

where $q/(1 + a)$ is the Y_t intercept and the "s-shape" is increasing when $a>0$ and $b<0$ and decreasing when $-1<a<0$ and $b<0$. Figure 14.39 describes this function under conditions of both growth and decay.

Gompertz Function [26]

The Gompertz function like the logistic assumes an "s-shaped" character. Figure 14.40 portrays this equation under conditions of both positive and negative coefficients. The equation for this function is

$$Y_t = qa^{bt}$$

and the Y_t intercept is qa with convergence at q. A growth "s-shape" is achieved by $0<a<1$ and $0<b<1$. A decay is achieved by $a>1$ and $0<b<1$.

Power Function

The power function is the last function discussed here. The equation for this relationship is

$$Y_t = at^b$$

In terms of the population-environment dynamic, the only condition in which this function is expected to be useful is when $0<b<1$. For completeness, figure 14.41 shows this equation under all conditions, both realistic and unrealistic.

The power function could be especially appropriate for data portraying the toxicity and forestry regeneration transitions.

Hypothesizing the Shape of the Transition

While there are other unbounded functions which could be utilized, the five described above are sufficient. All can be transformed to linear form with proper algebraic manipulation and then readily fit to empirical data. Note from figures 14.1 through 14.36 in this chapter, which described transitions, that several different functions could be used to portray each transition. Some functions can be excluded from consideration based upon logic while others are candidates but must await curve fitting before a final decision on the most appropriate equation can be made. For example, the demographic transition, where birth and death rates shift from high to low stability, would exclude all

26. Named after the developer of the function, an English actuary and mathematician, Benjamin Gompertz (1779-1865).

but a logistic function with $-1<a<0$ and $b<0$ or the Gompertz function with $a>1$ and $0<b<1$. In a similar way the original forest cover of the forestry transition that drops to a lower level, possibly as low as zero, can be fit with either of the above functions or with either variation of the negative exponential. The challenge of determining which function fits best in any circumstance is an empirical investigation which is beyond the scope of this chapter. Table 14.2 provides a summary of the possible functions useful in four of the transitions and our "best guess" based on logic and shape.

Other Approaches to Fitting Functions to Data

There are several other approaches to choosing the optimal function. The first is the application of differential equations, especially those with convergent periodic form, to the many phenomena in the global system which have a cyclical characteristic. Examples are the carbon, nitrogen, sulfur and fresh water cycles at the global level. At the regional level the El Nino current and individual water basin cycles are also evident. These cycles have a periodicity which might lend themselves to description by differential equations and undoubtedly affect the analytic representations of transitions related to them.

Another approach to curve fitting is to use a mathematical formulation which fits the data more closely but which is driven strictly by the character of the information rather than an underlying theory. An example of this method is cubic spline interpolation. Extremely close fits can be achieved but because they are bounded functions, the mathematics are applicable only over the range covered by the data and are therefore not useful for predictive purposes. The advantage to an approach such as cubic spline interpolation, however, is the possibility of uncovering some pattern such as a periodicity in the data which then could be studied further with more orderly functions.

Deriving Advantage from Functional Characteristics

In many of the functions whose parameters have been determined empirically, there are mathematical characteristics which might be helpful in describing and classifying transitions. For example, when a logistic function, shown as figure 14.39, is fit to a given set of data, the characteristics of that specific transition are uniquely represented mathematically. The amplitude of the transition is the difference between q and $q/(1 + a)$ and the steepness or maximum rate of change of the transition is at the inflection point where the second derivative is equal to zero.

TABLE 14.2. Transitions in relation to functions

Type of Transition	Possible Functions		"Best Guess"
Demographic	Logicstical Gompertz	(-1<a<0) (a>1)	Gompertz
Epidemiological (Infectious diseases)	Negative Exponential to the limit L Logistical Gompertz	(a>q) (-1<a<0) (a>1)	Negative Exponential to the limit L
Epidemiological (Degenerative diseases)	Negative Exponential to the limit L Logistical Gompertz	(a<q) (a>0) (0<a<1)	Logistical
Forestry (Original Harvest)	Negative Exponential Negative Exponential to the limit L Logistical Gompertz	 (-1<a<0) (a>1)	Negative Exponential to the limit L
Forestry (Regeneration)	Exponential to the limit L Logitstical Gompertz Power	(a<q) (a>0) (0<a<1) [yt=atb)	Exponential to the limit L
Toxicity (Industrial Production)	Exponential to the limit L Power	(a<q) (0<b<1)	Power (0<b<1)
Toxicity (Toxic Emissions)	Exponential to the limit L Power	(a<q) (0<b<1)	Exponential to the limit L (a<q)

The relative duration of the transition is somewhat more complicated to define but also can be determined by other mathematical properties. One promising approach to determining duration of transition is to apply some of the principles found in Chaos Theory. When Feigenbaum's graphical analysis methodology was applied to the logistic function portraying the North American demographic transition, points of convergence at both the beginning and ending of the transition were apparent. A threshold value on the transition curve, different from the inflection point, separates the function into two distinct components. Whenever the value of the function exceeds this threshold, stability occurs at the end of the transition. If the value of the function is less than this threshold, stability occurs at the beginning of the transition. This interesting mathematical property may have a real world counterpart. Perhaps there is a threshold point in a transition which defines irreversibility.[27]

Implications of Fitting Functions to Transitions

But what is to be gained by fitting an exponential or logistic or for that matter *any* function to transition data? The answer lies in the possibility of gaining additional insights from the process. First, comparisons within and across sectors can be more easily made if they are in functional form. The parameters of the functions can be compared rather like an index is used for some types of linear data.

Second, there may be insights gained simply by the process of fitting a function to historical data. Different mathematical functions often have very specific underlying characteristics which can provide useful ideas. An example of this can be taken from forestry. Each oak leaf has its own unique characteristics including the dimensions of the fingers on the leaf. If the combined length of fingers is plotted by frequency of occurrence, the resulting distribution is best fit by a normal distribution. This implies that the elements that go into determining finger length are additive. On the other hand, if the distribution had been fit best by say, a log-normal distribution, the implication might be that the elements determining finger length are combined in a multiplicative way. Simply because one distribution fits data better than another does not provide conclusive evidence of an underlying truth but instead may help pose hypotheses which then can be tested by more rigorous means.

The third potential benefit of applying functions to transition theory lies in the identification of lead indicators. Once transition data is successfully fitted to an appropriate function, then for a given condition and point in time, the

27. I am indebted to Dr. Sandra Lach Arlinghaus, first, for pointing this out, and then for showing me how it works mathematically.

future trajectory of a transition can be predicted more accurately. Identifying lead indicators is facilitated because, with an orderly function, only one, or at most, two parameters need to be determined to define the trajectory. This advantage is even more evident when several functions are occurring simultaneously.

Fourth, mathematical functions may facilitate investigation of several sectors interacting with each other. Earlier sections have discussed the special societal vulnerability associated with transitions occurring simultaneously. From a modelling perspective, this simultaneity is often very difficult to describe and analyze, which explains why less progress has been made in this area to date. However, being able to portray multiple transitions with specific functions could be helpful. There is no question that each transition interacts with the others to some extent. And to the analyst, this means that a reliable model must be structured as a set of simultaneous relationships. Describing transitions as functions facilitates this manipulation.

Finally, there is the question of how risk analysis can be related to transition theory. With a transition mathematically determined, it is possible to move up and down the curve in relation to time. In a similar vein, when risk analysis generates a function relating dose to level of toxicity, then the two can be combined to form a cross product. This yields a risk in relation to position on the transition curve. Policymakers then can discuss intervention options in terms of changes in the transition trajectory and at the same time have an estimate of risk reduction.

Chapter XV

Perceiving Population-Environment Dynamics: Toward an Applied Local-Level Population-Environment Monitoring System

Frank D. Zinn
Steven R. Brechin
Gayl D. Ness

Introduction: Three Stories

A New View

Agricultural officers had been trying for some time to persuade members of a small village in East Africa to change cultivating practices that were degrading the land. Young sons needed new land, and moved into fragile hillsides with traditional farming practices. One or two years of good crops were followed by extensive erosion, loss of soil fertility, and the move to even more fragile lands. During a fortuitous visit by a foreign remote sensing scientist, the village chief's son was taken for a helicopter ride (the chief refused the invitation), in which the scientist produced a video film of the village. That evening, on a portable TV screen, the villagers were shown their land from the air. It was a view they had never seen before, and they could not believe it was their village. The chief's son assured them it was, and pointed out specific huts. Fast reversals and reruns brought home the picture of erosion and degradation. From that point on the agricultural officer had considerable success in showing people how they could use new farming practices that prevent erosion.

Local Expertise

In a poor isolated Caribbean fishing village, an experiment was being conducted on a new mariculture technology. The village's population growth had pushed many sons and daughters to move out to the towns in search of work, but there were still too many people, and the fishermen knew they were overfishing the reef that is their livelihood. The new experimental technology showed great promise. It came from the outside, involving international scientific research centers, international aid agencies and the central government with its myriad technical agencies and political dynamics. The local fishermen were both helpful and hopeful. They helped the research team adapt the technology to the local situation that they knew very well, and clearly saw the prospects of rising income. Then the situation changed. International and national politics intruded and halted support for the experiment. The local fishermen wished to continue the project, but they lacked capital to continue without external help. Further, their advice on protecting the experimental materials in the face of oncoming storms was not heeded. The storm destroyed the experimental materials and the fishermen lacked the capital to start again. They are back to eking out a meagre existence, with little alternative but to overfish the reef and send their children away to the towns in search of work.

No Easy Solutions

The hills of a remote area were being cleared of forests for small agricultural homesteads. The homesteaders were sons of lowland farmers, forced to move into the hills to find their own land. The soil was rich and the harvests were good. The income generated by the farmers here was significantly higher than many others in the region. The news spread and others came in search of their own good land. The government also promoted resettlement there to relieve pressures on the land in more distant densely settled areas.

The demand for land increased, forcing the new immigrants to cultivate steeper and steeper slopes. Some farmers began to invade protected forest areas established years ago to preserve the area's water resources, including rivers that were navigable downstream. Erosion began on the steeper slopes, but quickly ate up more productive low-lying land. Farms failed, forcing movement to even steeper slopes. Farther down the slopes, heavy erosion, flooding and landslides became more common. Formerly dependable springs and streams disappeared or became unusable. As time went on, the distribution of people and activities became more uneven, and disparities of health and welfare began to widen.

Neither the central nor local government seemed capable of managing the rapidly deteriorating situation. The forestry department attempted to enforce the boundaries of the protected area, but had too few personnel and no alternative to offer the illegal squatters. One agricultural agency worked with farmers to promote increased production, while another tried to promote the adoption of soil conservation techniques. Health specialists attempted to deal with growing health problems caused by lack of clean water, but had only curative treatments to offer. Government agencies found themselves each dealing with only one piece of the puzzle, and no one agency had much success or knowledge about what the others were doing. The situation grew worse. Finally, in desperation, the national forestry office prevailed on the government to send in troops to remove the illegal farmers which, unfortunately, led to violence.

These stories are composites based on situations found in the research sites examined in this volume. They illustrate familiar aspects of modern population-environment dynamics. Population growth presses on the natural environment. Traditional technologies that were sustainable in the past, become destructive when used with larger and more densely settled populations. External attempts to help develop new technologies fail when they do not involve local people. Government's maze of conflicting policies and inadequate program implementation fail and government itself becomes associated with severe environmental destruction. Attempts to protect natural areas from encroachment not infrequently lead to skirmishes and bloodshed.

The stories illustrate some important points. First, identifying a "best course" for development involves making *tradeoffs* between potentially competing interests and goals. The idea of sustainable development has become fashionable in recent literature. It suggests that economic development must be promoted through the use of resources in ways that can continue to support future as well as present generations (Reid, Barnes, and Blackwelder 1988; World Commission on Environment and Development 1987). Though many governments officially support the notion of sustainable development, there exists little agreement on a precise definition of, or approach to, the concept. The concept is especially weak in its failure to recognize that trade-offs are required (Thrupp 1989; Tisdell 1988). In other words, sustainable development is a conceptual goal, and the reality is that planners must make hard decisions regarding how it is to be achieved in the planning process.

Second, *local participants* must be brought into development efforts to ensure their success. Most production is strongly affected by a wide range of local conditions. Climate, soil, land cover, labor resources, markets, and past practices all have distinctive properties with which local producers usually have extensive experience. This local expertise is needed to adapt generalized strategies and techniques to local conditions. Local commitment is also needed to mobilize the resources that the outside cannot provide.

Finally, any attempt to understand such a highly complex and dynamic situation, requires ongoing observation, or *monitoring* of many different aspects of both human actions and environmental conditions. The highly specialized monitoring of different technical agencies may be adequate to their specific problem, but not to the whole. The disparate monitoring capacities must be brought together in a system that is comprehensive and integrated in order to accurately perceive a highly complex situation. A comprehensive understanding is essential since solutions which do not address the whole situation are doomed to failure. In this chapter, we use these lessons as a basis for proposing an applied approach to understanding population environment relationships through improved monitoring at the local level.

Integrated Monitoring for Comprehensive Planning

Monitoring Population and Environment

Monitoring is the means by which we perceive changes in the world around us. The idea of monitoring the environment is hardly new. Environmental attributes like temperature, wind, rainfall, land use and water quality have all been recorded and monitored in many places for many years. We have also long tracked population attributes like growth rates, size and density, migration, mortality and morbidity. In addition, we have monitored our economies with sophisticated techniques like national income accounting and input output analyses and a raft of economic surveys.

Our monitoring capabilities and activities have increased in sophistication over the years. Like human populations and human productivity, this technical development has grown exponentially in the past century or two. Population censuses, surveys of social conditions and health statistics all spread slowly in the nineteenth century. In the second quarter of the twentieth century, national income accounting began its development, along with the large scale probability sample survey and aerial photography. All of these techniques exploded in the second half of the century, and were supplemented by a major technological breakthrough in satellite imagery, remote sensing with electronic rather than photographic data collection and storage capacities.

In some respects, remote sensing, either by photographic or electronic means, may be the most significant of these developments in monitoring technology. The ability to look down on ourselves and our environment provides us with a totally new way of observing the world. We can see spatial patterns that are practically invisible from the ground. We can gain intimate views of the trees, but cannot see the forest. From the air we can see the shape of the entire forest itself. When we track this over time, we can see the expansion or shrinkage of the forest. We can also examine heat flows, atmospheric conditions, oceans, lakes and rivers, and watch their changes over time. Electronic remote sensing offers, in addition, the endless possibilities of computer manipulation and image enhancement. This permits the examination of changes over time.

Forests, oceans, lakes and rivers are highly visible, but modern remote sensing with computer manipulation also offers the opportunity to include a range of far less visible conditions in a visual spatial display. Thus, we can show the spatial distribution of wealth, using national income data; of health, using life expectancy, infant mortality rates or the incidence of specific diseases; and of communications, using indicators such as school enrollment, ratios per capita or newsprint consumption.

One of the problems we face, however, is that these new monitoring technologies have tended to come in packages developed by specialized scientific disciplines. In fact, the great advances have been possible precisely because of the specialization of the scientific disciplines. This enables scientists within each discipline to tell us a great deal about that portion of the world that they observe. Few, however, can tell us much about relations between their specialized observations and phenomena monitored by other specialized disciplines. The result has been to isolate these packages of highly specialized knowledge and, not necessarily, to increase our understanding of the larger whole.

One outcome of this specialization, therefore, is that our efforts to plan for and achieve, sustainable development may be undermined. Knowing a great deal about isolated events is not sufficient to establish the understanding of the whole that is necessary to promote sustainable development. To be successful in this regard, we must be able to frame our goals and our problems in a comprehensive way. The process of framing, questioning, answering and taking action requires utilizing a monitoring approach that enables us to see the dynamics clearly and interactively (Drake, Miller and Schon 1983). We must create a framework that will allow us to bring together, in meaningful ways, data from many sectors and disciplines and, by definition, from a number of monitoring approaches.[1] We call this type of framework **Population-Environment Monitoring Systems (PEMS).**

As might be expected, this integration presents some of its own specialized demands. There are, for example, a number of methodological issues that emerge when we attempt to integrate data collected from multiple sources. A few of the more important are the need to: (a) use similar units of analysis; (b) coordinate sampling frames to assure data from the various methods sufficiently overlap in space; and (c) schedule data collection activities to coincide with relevant critical events or with relevant seasonal activity. Fortunately, these problems are not insurmountable, and a number of organizations have been fairly successful in developing multi-method monitoring approaches. To assist in this development, the new Geographic Information System (GIS) technology offers great help.

---

1. There are many different types of monitoring activities. Some of the more common are: (1) social surveys and censuses where individuals are interviewed on a wide range of perceptions, activities, and characteristics; (2) point sources monitoring where data is collected at a particular geographic point, and from which generalizations to surrounding areas such as found with temperature, air and water pollution, or soil surveys; (3) secondary data, such as government records and documents; (4) remote sensing data, such as airplane photographs or satellite imagery; and so on.

Geographic Information System -- A Solution to the Problems of Integration?

An exciting and potentially useful computer-based application is the geographic information system (GIS). GIS software provides users with the capability to analyze data and present it in a map format. GIS systems can work with data stored as geographical units like points, lines and polygons. These data can then be analyzed, overlaid and displayed.

Because GIS is spatial in its orientation, it can be used to resolve some of the methodological problems associated with the integration of data. For example, point source data (like a water quality reading) can be visually represented by a symbol; flow data, like migration rates, can be represented using line widths and arrows; and areal data, like population density, can be represented by shading patterns. In addition, given the necessary data, GIS software can be used to identify areas which have gone from forest to agriculture, or other changes over a time period.

GIS can also present data in a way that is easy to understand. Maps can be produced which show locations, coverage, change and other phenomena. Such displays are easier to comprehend than tables, or statistical measures of association. In addition, map presentations of "hard data" can be easily overlaid in the mind of the planner with local knowledge. Because of their spatial orientation, maps provide an intuitive view of the world, which can support the cognitive process.

One way, then, to pursue the development of local-level population-environment monitoring systems is to develop GIS technology and capacity within local planning units. One hurdle to be overcome in this area is that currently available GIS systems are, for the most part, large and expensive, requiring relatively sophisticated computers and extensive training. Simpler GIS approaches may be more appropriate, given a specific set of circumstances.[2] Before we turn to the simpler technology, however, we should review some of the major developments currently under way in integrated monitoring.

Cases of Integrated Monitoring Systems

There have been efforts, largely at the international and national levels, to develop multi-method approaches to monitoring. One of the more ambitious is the United Nations Environment Programme's (UNEP) Global Environmental Monitoring System (GEMS) and Global Resource Information Database

2. In fact, GIS systems might be simplified to the point that computers are not necessary. During a visit to a district-level planning office (BAPPEDA) in Indonesia, the authors observed maps of land use that were generated by photocopying a map onto graph paper, and manually shading squares.

(GRID) programs. GEMS was created in 1974 as part of UNEP's original mandate on overseeing issues concerning the global environment. GRID was created in the early 1980s to integrate the various data banks collected through GEMS (and other) programs on a range of environmental factors, such as water, air, land, oceans and so on, so that they could be better used by decision makers (Gwynne and Mooneyhan 1989, UNEP 1985).

Although GEMS is a UNEP program with a mandate for only environmental issues, it has been able to develop several multiple sector (or integrated) databases through collaborations with other United Nations agencies such as WHO, WMO, and UNESCO.[3] Although important information has been collected, it is limited and spotty in geographic, temporal, and subject coverage. Although still under development, it is not yet systematic and comprehensive.

Considerable improvement was made with the creation of the GRID pilot project. GRID is a computerized Geographic Information System (GIS) which allows one to overlay layers of data and display output on a map. For example, GRID was used to monitor elephant populations in Africa. For this project, 290 data points were collected from a variety of sources and plotted on maps. This project allowed for a comprehensive study of the African elephant. The GRID project has been useful in better understanding the elephants range and density to improve and target conservation efforts. This is the first time that comprehensive information on the African elephant population has been collected and presented in a visual geographic format (Douglas-Hamilton 1988). Conceptually, GRID is a very powerful and flexible system. It is, however, limited by the existence of only a few GRID centers and its sophisticated nature.

The Famine Early Warning System (FEWS) is another impressive monitoring system created by the United States Agency for International Development (USAID) in 1985. FEWS integrates data from other monitoring systems (satellite, and ground surveys) and most importantly, data that are from different sectors (earth sciences, economic, agricultural, health, etc.) (Walsh 1986 and 1988; Bass 1986). Recently a GIS has been added to FEWS to increase its flexibility and usefulness (Science 1988). The idea behind the FEWS experiment is to predict famine, before it happens, to give relief agencies some lead time in preparing for the emergency. It is directed at Africa's seven most drought and famine prone countries: Mauritania, Mali, Niger, Chad, Sudan, Mozambique, and Ethiopia. Both physical and socioeconomic data are collected from ground surveys and by satellite imagery. Integrated information on crop climate, crops yields, household income and so on are collected and analyzed. Frequent reports are produced

3. See for example: UNEP and WHO 1987, *GEMS: Global Pollution and Health*,UNEP, WHO; WHO 1983, *GEMS/Water Data Evaluation Report 1983*, UNEP, WHO, UNESCO, WMO; WHO 1982, *GEMS: Estimating Human Exposure to Air Pollutants*, UNEP, WHO.

on pre and post harvest conditions and are distributed to responsible individuals and agencies for their evaluation. Similar early warning systems are operated by FAO (United Nations Food and Agricultural Organization) and by European countries.

There are, of course, some difficulties with FEWS. Acquiring consistently high quality data from ground sources remains an important problem, as it does with any monitoring effort. Particular difficulties have been encountered in obtaining reliable data from government officials on the ground. It is not surprising that areas subject to severe famine are also poor and have very weak data collection and recording systems.

It might also be mentioned that, though FEWS was designed to serve USAID-Washington, there are presently efforts underway to make FEWS more adaptable and accessible to other users, i.e., for USAID country missions and host country governments. This has resulted in the development of software packages for personal computers. Though developed specifically around data related to predicting crop yields, FEWS is moving toward the local-level approach we advocate.

There are also integrated monitoring systems employed at national levels. In Indonesia, for example, the Ministry of Population and Environment (KLH) attempts to integrate population and environmental data through the use of a population environment balance sheet (*Kependudukan dan Lingkungan Hidup Daerah* or NKLD). NKLD is a reporting framework that was developed by KLH in order to encourage sectorial agencies to look at a broader range of variables and relationships. A set of guidelines, based on the UN Environmental Statistics Guidelines, was established for producing the report in 1987. These guidelines provide a framework in which population-environment dynamics might be considered, but the report generally describes population and environment factors independently.

Though each of these integrated monitoring developments has had a measure of success, they all have a similar limitation. They are designed to respond to international or national level planning information needs. They rely on expensive computer hardware and data collection techniques. They are, in short, highly capital intensive and generally deal on multi-national or national scales. What merits more attention, therefore, is the development of integrated systems for local level use.

The Dimension of Scale in Comprehensive Planning

A second important dimension of monitoring for planning, in addition to integration, is that of scale. Planning and monitoring takes place at many different political levels, and the framing of planning problems and objectives are different at each.

TABLE 15.1 A Typology of Scale and Planning

Scale	*Planning Function*
International	Coordination between nations, crisis support
National	Standardization of policy and targets within nations
Regional	Intermediate standardization and coordination of local units
Local	Implementation and adaptation of regulations and interventions within plans provided by higher levels

As table 15.1 indicates, the primary responsibility for implementing planning activities often resides at the local level. At the same time, these activities are guided by central policies, and local units are accountable to central levels. For example, international policies might regulate trade of lumber products between countries. National planners might establish minimum guidelines for the assessment of development projects on forested land in local land-use planning. Regional policies might coordinate development plans of local units. Local planners might develop land use plans and establish protected areas within their jurisdiction.

It is important to recognize that it is at the local level that plans are implemented, and that too much standardization of these activities can lead to ineffective plans or counter-productive results. There is a great deal of variation between localities that, in fact, requires adaptation within the framework of general policy goals. Different local units face different population and environment problems or, at a minimum, variations within the same problems. Ideally, the role of the local planner is to adapt interventions, or operationalize general approaches to achieve general policy objectives. For example, it may be that the national government has declared that some minimum number of hectares of forest land be protected in each district. It is then up to the district planners to select specific areas in their district for protection. Each district will want to protect forest land in a way that minimizes the negative impact on local economic activities.

In order for this planning approach to be successful, at least two conditions must be met. First, local people must both agree with the general objective, and be motivated to promote it. In our example, locals must understand the importance of protecting forest land, be motivated to protect and enforce protection, and have the capacity to carry out protection activities. Studies note that local-level planning and management can be more effective than the central level in cases where local people understand they have a stake in the health of their local environment and resources, and have the social capacity to take action (Brokensha and Rilley 1989; McNeely 1988; Dani and Campbell 1986; Uphoff 1986).

Second, both central planners and local implementers must have the ability to see phenomena in an integrated way. If the guidance required to establish reasonable policy is provided by higher levels, then the ability to plan and to see changes in the local situation will result in viable plans. Local planners, after all, have the capacity to see local changes in light of what they know about their jurisdiction. The application of local knowledge to the data collected will result in more accurate interpretation. Moreover, the application of local knowledge in the planning process will lead to programs which are feasible and effective. Finally, using local monitoring to see the results of interventions will promote learning about the efficacy of these intervention strategies. This dynamic (of seeing, learning, acting and seeing) is an effective approach to planning. Without the ability to see, the local planner is severely hindered.

Thus, for local comprehensive planning to be proficient, integrated monitoring must be synthesized into the process.[4] Unfortunately, a formal system of this type is not commonly found, especially in developing countries. Local level planners and decision makers usually don't make much use of the information they collect, and often just receive the instructions from higher levels on how to solve their local problem.

Framing an Integrated Monitoring Approach for Local-Level Comprehensive Planning

Attributes of a Local-Level Monitoring System

It is now well recognized that for many development plans to succeed, two conditions must be met. First, local participants must agree with the goals and must be committed to their achievement. If they feel a project is somehow "theirs", they are more likely to protect it from attack, to save it from neglect, and to adapt it to the most critical local conditions. Second, both central planners and local participants must be able to see how any given project fits with other conditions, and they must be able to see a whole range of conditions roughly simultaneously to decide on the priorities of specific projects.

4. Ironically, local planning activities traditionally operate in a PEMS framework. However, central governments have mandated planning activities, but have not provided support for local-level, comprehensive monitoring. As a consequence of the large and rapid changes in the local context brought by development and technology, this support is essential.

One of the implications of this set of conditions is that monitoring of population-environment dynamics must be for the benefit of the local planner--it must be driven by questions pertaining to the local condition and local planning activities.

Often, however, information systems are designed by central planners to answer their own questions. Local data are usually aggregated at higher levels, and provide little or no feedback to the local user. When local workers see little use in the data they collect, they are not likely to take much interest in its quality. In fact, when they do not know how data are used, they may well be led to falsify information to avoid upper level complaints. Central level planners and administrators often complain about the quality of data submitted by people in the field. One reason for quality problems in this circumstance is that local units are collecting data for which they have no use. There is little doubt that when local units use data in their ongoing activities, and those units have some reason or incentive for obtaining accurate data, then the quality of data is likely to be high (Drake, Zinn and Antonakos 1990).

This does not mean that local data collection must be kept separate from central needs. Quite the contrary. Local data collection should be organized with sufficient standardization to permit comparisons with other local units. This provides important information both for the central planners and the local participants. For example, if subunits at any level provide information that permits them to be ranked along with other subunits, both central and local planners can ask what conditions explain high and low performers, and then act on that information.[5] Local conditions may make it easier for some areas to achieve higher performance, which can lead planners to adjust goals according to local conditions. It can also be that performance levels are closely related to the amount of resources that come from the center. This kind of observation can be helpful to planners at all levels, because it permits them to make the argument for more resources to raise performance levels more widely. The goal here is not to replace central-level planning activities, but to incorporate or integrate monitoring into comprehensive planning processes at all levels.

This leads us to the specification of a local-level integrated monitoring approach. We propose the following as guidelines for the development of such a system. The interrelated guidelines are general, and would need to be refined to fit national as well as local planning needs and conditions.

---

5. The Indonesian National Family Planning Coordinating Board uses a very effective tactic of this type. At each level, from National to Province and District, subunits are ranked by acceptors recruited. This enables all administrators to search for the causes of high and low performance and to use the information they gain to increase performance throughout the country.

Driven by Local Level

It is important that a monitoring system be driven by local-level planning questions and information needs. By developing the system in a way that information collection is geared towards local use, there is an incentive for data to be carefully collected and verified.[6] This also implies that local people control the system and use it for their own purposes, but not without regard for higher level planning concerns and information needs.

Appropriate Technology

The monitoring system should be developed using the lowest reasonable technology. If computers are necessary, the hardware and software must be appropriate to the local environment. As computer technology continues to improve, hardware and useful software will become even more commonplace with important applications for planning purposes (Klosterman 1988).

Simple Methodology and Presentation

To conduct comprehensive planning, data must be collected from many sectors. It must also be integrated into one data management system for analysis. At the same time, the methodology used to analyze and display the data must be sufficiently simple so that the local users can understand the process. There is a tendency at the center to opt for the more high-tech processes, because they look good. It is especially important, however, to have a very clear sense of the capacities of local participants, and to select a technology that they can use easily.

Flexible and Adaptable

Since the system must support local planning in a variety of settings, it must be flexible and adaptable. The users must be able to add, modify or delete indicators in the system. Users should have the ability to adapt methodologies and output formats while still keeping data sufficiently standardized to meet the planning needs of higher levels.

Linked to Other Levels

The system should provide information to central levels to help in the formulation of policies and coordination of activities. By building in feedback

---

6. The problem of data quality is a challenging one which can often be ameliorated through better communication and/or feedback to the field.

loops within the network, local-level monitoring systems can provide information for policymaking at other levels, as well as help other local planners learn and develop solutions to their problems.

The PEMS Project

Presently, faculty at the University of Michigan, Michigan State University and Princeton University have joined together to establish a PEMS Project. The objective of the project is to collaborate with institutions in developing countries in the design and implementation of systems to monitor population and environment interactions at the local level to better achieve sustainable development. Specific cases, which are discussed in more detail below, have been successful to date largely due to the participatory approach taken in the design of these systems. Local institutions frame the population-environment problem of local concern. Assistance is provided to help identify and adapt methodologies and monitoring approaches which integrate information from pertinent fields and employ technologies appropriate to the particular setting. These systems can be used both to understand the problem and monitor the effects of interventions.

Indonesia: Center and Province

Central Level

The PEMS Project is working at two levels in Indonesia, with the Central government and with the University of Sriwijaya in South Sumatra. At the center, the Indonesian government recently created a Population and Environment Ministry, known as **KLH** (*Kependudukan dan Lingkungan Hidup*). KLH is a ministry of state, or a staff (as opposed to a line) ministry. Its task is to link together the wide range of government agencies dealing with some aspect of population and environment, and to induce each of the agencies to view their activities in a broader context.

KLH now produces a periodic report from each district on population-environment dynamics, called the **NKLD** (*Kemendudukan dan Lingungan Hidup Daerah*). It is a reporting system established in 1987, based on the United Nations Environmental Statistics Guidelines. The guidelines led district and provincial levels planners to consider a broad range of population and environment conditions in the same report. The NKLD has been largely descriptive, with little in the way of analysis.

The PEMS project began working with the Indonesian government in December 1989. In 1990, PEMS made a presentation on monitoring and GIS to the Indonesian Ministry and to the Central Planning Office (BAPPENAS).

Visits were made to two trial districts, where the Ministry was developing its new reporting system. In those districts, agreements were made on the data to be collected and personnel from both the district and the central government were trained in the use of a simple GIS system called QUICKMAP. In 1991, training was extended to district level officials in seven provinces. In each case the training is provided to five local officials from the Population Environment Ministry, and five from the local branches of the central planing office.

The Ministry is now introducing the use of GIS into the NKLD report at the district level. At the center the Ministry is developing software to produce reports at the district level. A prototype has been developed and is now being tried in the field. The new system is made up of two packages: PROFILE and QUICKMAP. PROFILE is used to enter basic population data from the local planning report and to print it out in the simple NKLD format. QUICKMAP is used to print those data in map form. At the moment however, the data included is limited to age and sex pyramids and certain land use information.

At this stage, the objective of the training and monitoring system is to create an awareness of population-environment conditions among local planners. The analysis performed does not yet go beyond university thematic maps and histograms. The monitoring system is not specifically linked to any planning activity, nor is any evaluation of data made by the system. This reflects the fact that KLH is a staff rather than a line ministry. It encourages local planners to observe a wide range of conditions in a single integrated framework, but it cannot direct that they do so. It is using the simple and descriptive mapping tool to attract the attention and interest of the officials of the technical agencies. Gradually, NKLD reports will display more and more population and environment data and linkages, and may provide some sample analysis from the mapping system. If it is sufficiently attractive, it will be taken up by other agencies. The coordination, however, lies ahead.

Provincial Level

The PEMS Project is working with faculty at Sriwijaya University in Palembang, South Sumatra. South Sumatra is one of the most resource-rich regions of Indonesia. The province is blessed with natural resources such as forests, oil, gas, coal, and other minerals. In addition, it is a site of much industrial production. It also contains extensive critical habitat such as coastal wetlands. PEMS's objective is to assist the University's Population and Environment Research Centers in building their institutional capacities in comprehensive integrated research and planning to provide policy advice to local and provincial governments in achieving sustainable development.

Maps of the Province and of Lahat district have been digitized, and data files are being assembled for each district in the Province, and for the subdistricts of Lahat. These data include a wide range of social, economic, health, educational, agricultural and land use data. These are the typical data files one finds in statistical yearbooks or in the collections of any specific government agency. They are also the kind of data that tend to be used and displayed nationally, but not locally. One aim of the PEMS project is to make those data readily available at the province, district and subdistrict level.

One of the first problems is gaining access to public data. University researchers are generally respected and can obtain much data, but there is also some suspicion and reluctance in some agencies to share data. There is also a perennial problem of data quality and coverage. Much local data that should be available is limited in coverage, and much more is of questionable quality. Thus researchers are constantly seeking ways to check the validity of the data they use in the PEMS system.

Local planners and university researchers already have a series of questions that PEMS can address. In Lahat district, for example, there is recognition of serious illegal forest clearing for coffee planting. Local planners and university researchers are working together to understand what drives this process. Two competing hypotheses have been developed, and a search is underway for data to test the hypotheses. Until 1983 the local forest reserves were protected by a traditional local authority system, the *Marga*. This has been dismantled by the central government. Villages were placed under an appointed headman, and the forests were placed under the protection of the forest service. With the new system, boundary enforcement is not implemented by customary law, but by bureaucratic control, which itself is highly underdeveloped. One suggestion is that the dismantling of the traditional authority system hastened the decline of forest protection and increased the rate of deforestation. The competing hypothesis is that deforestation is a more linear process, driven by population growth, from both natural increase and in-migration. Population data are being entered into the PEMS data management and GIS system, and data are being gathered on forest clearing over time. At this point, satellite imagery is being exploited. A search is being made for images from four points, 1978, 1982, 1985 and 1989. If sufficient cloud-free images can be located, computer enhancement can be used to examine the rate of deforestation before and after the 1983 administrative change. This will permit local researchers to determine whether the rate of deforestation is gradual, thus more likely the result of population growth, or whether it shows a marked increase, indicating an impact of the 1983 administrative change. The outcome can be fed to the provincial and national level planners for possible action.

Mexico

In Mexico, the PEMS team is working with Dr. Carlos Santos-Burgoa, Dean of the School of Public Health, which is located in the Mexican Institute of Public Health Research. Dr. Santos-Burgoa took part in early discussions on the development of PEMS. Currently the PEMS team is working on three projects in Mexico, which demonstrate the range of activities and the way in which PEMS can be adapted to meet different needs of the user.

One project provided the Institute's Center for Infectious Diseases with a digitized map of Mexico, showing state boundaries. Training was also provided to Mexican colleagues in digitizing maps and linking them to data files. The first maps were used to examine the spread of AIDS, using data from the early 1980s to the present. Following this assessment, the integrated data and map system are being used by staff from the Center to examine the spread of a number of other infectious diseases.

A second project provided a digitized map of 13 towns in the state of Mexico. In each town the individual blocks were digitized, and the map package was integrated with a data set from a large sample survey done in 1989. The survey contains a great deal of information on individual health conditions and use of government health assistance. It also adds current data on location and availability of health facilities in each of the 13 towns. Previously, the survey data permitted only statistical analysis. With the mapping package there is now considerable examination of the spatial distribution of these conditions.

In the third project the PEMS team is assisting to develop a management information system for a large multiple-department project on management, planning and programming for local health clinics. The focus of this project will be on the quality of care. The PEMS team is undertaking an information needs analysis and the project is considering the use of GIS and other graphic display techniques to meet the needs of the local health clinics.

Zimbabwe

Since early 1990, PEMS has been working with the Zimbabwe Department of Natural resources to evaluate a small integrated rural development project in Mukarakate, approximately 100 kilometers east of Harare. The area is an equilateral triangle bounded by two rivers that merge on the north, and a road running across the south end. It is a mere 72 square kilometers, containing 138 families and 950 persons.

Many rural areas of Zimbabwe suffer from excessive soil erosion, which locks farmers into a self-sustaining cycle of poverty and environmental degradation (see McIntosh, in this volume). The pilot integrated development project is testing a series of interventions--such as soil conservation, agroforestry, animal husbandry--to find a way to turn the vicious circle into a virtuous one. The project is organized locally under the leadership of a community committee, which meets to identify local problems and plan collective activities to deal with these problems.

PEMS is being used as a device for evaluating the project and, in the process, for developing a monitoring system tailored to the needs of the local population. The first task is to map the area, locating individual family plots and open public lands. A nearby government clinic serves the area, and has records of many of the households. Data from the clinic are being organized for the PEMS data management system. In addition, a census is being planned to collect demographic, health, and economic data to be included by household. A geographic survey is being planned to record soil conditions and land cover throughout the area. These data will also be integrated into the data management system identifying land cover and soils by individual holdings and open lands. Finally, rain gauges and river gauges are being established at key points, using the local committee for record keeping and maintenance of the gauges. These data, too, will be added to the basic data management system.

When the basic data are established, periodic six month household surveys will collect data on population changes, health and income. Other periodic surveys are being planned to follow on the economic and geographic surveys. In some cases, aerial photography can be used to supplement ground based data collection. All of these will be used to examine changes in living standards and environmental degradation and to evaluate the integrated rural development project. GIS will provide one means of generating visual displays of the changes. If the evaluation process proves useful it will be extended by the Ministry of Natural Resources to other externally funded development projects in Zimbabwe.

One set of problems for this project is organizing the data collection, and mobilizing the specific agencies necessary to do the work. The most demanding part of this is bringing together the separate agencies of government whose problems touch one another, but whose officials seldom communicate. The local village planning committee will be the primary locus for mobilizing the information, putting it into usable shape, and defining the problems that are to be examined. Through the local committee, we can keep the focus on immediate problems and can tap intimate local knowledge of the population and environment conditions. More difficult will be to bring together the various government agencies responsible for serving this population and to induce them to work together.

Conclusion

The work to date indicates that it is a relatively simple matter to introduce appropriate technologies into planning and policymaking offices at a local level. In fact, in Indonesia it is clear that a simple GIS approach is far more useful and more popular than many of the high-tech systems that have been promoted by other donors or by the central government. Single variable displays are easily produced and understood. More effort is required, of course, to teach people the more complicated methodologies required to indicate relationships between variables. This, and the quality of available data, remain the most significant hurdles to be overcome in PEMS sites.

In effect, organizing the monitoring activity on the ground is the most difficult part of the project. But it is at that point that the real success of the project will be found. If local people can be given a new way to look at their population environment relationships, they may also be given a greater capacity to take into their own hands the management of those relationships. It is here that the "local action" of the common slogan will work. Our aim is to develop a technology that will truly permit local people to think globally, and to see themselves as part of that global environment, so that their local actions will be well-informed, and will lead toward a more healthy ecosystem, or toward more sustainable development.

It is commonly noted that generating sustainable development, saving the earth for future generations, maintaining a responsible stewardship, or protecting the viability of the earth as we know it will require a substantial change in the way we live, work and play, reproduce and consume. The "we" in this case is highly variegated. It involves collectivities from world-wide organizations to regional, national, and local governments, to small villages and families, as well as individuals. All will have to change patterns of behavior in critical ways to provide for future generations. International and national policies will be required to control large-scale producer organizations. But much of the change will need to occur at a much smaller scale: by individuals, local groups, and agencies.

Change at this level is well captured by one of the most common of the slogans for environmental protection: think globally and act locally. Acting at any level depends on perceptions, which in turn are affected by the monitoring technology being used. There is now extensive work under way to develop monitoring technologies that integrate the wide variety of data needed for effective environmental protection. Most of this development, however, as we have seen, requires large scale sophisticated electronic equipment.

The PEMS approach, on the other hand, attempts to develop a simple, portable system that can permit local groups to integrate different data sources into one system, which also permits graphic and mapping display. This will allow local action to be based on a greater local monitoring capacity. Also more regional level action can be pursed by higher level agencies by analyzing the collected data of a number of local areas. This gives PEMS a useful "nested" quality to its applicability with the potential to assist planning at many levels.

Section V

Summary, Conclusion and Next Steps

Gayl D. Ness
William D. Drake
Steven R. Brechin

This final chapter contains an analytical summary of the volume that is organized around three main themes: a) issues of **scale and complexity**, b) the concept of the **ecosystem** and the problem of **boundaries**, and **c) human institutions** and their deliberate interventions in the population-environment dynamic. Rather than simply restate what is said above, the summary attempts to draw general observations that emerge from the chapters. Following the summary is a discussion of a research agenda for the future. The agenda will follow the aforementioned three themes. In each case, we shall focus attention on taxonomies, empirical observations, and causal relationships. We begin our summary by first reconsidering the paradigm set out in figure 16.1 in chapter 1.

Paradigm, Scale, Boundaries, and Institutions

Paradigm

It is clear that the population-environment dynamic is not as simple and straightforward as presented in figure 16.1, which shows an unspecified two way relationship between two equally unspecified elements: population and environment. The chapters illustrate that the overall dynamic is extremely complex, dependent upon a number of interrelated mediating factors, and displaying an identifiable range of outcomes. From our review, we can begin to specify population, environment, and the dynamic relationships with a bit more detail. Our attempt is portrayed graphically in figure 16.1.

Figure 16.1 provides a more detailed description of population and environment which is essential for determining and understanding the interactions. Population can be defined demographically by six basic variables: births, deaths, migration, size, composition, and geographic distribution. All but age/sex composition have been discussed by our authors.

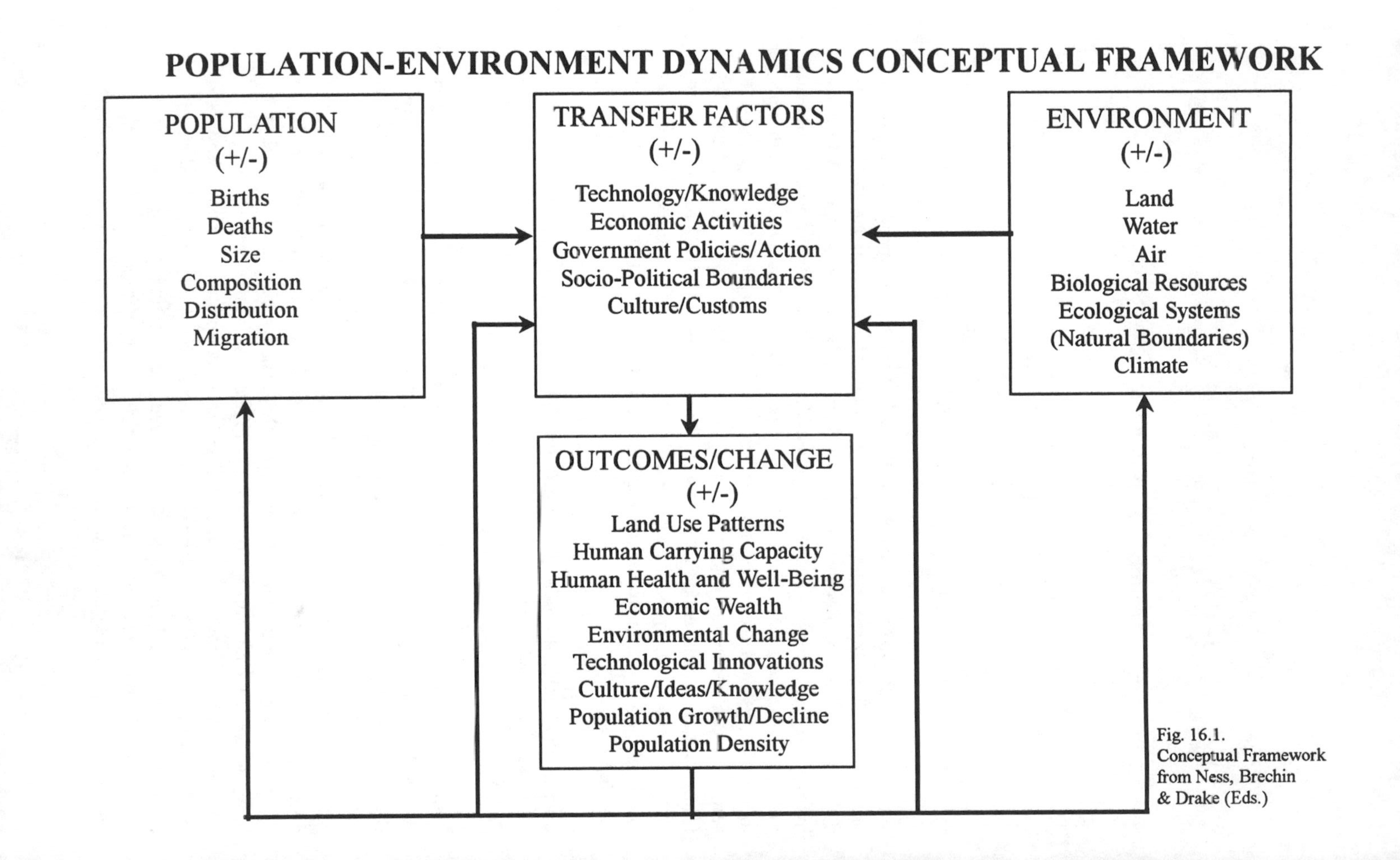

Fig. 16.1.
Conceptual Framework
from Ness, Brechin
& Drake (Eds.)

In addition, the writers have elaborated other social dimensions of variance by which populations can be defined and described. Populations are organized in different ways: by race, class, language, religion, nation, and geographic region. They also show different compositions of individual skill and different types of economic or productive activities. All of these conditions reflect an impact of environmental conditions and, in turn, have a profound impact on the environment. Most of these dimensions of variance have been used in our chapters to examine a specific case of population-environment dynamics.

The environment has also been more fully elaborated in our chapters. We are dealing with a total global ecosphere, divided into ecosystems, defined somewhat arbitrarily by the specific intent and technology of any given observation. Both ecosphere and ecosystems include the land, water, air, all biological resources (flora and fauna), and climate. Ecosystems may be identified by their ecological or natural boundaries, forming three dimensional spaces of highly interactive elements that can be conceived of as living organisms. When we deal with the human population, however, those systems are increasingly bounded by social constructs, and political boundaries. Just as ecological boundaries, these political lines define distinctive patterns of density and mark transition points between patterns of high and low resource or energy transfers. Though these boundaries are conceived in the minds of human beings, they are no less real in their impact on the population-environment dynamic. If anything, they may be, in fact, more dynamic and variable, for as has been noted often in sociology, anything that is socially defined can be socially redefined.[1] Today it is becoming increasingly clear that human society cannot separate itself very far from the sustaining ecological processes. Likewise, the sanctity of the nation state, codified in international law, is giving way to the recognition that state boundaries do not control pollution, and states by themselves may not be capable of protecting the global environment. In effect, the observation of global change is creating a new set of social definitions of the world's working boundaries, political and cultural, as well as ecological.

The two most distinctive elements emerging from our analysis and shown in figure 16.1, however, are the intermediate or mediating variables, and the concept of a distinctive outcome. The absence of arrows directly connecting population and environment is quite deliberate. That connection, we argue, works only through the mediating conditions. The *mediating* conditions include: government policies and actions, political boundaries, the tools, techniques and machines used by people, culture, and all other human institutions and creations. These intermediate variables help to define or shape the relationship between population and environment and to produce specific

---

1. The events in Europe over the past three years remind us of how quickly and dramatically the social definitions known as political boundaries can change.

outcomes, such as environmental degradation, population growth, and land use patterns. The human carrying capacity of the land, for example, must be viewed as the outcome of a complex set of environmental, socio political, and technological conditions. The mix and character of these conditions will define how many people will be supported by the land, at what level of well-being, and at what costs to both current and future resources of a specific set of ecosystems. In turn these outcomes work through numerous feedback loops to influence the intermediate variables as well as the basic population and environmental conditions. Undesirable outcomes may imply population decline, or may spawn new policies and actions either to promote population growth, or to maintain an equilibrium and protect future resources of an ecosystem. Technological improvements such as modern medicines or better agricultural practices, may greatly reduce death rates throughout the world, which then produce a new stress on the ecosphere and on specific ecosystems. All of these outcomes can be described as various manifestations of the health of an ecosystem, discussed below, either in terms of current vital signs or its capacity to recover from stress.

Finally, to some, our framework, presented as figure 16.1, may resemble human ecology's traditional POET model (Population, Organization, Environment, and Technology) introduced several decades ago in sociology and human ecology.[2] In some ways ours is merely an elaboration of that earlier paradigm. In particular we have attempted to specify in greater detail what we mean by population, by environment and by their relationships with other variables. There are, however, some important differences. We have placed technology and organization into the same construct. This was done for convenience and could be separated out again. We see, however, technology and organization (or human institutions) as playing a crucial role in mediating population and environmental factors. Instead of a filter, the POET model tends to represent technology and organization more as separate and equal variables. The result of the interaction between population and environment, and the mediating variables is represented by specific outcomes and changes. The POET model does not clearly specify mediating conditions or the outcomes. Most importantly perhaps, is that our framework is not based upon a notion of balance or equilibrium. Instead our feedback loops represent cause and effect and responses that may or may not be functional to the entire system or its components.

There is much more to be said about the lessons learned on population-environment relationships from this volume and on where we should go from there. For further discussion of these topics we turn now to the analytic summary.

---

2. For a good summary and critique of human ecology and the POET model from an environmental sociology perspective see Humphrey and Buttel 1982, pp. 41-52.

Scale and Complexity

The chapters reflect a progression from larger to smaller scale, from broad conceptual issues and long historical trends to large scale then to small scale systems. Broad perceptions and long time spans imply abstract constructs, relatively homogeneous conditions, and smooth trajectories of change. Microsystems imply concrete actors and actions, heterogeneous people, groups and environments, and the immediate results of actions, with often radical changes in short periods. That is, the progression in scale is one from greater simplicity to greater complexity, and the potential for relatively more rapid change.

Undoubtedly, scale affects the detail observed, and thus the complexity of the object. Deep among the trees we can examine individual plants in great detail, but it is difficult to see the forest. When we can see the shape and extent of the forest, we necessarily lose sight of its detail. Linking observations at various scales poses difficult conceptual problems in any analysis. Every general characteristic of a population or environment is made up of many minute details, most, but not all of which are accurately represented in the general condition. There are also minute details or actions that diverge widely from the general. Thus, understanding both the general and the specific requires that they be linked together in observations. This implies a capacity to make observations at different levels of aggregation, and to integrate those observations into a large whole. We cannot know, for example, whether the migrations and family sizes observed in Buen Hombre, Lahat, Locknevi or Tannougou are local deviations from or accurate representations of national level population dynamics unless we can aggregate observations from all the Buen Hombres--of the Dominican Republic, the Caribbean or the World--in one data set.

Such aggregate data collection is, of course, precisely what nation-states and the emerging world community have been doing with increasing skill and energy over the past century. It is, in fact, the exponential rise of this new capacity for observing human and natural conditions that has led to worldwide concern for the potentially destructive nature of the current population-environment relationship. Censuses and large scale sample surveys have permitted us to see how whole collections of individual or small group actions add up to a distinctive character of the population-environment relationship. This reflects a general rule to which we shall return often. Monitoring affects behavior. New monitoring technologies imply new ways of seeing, which in turn imply the identification of new problems, and new forms of population or environmental stress. These stresses, as Kartomo has suggested for Indonesia, often lead to actions specifically directed at addressing or adapting to the new stresses.

Viewed from different levels of scale, the population-environment dynamic takes on different characteristics. Far from being mutually exclusive or contradictory, this progression through scale can give us a fuller and richer picture of the entire population-environment relationship. The progression, or the impact of scale on complexity is evident in our three basic elements: *population*, *environment*, and the *population-environment dynamic*.

Human Populations

Populations can be viewed in the abstract as the total world population, which until recently has shown only very slow rates of change. As we saw in Ness' chapter 3, the human population grew very slowly until about 1700, then experienced increasing growth rates until just a decade or two ago. The absolute numbers rose only very slowly throughout history, but have shown exponential growth over the past three centuries.

As we examine specific situations, disaggregating the total into regional or geographic units, those populations become differentiated along a number of dimensions. Even the overall growth rate is differentiated by time and place. Different world regions, marked by rather different population and environment characteristics, share a common role in the overall pattern. All grew very rapidly in a short period of time, but they did this at different times, and with different trajectories.

When we move to smaller scale, we find even more variance and more dimensions of variance. The quality of life varies, marked dramatically by death and disease rates, standards of living and levels of energy consumption. Populations are also differentiated by resource control, or social class, by ethnic identities and languages, and by the authority systems that affect their lives and to which they are variably connected. At the micro level of individuals and small groups, populations are differentiated by such things as families or kinship groups of different size and resource control by the specific character of their activities, or the work they do, and by the skill they bring to that work. Not only do we see greater detail, we can also see more rapid change.

None of the characteristics that mark these dimensions of variance is written in stone. All change over time and space. Numbers and rates of growth change, for both the whole and for small units. The quality of life, energy consumption, resource mobilization, even ethnic identities and the structure of families change. The character of work and the skill people bring to that work, or the resources they control can vary with time. Perhaps the most important type of variance in all of this is the rate of change for any of the characteristics. It is both a fundamental cause and a significant consequence of other conditions. We shall return shortly to this issue of the rate of change.

There is also a much more immediate and conflictual character to the interactions among populations that we can see as we move from the larger to the smaller groupings. Historically, western expansion caused the demographic collapse of the native American populations in a process that stretched over a century. We are reminded of the underlying potential for conflict in migrations today. For example, Fulani herders intrude upon subsistence agriculturalists in Benin, bringing increased pressures on the land and raising old patterns of conflict to new levels of intensity. Recent migrants intrude upon South Sumatran villagers, changing the forest use almost daily. Population growth and resource constraints force some people out of specific environments while others remain. The conflict is real and immediate with daily negotiations between groups and individuals. South Sumatra is not alone in experiencing bloodshed over issues of population growth, migration and changing land use. Teitelbaum and Winter remind us that these conflicts have also had an enduring place in the way people think about population-environment dynamics.

Environments

The environment can also be viewed as highly abstract and homogeneous conditions, land or forests that imply both opportunities for and constraints on a population. From a distance, a landscape may appear homogeneous and changing only gradually over long time spans. On closer inspection, land cover, quality and crops can vary radically within relatively small areas. Further, the heterogeneity of the smaller landscape has both natural and human dimensions, as precipitation, watersheds, or soil quality vary in complex interaction with irrigation systems, planting technologies and population densities. Thus, macro and micro perspectives can generate very different perceptions of the environment. Our most recent developments, from satellites to electron microscopes, have permitted us to see the entire earth system from outer space at the same time that we can see more fully the molecular make-up of all life. Drawing these diverse observations together to form an integrated picture of the earth's system has become the urgent and pervasive task of a burgeoning new interdisciplinary field, Global Change.

As with population there is also a qualitative variance in the environment, which can be seen in both large and small scale. It can sustain more or fewer people at any one moment. World population density has grown radically along with population growth, but it has consistently been much higher in Asia and Europe than in Africa or the Americas. Further, the environment can be degraded or enhanced, reducing or increasing its capacity to support both human and other populations in the future. Recent increases in the world community's ability to observe these varying and changing capacities has led us to change radically our definition of the environment.

As Teitelbaum and Winter noted in chapter 1, nineteenth century economists saw the environment as a limitless expanse of resources to be exploited for human benefit. If such exploitation proved harmful to people or other species, it usually went unrecognized because the exploited essentially had no voice. The new observations of degradation have led us to see environments as fragile systems in which today's exploitation can mean a future of barren landscapes. This leads us to search for long term sustainability in our use of the environment.

The Population-Environment Relationship

The population-environment relationship itself can be seen in a number of ways. Environments viewed simply as resources have led to the view of populations which can exercise adaptive strategies. Thus populations in general can be seen to generate technologies and forms of organization aimed at increasing the carrying capacity of their environment, exploiting resources, permitting population maintenance or even growth. This has happened on a global scale over a millennium, bringing high levels of population density and resource consumption that may now threaten the very survival of the human species and perhaps even the entire planet as a living system. At the micro level we have also seen specific populations outgrow their environments, implying a successful adaptation, but also thereby forcing some members out of that specific environment. This, too, has occurred over long time periods in the past, but it is also a very immediate action that can be observed in specific locations on a near daily basis. It occurred over almost two centuries in Southeast Asia, in one or two generations in late nineteenth century Sweden, and in Benin and South Sumatra we see it happening as we make field observations.

A population's perception or definition of the environment itself can change, however, with dramatic effects on the population-environment relationship. In this sense, the view of the environment that Teitelbaum and Winter attribute to the nineteenth century must be seen as distinctive to a particular time and place. The rise and spread of industrial capitalism saw an emphasis on efficiency, deriving from a pervasive utilitarian perspective (Hays 1959). It was associated with the view of the environment as resources to be exploited. Nature was to be tamed and overcome to the greater glory of mankind. That view has changed dramatically only in the past few decades. Since the movements of the 1970s, the environment has come to be seen as an ecosystem, implying a functional interdependence of parts in a living system. This has entails a forward perspective, which has led to consideration of the sustainability of the planet (Pursell 1973; McCormick 1989).

The distinction between environment as resources and environment as ecosystem has led to a new perception of a qualitative variance in the population-environment dynamic. Interactions can enhance the health and well-being of a population in a productive system that can be sustainable for the foreseeable future. But they can also be viewed as destructive of both populations and environments, producing lower living standards, and certain deterioration for the future as well. We suggest from these observations that an ecosystem in which relatively high living standards are achieved today and protected for the future is a more healthy ecosystem than one in which we have lower living standards either today or in the future. Some of the chapters illustrate the emergence of deliberate actions designed to sustain a more healthy relationship into the future. That is, the population-environment relationship itself comes to be the object of deliberate future oriented actions, or public policies. We have seen such policies both in the forest protection of traditional microsystems, and in emerging larger scale national and international systems.

Drake also shows us that the relationship between population and the environment can be conceptualized essentially as a story of transitions, or the transformation from one equilibrium to another. Populations change to adapt to environments and environments change as populations adapt. The most well known transition is that in which a population fills an empty environment, growing slowly at first, then accelerating, and finally slowing the growth rate to reach some form of stability. Populations may also decline, however, rather than reach stability, as they are crowded out by others that adapt more effectively to the environment. We have seen these waves of succession occur over millennia and among many species of plants and animals, and among different socially defined groups of humans as well.

Our modern age, however, has seen a radically new set of transitions associated with the rise of the human species and the use of fossil fuels. This has led Drake to propose the concept of a family of transitions that marks our age as distinctive. In these transitions, the human species has shown immense capacities to dominate its environment, and to alter it sufficiently to threaten future life for the species. The transitions themselves have become differentiated by what can be called their impact on the health of the ecosystem. It is quite possible that certain elements of the current family of transitions will lead to a less healthy system, marked in the extreme case by the collapse of the human species.

Two aspects of this family of transitions have gained great salience: their speed and interconnectedness. Drake suggests that the slower speed of past transitions permitted adaptive feedback mechanisms to work well enough to support a healthier ecosystem (one that could support present as well as future populations). The high speed of the current transitions places heavy demands on the feedback mechanisms and may delay effective adaptive behavior until irreparable damage is done to the environment, destroying its future supportive capacities.

Further, different parts of the family of transitions are differentially connected to one another. Close interconnectedness appears to increase the stress within the transitions themselves. The modern scene is especially stressful because many members of the current family of transitions are closely interconnected. The demographic and epidemiological transitions, for example, are tied to the agricultural and toxicity transitions. This makes it more difficult for any single transition to work itself out easily and to find an equilibrium that supports both current and future populations.

Thus the idea of transitions as a central characteristic of the population-environment dynamic leads to a specific series of questions on the characteristics of those transitions. It also leads to new questions about the causes and consequences of those transition characteristics. That discussion will follow shortly, but first we must make two other summary observations.

The Boundaries of the Ecosystem

The concept of the environment is a messy one in part because its boundaries are always somewhat arbitrarily defined. When we deal with the atmospheric changes that threaten to produce global warming, the earth's atmosphere provides the boundary. The entire planet is the unit of analysis. Here is an ecosystem where dynamic interactions are found between physical elements, gasses and living systems, which produce an equilibrium of earth's temperature within a remarkably small degree range, persisting over millions of years. To the social scientist this is a relatively easy system to grasp. Its boundaries are clearly marked, it has a relatively manageable number of sources and sinks for the gasses that are critical for maintaining the equilibrium, and many of the physical and chemical theories, are both simple and possess considerable explanatory power. Even here, however, there are conceptual issues of the boundaries that make the idea of the environment theoretically messy. Should the sun be considered an exogenous, or an integral part of the ecosystem? That is an issue that is often solved in a somewhat arbitrary and ad hoc fashion. This makes the boundaries of the system being observed somewhat arbitrary. The boundaries are in large part conceptual constructs. The particular construct we create, however, raises questions of the implications of the decision.

The issue can often be avoided simply by focusing on a somewhat artificial or constructed category of environments or activities. The fact that these categories are constructed, however, does not mean that they are without profound consequences. One can deal with the human habitat, as inferred from Teitelbaum and Winter's historical thinkers. Conceptual boundaries can also cover all agriculture, or tropical forests, as in the analyses of Ruttan and Grainger. When we move to geographically and historically specific situations, however, either in large or small collectivities, the problem of boundaries and the definition of the ecosystem becomes more acute. In nation-states we deal with boundaries that are political as well as natural. Within the political boundaries there are interactions as distinctive as those found within natural boundaries, where ecosystems are most commonly marked by variations in interactions, which are often reflected in the density of specific plants or animals. Ness argued that the political boundaries are of greater importance than the natural ones in determining the changes in rice production or human reproduction in Southeast Asia. At the micro level, Brechin saw the boundaries of a protected area violated or altered from a combination of international market mechanisms reflected in increased prices for coffee (a result of inelastic demand and reduced supply from damaged crops in Brazil), and as the result of a growing nation-state intruding into traditional authority systems that had defined those boundaries in the past. This change of boundary definitions, coming from changing political institutions, and market intrusion were found to have a marked impact on the way the population interacted with its environment.

The theoretical messiness of the concept of the ecosystem also has roots in what we can call the technology of observation. The ecosystem as a unit of analysis has always been defined in an ad hoc fashion, by whatever conditions, living or non-living, the observer chooses to recognize. That choice, of course, is very much the product of our technical capacities to observe specific things. The entomologist constructs ecosystems quite different from those of the atmospheric scientist. In this sense, ecosystems have an arbitrary quality, determined largely by the monitoring technology being used.

But if ecosystem boundaries are shaped by distinctive capacities to observe, they are also open to redefinition as we construct new monitoring technologies and learn how to see other things, or different connections. Recently, for example, the entomologist and the atmospheric scientist have come together as one deals with the source (termites) of a greenhouse gas (methane) important to the other. In the same fashion agronomists, chemical engineers, health scientists, and atmospheric scientists are brought together to consider multiple actions that have important implications for other ecosystems.

That is, ecosystems, as any other object, are in part socially defined, and anything that is socially defined can be socially redefined. We have seen that a large part of the driving force behind that social definition and redefinition is the technology of observation we bring to the world. As we develop new tools of observation, we see new things and those perceptions change our notions of the boundaries of the systems we observe.

If this makes it difficult to define the ecosystem and its boundaries with any long lasting precision, then we must simply live with a messy theoretical system. The relationship between any population and its environment is a rich and complex movement of things. This becomes amply evident as we progress through the chapters of this volume. They remind us that however deeply we may get into any specific set of observations, we will do well to remember that what we see is in part a function of the ecosystem boundaries implied by the deliberate act of observation, i.e., we should always be attentive to what we are not seeing. Especially when we are trying to understand population-environment relationships, we must remain open to the possibilities of important conditions and connections that can be seen only with the use of different monitoring tools, which imply different boundary constructs. That is, our efforts at understanding must be interdisciplinary.

Institutions and Interventions

Most of the studies in this volume illustrate collective human actions that are deliberately focused on the future and on some broad or specific goal of managing the population-environment dynamic. That is, we are witnessing public policies at work. Policies governing agriculture, forest use, energy production, human fertility, mortality and health, and trade have all been formulated and have had a marked impact on the relationship between population and the environment. Policies can work, deliberately or even inadvertently, either to promote or decrease the life sustaining capacity of the environment. The recent success of the human population in dominating its environment has given rise to new and urgent concerns for policies that support the environment. Environmental and population policies aimed specifically at a sustainable relationship between the two have become the focus of much new collective action.

The chapters also illustrate remarkable changes in these policies. The historical emergence of the state with its policies to stimulate economic development, thus dramatically changing the character of the environment and its carrying capacity is perhaps the most important of these recent changes. Here, too, however, we have seen more detailed changes as we move from the broad historical to the more immediate settings. The past policies promoting the unabashed exploitation of the environment, for example, have given way only in the past few decades to policies more concerned with the protection of long term sustainability, however incomplete its definition. Perhaps the most remarkable change, however, has been that from pro- to anti-natalism in population policy, which has occurred only in the past four decades. This recent revolution reverses centuries of public policy in support of larger populations. These policy changes both derive from and produce increased technical and organizational capacities to monitor and to manage ever larger portions of the population-environment dynamic, implying an increased centralization of control over larger and larger populations and territories.

In the progression from macro to micro systems, these broad and radical changes become highly differentiated, moving in different directions and at different speeds. The move from pro- to anti-natalism, for example, began in

Asia, then moved through Latin America and is only now emerging in Africa. Further, even in Southeast Asia where on the whole, there has been a rapid regional movement, some countries, such as Burma, remain staunchly pro-natalist (see Ness). Others, such as Malaysia have moved from pro-natalism to anti-natalism to an ethnically mixed set of policies. McIntosh examines both population and agricultural policy changes in Zimbabwe, which currently leads all of Sub-Saharan Africa in these policy changes. Martine has shown how Brazil's national agricultural policies have changed over time, with marked impact on the population-environment relationship, whose outcome is deleterious for the poor, highly beneficial for the rich, and generally adverse for the entire population. Such differentiated policy conditions can be found throughout the studies in the volume, forcing us to recognize the importance of human political institutions for mediating the current population-environment relationship.

As we move from macro to micro levels another aspect of institutions becomes important. If the state implies increasing centralization of control over the population-environment relationship, its control is also subject to local limitations. Local institutions can impede or support state policies. In the most successful policies, this has led to an increased emphasis on local participation. Policies established and promoted by large central governments may be capable of mobilizing massive resources, but their implementation, and their adaptation to varying local conditions requires that small scale groups of people gain an active role in carrying them out. Whether local groups and conditions support or obstruct these central policies becomes a major problem, both for scientists seeking to understand processes, and for political leaders attempting to control them.

At the same time, this concern can be found at a different scale. Ruttan notes the importance of macro-micro linkages beyond the nation-state and its local areas, i.e., at the international level. Ruttan argues convincingly that in order to continue improving agricultural production worldwide, serious adjustments need to be made to our present organizational efforts. The institutional arrangements that gave rise to the Green Revolution are insufficient for future advances. Agricultural innovation and increased production will require more attention to critical differences found among scattered micro-settings. Consequently more decentralized agricultural research programs will be necessary, but under a larger umbrella of coordinated efforts.

Though the studies presented here illustrate the power and importance of human institutions, they do not in themselves constitute a coherent theory of institutions, policies, and the population-environment relationship. Here we can present a series of propositions inferred from the chapters in an attempt to develop a more systematic discussion of human institutions and population-environment dynamics.

The Health of the Population-Environment Interaction

First, we argue that it is possible to conceive of the population-environment dynamic as an outcome that can be assessed at any one time as more or less "healthy". For the human population more healthy conditions are those in which common indicators of death and disease are reduced, and in which human individuals have greater opportunities to develop the talents they possess. For the moment, the conditions under which North Atlantic, Oceanic, and Japanese populations live is more healthy than those under which, for example, South Asian populations, live. As a first approximation, measures like infant and maternal death rates, levels of education, and equality of wealth distribution serve as rough indicators of this variance.

This formulation need not raise questions of the conflict between human and other species, or between the future and current quality of life. To avoid this conflict, however, we must conceive of the environment not as resources to be exploited but as an ecosystem. This assumes that the human population is merely a part of a larger system and that human health is not possible without the health and sustenance of other species in that ecosystem as well. As previously noted there is a rising human perception of the environment as ecosystem. However, continued political support for environmental exploitation attests to the fact that this perception is not yet universal and perhaps not even dominant in public policy. Nonetheless, the concept of the population-environment dynamic outcome as an ecosystemic condition permits us to conceive of the health not only of the human population, but of the entire ecosystem of which humans are a small part. That is, we assume that we can conceive of an outcome of the population-environment dynamic in which human activity is not destructive of the ecosystem.

This assumption provides a criterion against which to assess the health of the current population-environment relationship: it implies current standards of living which are high, but do not compromise future levels. If we accept this assumption, however, we are left with the question: what kind of institutions produce such an outcome?

The Primacy of Political Institutions

Of all forms of human institutions, political institutions today are of primary importance in determining the character of the population-environment interaction. Political institutions are those that protect a society from external attack, maintain internal order, articulate interests, establish collective goals, and mobilize resources to achieve those goals. Such institutions are typically called *government*, though that term often does not exhaust the institutions that perform political tasks. Furthermore, these tasks need not be played by differentiated institutions that we can identify as "government."

The primacy of politics is especially evident in the character of past and present demographic transitions. From the eighteenth through the early part of the twentieth century, mortality declines and subsequent fertility declines were driven primarily by technological changes and market forces, by expanding trade and economic development, which were associated with changes in reproductive norms. To be sure, governments in the nineteenth century played increasingly important roles in mortality reduction through public health activities, but until mid-twentieth century these were probably less important than economic growth and trade expansion. Similarly, and more strikingly, the fertility declines that came in the western world, and in Japan before 1950, often came *against* the stated interests and policies of government.

On the other hand, the mortality reductions that followed World War II came very much from both national and international political institutions that created networks for the distribution of the new mortality reducing technology developed in association with the war. The same is true in the case of modern fertility decline. In the Third World, where governments have taken strong stands, creating effective fertility limitation policies and programs, fertility has declined far more rapidly than it ever did in the past of currently industrialized countries. China and Thailand are perhaps the two most dramatic examples, but there are others as well. Taiwan, South Korea, Indonesia, some of the states of India, Tunisia, Zimbabwe, Mexico, and Costa Rica can be cited as important examples where government activities have had a strong depressing impact on fertility.

There is also, of course, rapid fertility decline driven by market forces and normative change as well. In Hong Kong and Singapore, government action complemented more powerful economic changes that were producing both mortality and fertility reductions. In Brazil, fertility has declined in the face of government opposition to modern contraceptive technology. In that case, however, it is clear that the urbanization transition has been a major cause of the spread of fertility limiting technology. Thus, rapid economic development and urbanization can also drive fertility down, even in the face of government opposition. But even these development forces work more swiftly, and with less cost to the health of mothers and children, when governments play a strong and supportive role in fertility limitation. Finally, even where there is little or no economic development and urbanization, as in Sri Lanka, Kerala (India), and most dramatically in Bangladesh's Matlab Thana, political institutions can still produce substantial fertility reductions.

Political institutions are also of great importance in driving environmental protection or degradation. The environmental disasters of Eastern Europe and the former Soviet Union are the most recent and dramatic illustrations of the high cost of certain forms of political action. But the examples can be extended much further as well. First and Third world government agricultural policies, commodity and input subsidies, support for research and development, and even legal constraints or supports for private sector research and development are all demonstrably important in determining agricultural outcomes. The same can be said for other environmental outcomes as well. Japanese energy efficiency and US inefficiency are both closely tied to government policies that range from fuel prices to research and development policies.

Thus, political institutions are of critical importance in determining the health of the current and future population-environment interactions. It is now necessary to identify the specific conditions of political institutions that determine the health of these interactions. Such identification is rendered difficult by the great importance of history, time, and ecological conditions in determining how political institutions work in any specific locality. Nonetheless, we can attempt some generalizations from the studies in this volume as we propose a research agenda for the future.

Future Research: A Transition Perspective

The first step in formulating a theory of transitions was proposed as a method of helping to deal with the complexity of population-environment dynamics (see Drake's chapter). Although we are some distance from formulating a theory of transitions, identifying this pattern of change provides us with a distinctive *perspective* on the population-environment dynamic. This perspective suggested that for many sectors in a society, we can identify a critical transition, in which conditions move from one equilibrium to another through a period of rapid change. We can also observe a number of other characteristics associated with those transitions. A fuller understanding of those transitions and their common elements can help in the study of population-environment dynamics. That is, this perspective of transitions can be used to address the basic problems identified above in the three major themes: scale and complexity, boundaries and ecosystems, and human institutions. Thus we use the transition perspective to suggest an agenda for future research.

Scale and Complexity

The first task is to encompass the full range of population-environment dynamics at all scales while keeping their great complexity manageable. This can be done by focusing on the critical observable transitions of our era. To deal with these transitions more systematically, a number of basic steps can be identified.

Developing a Taxonomy of Transitions

Nine transitions were presented by Drake, providing examples from different sectors of society. Although these examples cover a wide range of conditions, they are by no means exhaustive. Therefore, an important next step is to develop a taxonomy of types of transitions and a nomenclature useful in their classification. For example, we have already suggested two broad categories: 1) transitions typifying different sectors in society, such as the demographic or urban transition, and 2) those representing the means of change, such as the toxicity transition. Within each broad category, there are also logical subgroupings. For example, the toxicity transition is actually a grouping of transitions representing many different types of toxins or pollutants emitted into the air, ground water, surface water, and the earth.

As this taxonomy is developed, care must be taken to distinguish between the ever-present changes occurring in society and the specific change represented in a transition as defined here. The key aspects of transitions as we define them include an initial condition of relative equilibrium, moving from slow to rapid change, and returning again to relative stability, generally at a different level. Second, a transition period is accompanied by an imbalance and/or change between key variables that is, in some way, unusual. Returning to relative stability after the transition does not necessarily mean that the determinants or outcomes of the new equilibrium are the same as before the transition. For example a transition in the forestry sector may result in more forests or a desert. What makes a transition take different forms?

Obtaining Evidence of Transitions at Different Levels

Some readily available data on transitions were presented in Drake's chapter. These data portrayed transitions at the regional, national or international level and were obtained from secondary sources. For example, data on the deforestation transition were arranged by region of the world over the period 1850-1980. We suspect that transition data of this type, which are aggregated to regional or higher levels of government, often mask the degree of change occurring at the local level. The volatility of change probably is much greater locally, but as local transitions are combined with others occurring at slightly different time periods, the data appear to show a more gradual change spread over a longer time span. For example, as a region is deforested, logging in any single local area occurs over a relatively short period and results in a rapid local transition. As one area is logged, foresters move to another locale to repeat the process. Then when deforestation data are gathered for the entire region, logging rates per unit of time can appear modest, simply because all local areas are not logged simultaneously. In short, for some sectors, it is quite probable that local transitions are much more volatile than regional data suggest because time lags and changes in scale dampen *apparent* change as data are aggregated. It is also possible that there are other phenomenon whose rates of change are masked by aggregated data. Therefore, a next step in developing transition theory as a useful analysis tool, is to obtain data representing transitions at all levels of aggregation in order to explore whether this masking effect exists and, if so, how to better interpret regional and national level data. Of course, it is also possible that some sectors experience transitions that behave in the same way at all levels of aggregation. Each of the types of transitions identified in the taxonomy should be explored empirically to determine how universal the masking effect is. Here the Population-Environment Monitoring Systems (PEMS) described in an earlier chapter could be very useful. By focusing on local-level population-environment dynamics through integration of information from multiple sources, PEMS can help to fill in the knowledge gap on local transitions. In addition to affecting the perception of the rate of a transition, this aggregating problem also touches on the issue of boundaries and ecosystems, to which we turn later.

Obtaining Evidence on Regions of Multiple Transitions

One of the key arguments for viewing population-environment dynamics as a family of transitions is that transitions often occur simultaneously in several sectors. Therefore, we must obtain empirical evidence on *geographic regions* that are experiencing multiple transitions at the same time. In this case, an attempt should be made to identify changes in all the important sectors for each of several local settings. These locales then could be classified by the degree to which they indicate transitional volatility. If earlier contentions about heightened societal vulnerability during transitions are valid, then there should be indications of such risk.

Because of the field-level difficulty in implementing such studies, it is unlikely that the status of all sectors could be determined for a given region in any single project. Furthermore, because the rich cultural and other local contextual characteristics of any location are so varied, such geographically bounded studies should be replicated many times. It is interesting to note that this type of study is similar to case studies done in many fields, including geography, natural resources, public health, anthropology, and public policy. Consequently, another possible approach to this inquiry would be to draw upon already existing studies in these fields, but recasting them in terms of transition theory. Because of PEMS basic mission to integrate data from various sectors at the same time, the PEMS system could help to provide the information required to better understand the effects of multiple transitions within a particular local region.

Identifying Causes of Transitions

Perhaps the most important and potentially beneficial next step in the development of a theory of transitions is identifying their causes. Understanding why a transition happens is fundamental to knowing how to modify its trajectory. If it is true, as we believe, that societal vulnerability is often higher during periods of rapid change, then in some circumstances, it may be prudent public policy to slow the transition if it is possible. At other times, it may be more helpful to predict the timing of rapid change in order to plan for the infrastructure needed to accommodate new conditions following the transition.

The impact of such studies on policy is well illustrated by the case of the demographic transition, and especially of the modern fertility decline that will end the transition in the Third World. There is an extensive debate over the causes of fertility decline, with direct implications for policy. The "demand side" theory posits that when major socioeconomic change occurs and the value of children declines, people will find ways to control fertility by themselves. However, it is unclear at this point whether fertility reduction is caused by improvements in education and affluence or by the clear identification of opportunity costs to the family in having additional children. Education and affluence are traditional hypotheses, but recently there is mounting evidence favoring the opportunity cost hypothesis. It is quite probable that public policy which attempts to modify the trajectory of the demographic transition would differ depending upon the extent to which the competing hypotheses are true.

Studying Linkages Among Sectors

In Drake's chapter it was posited that there are linkages or interactions between different sectors of society, which often alter transition trajectory. These interactions may amplify or dampen a transition. An example of an amplifying interaction is the recent urbanization transition in Mexico and Eastern Europe, which has occurred at the same time as the severe toxicity transition brought on from uncontrolled industrial production. Existing infrastructure has been overwhelmed, thereby amplifying adverse effects, which then become apparent in the epidemiological transition.

An example of a dampening effect among transitions occurred in the United States near the turn of this century. At the same time that a major urbanization transition was occurring in many cities, there was a significant educational transition taking place. Primary and secondary school coverage was being extended rapidly in both urban and rural areas. Consequently, rural-to-urban migrants were better equipped to deal with their new environment and urban regions were less severely stressed than might have otherwise been the case.

As our society has modernized, there has been a tendency for sectors to become more intertwined with one another. We have seen that this interrelatedness often extends across both sectors and scales. The opposite can occur as well, as intersectoral connections can also be reduced. An example of a weakened association is the traditional link between agriculture and forestry. It can be argued that the deforestation transition is historically related to the agricultural transition because expansions in land under cultivation often came at the expense of forests. Now, however as increases in agricultural production are derived mostly from intensification of land already under cultivation, this linkage between agriculture and forests is less apparent. It is therefore useful to keep in mind that the strength of relationships among sectors is by no means constant over time. In some cases connectedness between sectors is increasing and in others it is decreasing.

In summary, interaction among sectors can be either positive or negative and can either amplify or dampen each other. Furthermore, the degree of their interaction can strengthen or weaken over time. Thus, another dimension in the development of a theory of transitions is to explore these linkages, perhaps on a case-by-case basis. We contend that an important new arena in the study of public policy is to better understand when and how different sectors interact so that policy initiatives can be structured to take advantage of positive synergisms when they exist. What are the circumstances under which these associations amplify adverse effects and when do they dampen impact? Which linkages between sectors are increasing in strength and which ones are diminishing? Are there circumstances under which public policy can have a positive societal effect by encouraging joint development which either dampens adverse impact or reduces rates of change so that natural corrective processes can operate better?

This area of investigation, crossing traditional sectors, offers a useful connection between transition theory and risk analysis. Transition theory provides the calculus for looking across sectors and risk analysis offers the means for comparing alternative intervention strategies, which also span more than one sector.

Describing Transitions with Mathematics

The word "theory" has been used here to mean a sort of broad systematic strategy for linking similar observations concerning the behavior of different types of transitions in order to seek useful similarities among them. This real-world approach to theory shares with its mathematical counterpart the necessity of developing a logic for connecting related ideas. The mathematical approach typically consists of creating theory as a logically connected body of theorems. The real-world approach often does not prove theorems but seeks a similar style of connectedness.

Thus, an additional step in the research agenda would be to find a way to express real-world transitions in the language of mathematical theory. Mathematical models provide an entry point to such theory just as empirical evidence can provide entry points to real-world theory. Because curves of different shapes describe different transitions, a model that examines the geometric dynamics of the set of curves could be helpful. A tool which might be useful in this endeavor is Feigenbaum's graphic analysis from chaos theory--a style of geometric analysis that starts from a condition of geometric equilibrium, then shifts to, possibly, a state of slow to rapid change, then settles down to a new equilibrium. One curve might be well behaved (from the standpoint of geometric dynamics) while another, of just a slightly different shape, might be chaotic. Both fit data from the real-world transition under consideration. Mathematics could offer a way to analyze fluctuating (non-monotonic) data using techniques which themselves mirror the fundamental characteristics of transitions. This could provide reasons to elect otherwise equal choices by selecting the alternative most similar to real-world conditions.

The Appendix to Drake's chapter presented several traditional mathematical models that might be useful as entry points in portraying transitions. To begin, these functions should be fit to empirically obtained data. At least three benefits would be attained. First, if reasonably close fits are achieved, there may be new insights obtained on the character of the transition phenomenon derived from knowledge about the mathematical function that fits the data. Second, comparisons across different scales within a sector and among different sectors would be facilitated. Third, providing that unbounded functions are used, predictions about future rates of transitional change could be made by extending the time variable into the future.

Boundaries and Ecosystems

The *transition perspective* can also be useful in helping deal with the messiness of ecosystems and boundaries. We noted above that the concept of the ecosystem is theoretically messy, because its boundaries are basically defined by the specific monitoring technology, and the specific interest the observer brings to any observation. Further, we noted that some ecosystems have natural boundaries, but there are increasingly important socially constructed boundaries, especially in the form of the administrative boundaries created by the modern state.

The problems of building a taxonomy of ecosystems and boundaries, and of generating systematic observations of population-environment dynamics with that taxonomy, become acute when we try to decide what to observe. Which of the hundreds of classifications of natural or socially constructed ecosystems will be most useful? Which of the scores of human ecosystems shall we examine?

It is here that the transition perspective can assist us. We can use specific transitions as the focus of analysis, and let the scale of these units define ecosystems and their boundaries. This will not solve all of the problems of ecosystem identification, of course, but it will help us to identify those ecosystems that are especially relevant for the transitions that mark so much of current population-environment dynamics.

The cases provided in this volume illustrate specific transitions and from these we can suggest questions that should be addressed by future research. Since the cases in the book have both natural and socially constructed (administrative or state) boundaries, they permit us to examine a wide variety of ecosystems. Finally, recalling that the transition perspective raised the issue of the connectedness of different transition in what is called the modern family of transitions, we can identify ecosystems first by single transitions, then by multiple transitions. Finally we can turn to an increasing problem implied by the rise of the state, where multiple ecosystems, experiencing transitions at different speeds, are contained within a single boundary. We shall see that this is also a problem that characterizes the emerging world community.

Single Transition Systems

We can start with the well-documented demographic transition, which has been completed in the past and is currently underway in many Third World countries. We can then also examine the agricultural transition, following the discussions of Ruttan, and deforestation transitions, following Grainger, Brechin, and Agbo.

Low and Clarke describe two different fertility-migration transitions in nineteenth century Sweden. The boundaries within which these transitions occurred are partly administrative, created by the state and assigned to the church for registering the population. In part the boundaries are also natural ones, since they involve two different river valley systems with different patterns of crops and external trade. In these two ecosystems, we found that a greater diversity of production in one valley cushioned the population against the impact of larger regional or world price fluctuations, and slowed the process of fertility decline.

By contrast, in Benin the recent juxtaposition of an administrative forest-protection boundary in close proximity to a natural ecosystem boundary, of the rock outcropping, provided little opportunity for fertility adjustment during the period of rapid population growth. The result was increased land pressure, increased out-migration and reduced standards of living.

It is possible to suggest from this contrast that administrative boundaries that emerge slowly over a long time period under the gradual extension of state power tend to fit natural ecosystems more closely than do the more recent administrative boundaries imposed by the rapid growth of the state in Third World countries. Evidence can be found in the contrast between meandering European boundaries and the long straight lines (drawn in the sand by colonial rulers) that mark many boundaries in Africa. This leads to a series of empirical questions. To what extent do administrative boundaries match natural ecosystem boundaries? Has this matching, or coterminality, changed over time and does it vary systematically over world regions? Further, does the extent of the administrative-natural ecosystem matching affect the pattern of the demographic transition, or of any other population-environment transitions?

Teitelbaum and Winter described the variety of ideational patterns that accompanied the demographic transition in the Western world, primarily in Western Europe. In the following chapter, Ness also showed that this "western" transition of the past has been characteristic of all of the currently industrialized world. On the whole, the speed of that past transition, both in mortality and fertility declines, was markedly slower than the transitions we see today. This can lead to questions about the character of that greater European ecosystem that could produce slower transitions than we see today. We can also ask what intervening conditions, such as new medical, public health, and contraceptive technologies, are responsible for the difference in the speed of transitions in these great ecosystems.

Even within Western Europe, however, we can find rich variation in the decline of fertility (Coale and Watkins 1986). Thus in addition to seeing Western Europe as a single ecosystem, distinguished by their state systems from much of the rest of the world, it is also possible to reduce the scale and to disaggregate Western Europe into a large number of smaller ecosystems. In this case administrative boundaries define distinctive cultural or linguistic groups, or ecosystems writ small. Ness found the same thing with the ongoing fertility transition in Southeast Asia. Burma and the Philippines stand in a very different position from Thailand, Indonesia, Malaysia or Singapore in the general decline of fertility.

This can lead us to ask about the relationship between administrative boundaries and culturally or linguistically defined ecosystems. Are the administrative boundaries typically laid down respecting such cultural or linguistic differences, as happened in Eastern Europe after 1918? Or do administrative boundaries lead to a kind of cultural or linguistic homogenization of a population, which produces a more uniform pattern of the demographic transition? Thus the historical analysis of emerging ethnic identities in areal units defined as ecosystems by the distinctive trajectory of the demographic transition can provide important insights into the dynamics of the population-environment relationship.

Ruttan begins with a discussion of recent agricultural transitions, in which increases in output have come less from the extension to new land and more from the intensification of production on a more stable land area. He asks to what extent those ecosystems can support continued agricultural increases. By inference the North American corn and wheat belts combined can be treated as a single ecosystem with both natural and administrative boundaries. The speed and character of the transition in this greater area can be seen to be determined by natural factors such as soil and water, but also by administrative boundaries and political conditions that affect the price of both inputs and outputs in the system. Martine examines the same transition in Brazil, finding it less productive and less benign especially for the poor than the transition found in North America. He attributes these negative characteristics of the transition to government policy.

Grainger attempts to model the deforestation transition, with special attention to tropical rainforests. Brechin works in a micro unit of this larger tropical rainforest ecosystem. For both authors the relevant systems have both natural and administrative boundaries. Both see natural topographic and agronomic conditions shaping the speed and character of the deforestation process. In both, population growth is one, but not the only, critical element in shaping the transition. Both also see that it is the intermixture of population growth, technology, and political-economic forces that drive the transition. Further, in both cases the newly placed administrative boundaries define ecosystems by promoting, permitting, or limiting specific population-environment relations within the system.

These provide useful illustrations of research that can address fundamental issues of ecosystems and boundaries and in doing so advance our understanding of the population-environment relationship. The point of these is to use transition-defined ecosystems as the units of observation. From this we can ask important questions about the relative impact of natural and administrative boundaries on the population-environment adjustments that take place within that system.

Multiple Transition Systems

The transition perspective emerged out of the observation that the modern world is characterized by a *family of transitions* that appear to be occurring at roughly the same time. This leads to the identification of another set of ecosystems, defined by the specific mix and speed of transitions found in a given area.

Ness observed demographic and agricultural transitions occurring throughout Southeast Asia. The critical point of these observations is that these two transitions did not occur with the same speed, quality or strength in all of the countries of the region, despite the fact that the technology and capital resources for both transitions were equally available. Kartomo shows how stress and response led Indonesia to experience both transitions at roughly the same time. By contrast, Thailand saw the demographic transition move considerably ahead of the rice transition. The Philippines show the opposite, and in Burma both transitions are substantially retarded. McIntosh sees the two transitions occurring, at least potentially, simultaneously in Zimbabwe. In the microsystems of South Sumatra, Benin, and the Dominican Republic we see a number of transitions move with varying speed, but the small size of the arena and the short time span of the observations precludes a systematic assessment of transition speeds and trajectories.

In these arenas marked by multiple transitions we can ask a more probing set of questions than can be asked in one-transition systems. What is the impact of boundaries (natural or socially constructed, gradually or rapidly formed) on the character of the ecosystem, and especially on the interconnectedness of its family of transitions? Addressing these questions will be greatly facilitated by the development of a taxonomy of transitions, which should be a higher priority.

The State and its Multiple Ecosystems

The modern state has created a massive new network of boundaries that sometimes cut across natural ecosystems, and sometimes draw many different ecosystems into one interdependent political system. How can we use transitions to shed some light on the implications of the boundaries that contain multiple ecosystems? Considering demographic, epidemiological and productive transitions together, we can make a number of observations that suggest areas of future research.

The difference in the speed of these three modern transitions in different global regions lies behind many so called North-South tensions in the world today. The North can be seen as a single ecosystem in which relatively weak political boundaries tie together populations that have come relatively slowly and at roughly the same time, through the demographic and epidemiological transitions. It also marks a region of slow, but possibly increasing speed of the economic or productive transition.

The South, by contrast, is undergoing these three transitions, and in many cases they are occurring much faster that they did in the North. If they all occurred roughly simultaneously, as we can say they did in the Newly Industrializing Countries (NICs), the tensions with the North are dampened, or at least transformed into economic tensions that submit relatively easily to negotiation. If the transitions occur at quite different speeds--as we see with the epidemiological transition vastly outpacing the demographic or productive transitions in Africa, the result is heightened tension.

In effect, what we see is that political boundaries--North and South--that contain multiple ecosystems with vastly different rates of transition, provide for considerable political conflict and tension. There is a parallel as well at less than global levels. Part of the internal tensions of the former Soviet Union that have been so dramatically exposed in recent years, derive from the fact that its political boundaries encompass different ecosystems experiencing very different rates of various transitions. Nor is the former Soviet Union alone in this internal tension. India, China, Canada, Brazil, Mexico, and the United States can illustrate the same internal asymmetry in rates of transition. Indeed, the deviant cases of modernization in the new nation state may well be those rare cases of a highly homogeneous population that experiences a wide variety of modern transitions at roughly the same time. Japan, Singapore, Taiwan, and possibly Hong Kong and South Korea may provide good illustrations of this highly uncommon phenomenon.

The critical role of the nation-state in the modern population-environment relationship presents us with another vexing conceptual issue. The boundaries of these systems vary in what can be called their *permeability* (Ness and Brechin 1988). Some boundaries are more open than others to the flow of people, ideas, and goods. Further, this permeability varies along a generalization-specificity dimension. Modern communications permit ideas and expressions to flow more easily than people or goods across all boundaries. At the same time, languages define larger transactional ecosystems, making political boundaries within those systems more permeable to some ideas and expressions than to others. Immigration and trade policies define at least a portion of boundary permeability to the movement of people and goods. The widespread observation of smuggling and illegal immigrants only testifies to the fact that political boundaries can seldom be completely sealed off. They can, however, vary in their openness on many different dimensions. Finally, all of these dimensions of boundary permeability vary over time, sometimes dramatically. Europe in 1992, Hong Kong in 1999, and the coming of a North American Trade Agreement illustrate one type of change. Wars, revolutions, and violent conflicts represent another.

Thus using transitions--single or multiple--to define ecosystems, and then examining the overlay of natural and political boundaries around single and multiple ecosystems, can take us some way forward in understanding the impact of different kinds of boundaries on ecosystems. It can also lead us to ask a series of questions about the social arrangements, or institutions, within specific boundaries that affect the speed, pattern, or interconnectedness of different transitions. Finally, transitions can lead us to ask what are the conditions that determine how multiple ecosystems and different transition speeds are either translated into violent social conflict and tension, or are resolved with less than violent conflict. But this set of questions takes us to our final consideration, of human institutions.

Human Institutions

From the review of population-environment relationships provided above in the summary, several research questions emerge on the role of human institutions. They include: (1) Defining the outcome of the population environment dynamic as ecosystem health. (2) The role of normative structures; and (3) Understanding the character of political institutions.

Defining Ecosystem Health

Above we argued that the population-environment relationship can be conceived of in ecosystemic terms, and that it is possible to conceive of a dimension of health in that system. Recently, Dryzek (1987) has developed the concept of ecological rationality, which we find similar to that of health. He defines this as the capacity of *human and natural systems in combination* to cope with the particularity of environmental problems. He further identifies five criteria of ecological rationality, including negative feedback, coordination, robustness, flexibility, and resilience. Dryzek goes on to develop a taxonomy of nine forms of social choice, reflected in coordination across actors, and seven prevalent mechanisms by which those choices are made. This provides a useful set of ideas, but is less successful in operationalizing the concept of ecosystem health.

If we are to make progress in defining the health of an ecosystem, the first task is to operationalize the concept by developing effective indicators and accounting systems. There are, to be sure, many existing indicators that can be used. We now have a wide variety of economic, health, epidemiological, and social indicators for the human population, as well as measures of various forms of environmental conditions. The latter, however, tend to be dominated by negative measures. That is, we assess degradation, or assaults on the environment more effectively than we measure its health. In any event, this great plethora of measures is for the most part highly atomized by academic and scientific specializations. It is necessary to tie these various measures together into an integrated accounting system that permits us to see the ecosystem as a whole and not simply through a subset of its parts.

The attempt to develop such an integrated ecosystem accounting technique is what is proposed in the PEMS project. That project, however, aims at developing a device for small systems. A parallel is also needed for larger national and international systems. Perhaps the single most important step in this direction lies in changing the system of National Income Accounts. These Accounts represent one of the most powerful research techniques in the social sciences, and they have had an immense impact on public policy. Unfortunately, the accounts persist in the older western view of the environment as an unlimited resource to be exploited. What is needed is to include natural resources in the capital stocks of the National Accounts. Repetto et al. (1989) has amply demonstrated that this is possible, and the World Bank (1990) is making similar efforts. Until this technique achieves more universal application, however, the National Income Accounts will continue to be part of the problem of environmental degradation rather than its solution.

Aside from the System of National Accounts, more innovations will be necessary. An integrated ecosystem account would include not only human productive activities with natural resources defined as capital stocks, but also a variety of natural and human indicators of well-being. The work on social indicators can provide useful measures for the human population, but we must also find a way to include measures of biodiversity, and the well-being of the natural environment in this integrated account. If we can develop such an ecosystem account, we can more effectively address the question of the human institutions that affect ecosystem health.

We need not wait for the development of such an account before raising these questions, however. It is not difficult to identify cases of massive ecosystem destruction or of ecosystem protection and restoration. A useful step, therefore, would be to identify areas of ecological disaster, protection, and restoration and to use these for systematic studies of human institutional correlates. Focusing on a specific set of transitions--epidemiological, agricultural, and toxicity transitions, for example,--can provide a strategy for identifying these areas.

The Character of Political Institutions

We have already noted our belief in the primacy of human institutions in determining ecosystem health. Here we can draw on the observations in this volume to suggest some propositions about specific political characteristics and ecosystem health.

Earlier Ness and Ando (1984) suggested that the strength of what they called the political administrative system was a critical determinant of the success of a populations' adjustment to the stress of rapid growth. They argued that this adjustment, occurring in the Third World today, is a political ecological adjustment, in which the stress of rapid population growth is reduced through the reduction of human fertility. They called the adjustment a political ecological one because they saw political institutions leading and determining the success of the adjustment. The strength of the political administrative system was marked by three conditions: strong central decision making capacity with effective mobilization of local participants, effective socioeconomic monitoring, and a capacity to promote socioeconomic development. They used a grounded scale to operationalize the concept for twenty one countries from roughly 1950-80, and found that the strength of this system determined the timing of the anti-natalist policy decision, the strength of the implementation of that policy decision and the reduction of human fertility.

These three dimensions of political administrative strength have parallels in Dryzek's ecological rationality.[3] The strength of political decision making can reflect Dryzek's coordination and robustness. Monitoring is reflected in his negative feedback, or the capacity to perceive destructive behaviors and processes. The capacity to promote development is reflected in Dryzek's coordination, flexibility, and robustness.

Our chapters can illustrate both positive and negative aspects of these dimensions of political administrative strength or ecological rationality. Indonesia's commitment to development, and its capacity to monitor demographic and socioeconomic change led to positive responses to pressures through its population and rice programs. On the other hand, failures on these dimensions in Indonesia's forest protection and toxification can be said to be related to weaknesses in ecosystem health. Zimbabwe's newly achieved independence is associated with greater system strength and with more effective solutions of its agricultural and population problems. Benin's adjustment of the protected park boundary represents some flexibility, but in its incapacity to promote socioeconomic development and to monitor socioeconomic conditions is associated with deteriorating ecosystem health. These are highly impressionistic assessments of both political administrative system strength and ecosystem health, of course, but they suggest that the concepts are useful; and that they can be operationalized for the more systematic observations of covariance that are implied by a research agenda.

There remain at least three problems, however. These concern local participation, equality, and culture or value systems. Ruttan notes, for example, that effective *extension* systems are required to make health delivery systems as effective as agricultural development systems. Ness and Ando's political administrative system strength only indirectly suggests that extension to and mobilization of local participants is an important ingredient in the capacity of a central political system to make effective decisions. Thus, the question of local participation is surely an area that needs further development, both of ideas and of operational measures.

Equality presents another conundrum. Martine notes how Brazilian government politics favored the wealthy at the expense of the poor and thus led to a overall weakening of system health. Much has been written about equality and both development and population planning, but there is little effective resolution of the issues. The ecological disasters of the former USSR, Eastern

3 Stinchcombe (1974) has presented a similar formulation of conditions that affect what he calls the efficiency of industrial administrative systems. It is possible that similar formulations can be generated for the effective function of systems from the individual up through the global levels.

Europe, and China, however, suggest that an overwhelming aim of industrialization without a concern for equality and broad human welfare can be deleterious to system health. It can also be argued that equality implies a greater voice for local interests, thus increasing the capacity of a system to perceive negative feedback of environmental assaults. Here is another area of issues ripe for systematic research.

Finally, there is the area of culture or value systems that is almost totally absent in Ness and Ando's approach, and is not clearly articulated in Dryzek's. There are, however, strong suggestions that value systems have an important impact both directly on ecosystem health and indirectly on the political institutions that affect that health. Dunlap (1988) has shown how changing popular perceptions of the environment have supported wildlife protection efforts in the U.S. Other studies (Inglehart 1990, Harris Polls 1987, Kempton 1991) have found similar dramatic changes in popular perceptions and values throughout the world. Two general researchable issues arise from these observations. First, how do values affect individual behavior? Does the rising concern for the environment lead to more conservationist behavior at the individual level? Second, how, or to what extent, do such popular values get articulated in the larger community, in national and international policies? This is partly a question of values and political systems, and partly a question of the political economy, or the economic interests of various negotiators in the game of large scale national and international environmental policy.

The research agenda is long and complex. Much work remains to be done in developing useful concepts, in operationalizing those concepts, and in making the detailed field observations implied by any research agenda. This exercise in attempting to think about the population-environment dynamic has, we believe, helped to lay out some of the details of that agenda. If the work ahead is difficult, the stakes are very high. They may well include more than mere symptoms of ecosystem ailments. They may include the very existence of the planet's ecosystem as we know it. Let the work go forward.

References

Abernethy, V. 1979. *Population Pressure and Cultural Adjustment.* New York: Human Sciences Press.

Adamchak, D. J. and M. T. Mbizvo. 1990. "The Relationship Between Fertility and Contraceptive Prevalence in Zimbabwe." *International Family Planning Perspectives* 16(3):103-6.

African Research Bulletin. 1-31 January 1991. Political Series 28 (1).

Agbo, V. and N. Sokpon. *Population and Environment Interdependence in Northern Benin West Africa: A Case Study on the Rural Commune of Tannougou.* University of Michigan School of Natural Resources and National University of Benin, School of Agricultural Sciences.

Allen, J. C. and D. F. Barnes. 1985. "The Causes of Deforestation in Developing Countries." *Annals of the Association of American Geographers* 75:163-84.

Altieri, M. A. 1987. *Agroecology: The Scientific Basis of Alternative Agriculture.* Boulder, CO: Westview Press.

Arnold, G. March/April 1990. "The Land Dilemma." *African Report* 35:58-61.

Attolou, A. 1986. "Evaluation en cours d'execution dans les zones pilotes de Ouake." *MDRAC-DEFC Projet UN* Boukoumbe et Malanville.

Barker, R., R. W. Herdt and B. Rose. 1985. *The Rice Economy of Asia.* Washington, D.C.: Resources for the Future, Inc.

Bass, T. May 1986. "Famine--An Early Warning System." *Science Digest* :57-62; 82-84.

Baumer, M. 1987. *Le role possible de l'Agroforesterie dans la lutte contre la desertification et la degradation de l'environnement.* Wageningen CTA Cop.

Becker, G. and R. J. Barro. 1988. "Reformulating the Economic Theory of Fertility." *Quarterly Journal of Economics* (103):1-25.

Becker, G. S. and H. G. Lewis. 1974. Interaction Between Quantity and Quality of Children. In *Economics of the Family: Marriage, Children and Human Capital,* ed. T. W. Schultz, 81-90. Chicago: University of Chicago Press.

Bell, R. H. V. 1984. The Man-Animal Interface: An Assesment of Crop Damage and Wildlife Control. In *Conservation and Wildlife Management in Africa,* eds. R. H. V. Bell and E. E. McShar-Caluze, Washington, D.C.: U.S. Peace Corps. [Proceedings of a workshop organized by the U.S. Peace Corps, October 1983.]

Bendix, R. 1977. *Max Weber: An Intellectual Portrait.* Berkeley: University of California Press.

Bhatia, B. M. 1988. *Indian Agriculture: A Policy Perspective.* New Delhi, Newbury Park, CA: Sage Publications.

Biro Pusat Statistik. *Penduduk Indonesia: Hasil Sensus Penduduk 1990.* Jakarta, Indonesia: 1990.

Biro Pusat Stastik. 1987. *Statistik Indonesia*. Jakarta, Indonesia.

Biro Pusat Statistid dan Asosiasi Exportir Kopi Indonesia. *Statistik Kopi 1989.* Jakarta, Indonesia: 1989.

BKKBN and Community Systems Foundation. October 1986. "KB-Gizi: An Indonesian Integrated Family Planning, Nutrition and Health Program: The Evaluation of the First Five Years of Program Implementation in East Java and Bali." Ann Arbor, MI: Community Systems Foundation.

Blaikie, P. and H. Brookfield., eds. 1987. *Land Degradation and Society.* London: Methuen.

Blurton Jones, N. 1986. "Bushman Birth Spacing: A Test for Optimal Interbirth Intervals." *Ethology and Sociobiology* (7):91-105.

Blurton Jones, N. 1987. "Bushman Birth Spacing: Direct Tests of Some Simple Predictions." *Ethology and Sociobiology* (8):183-203.

Blurton Jones, N. 1989. The Costs of Children and the Adaptive Scheduling of Births: Towards A Sociobiological Perspective on Demography. In *Sexual and Reproductive Strategies*, eds. A. Rasa, C. Vogel and E. Voland, Kent: Croom Helm.

Blurton Jones, N. and R. M. Sibley. 1978. "Testing Adaptiveness of Culturally Determined Behavior: Do Bushman Women Maximize Their Reproductive Success by Spacing Births Widely and Foraging Seldom?" S.S.H.B. Symposium 18: Human Behavior and Adpatation, 1978.

Board on Science and Technology for International Development (BOSTID) and Institute of Medicine (IOM). 1987. *The U.S. Capacity to Address Tropical Infectious Disease Problems*. Washington, D.C.: National Academy Press.

Bongaarts, J. 1982. "The Fertility Inhibiting Effects of the Intermediate Fertility Variables." *Studies in Family Planning* 13(6/7):179-189.

Bongaarts, J. 1978. "A Framework for Analyzing the Proximate Determinants of Fertility." *Population and Development Review* 4(1):105-132.

Boohene, E. and T. E. Dow. March 1987. "Contraceptive Prevalence and Family Planning Program Effort in Zimbabwe." *International Family Planning Perspectives* 13(1):1-7.

Borgerhoff Mulder, M. 1988. Kipsigis bridewealth payments. In *Human Reproductive Behaviour: A Darwinian Perspective*, eds. L. Betzig, M. Borgerhoff Mulder and P. Turke, Cambridge: Cambridge University Press.

Boserup, E. 1965. *The Conditions of Agricultural Growth: The Economics of Agrarian Change Under Population Pressure*. Chicago: Aldine.

Boserup, E. 1981. *Population and Technology*. Oxford: Basil Blackwell.

Boxer, C. R. 1961. *Four Centuries of Portugese Expansion, 1415-1825: A Succinct Survey*. Johannesburg: Witwatersrand University Press.

Brandstrom, A. and L.-G. Tedebrand. 1986. *Society, Health and Population During the Demographic Transition, Report No. 4* Almqvist and Wiksell International, Stockholm.

Bratton, M. January 1987. "The Comrades and the Country Side: The Politics of Agriculutral Policy in Zimbabwe." *World Politics* 39(2):174-203.

Brazil's Agricultural and Demographic Censuses. 1990. Sao Paulo, Brazil: Government of Brazil.

Brechin, S. R. 1984. *Social Impact Assessment and Diffusion Theory: Convergence on Consequences* Unpublished Draft Paper.

Brechin, S. R. and P. C. West. 1990. "Protected Areas, Resident Peoples, and Sustainable Conservation: The Need to Link Top-down with Bottom-Up." *Society and Natural Resources* 3:77-79.

Brechin, S. R. and P. C. West. 1982. "Social Barriers in Implementing Appropriate Technology: The Case of Community Forestry in Niger West Africa." *Humboldt Journal of Social Relations* 9(2).

Brenner, M. H. 1987. "Relation of Economic Change to Swedish Health and Social Well-Being." *Social Science and Medicine* 25(2):183-195.

Britton, J. C. and B. Morton. 1989. *Shore Ecology of the Gulf of Mexico*. Austin, TX: University of Texas Press.

Brokensha, D. and B. W. Riley. 1989. Managing Natural Resources: The Local Level. In *Changing the Global Environment: Perspectives on Human Involvement*, eds. Botkin et al, San Diego, CA: Academic Press, Inc.

Browder, J. O.,ed. 1989. *Fragile Lands of Latin America: Strategies for Sustainable Development*. Boulder, CO: Westview Press.

Brown, L. 1988. "The Changing World Food Prospect: The Nineties and Beyond." *Worldwatch Paper*. Washington, D.C.: WorldWatch Institute. October, 1988.

Bunkley-Williams, L. and E. H. Williams Jr. April 1990. "Global Assault on Coral Reefs." *Natural History* :46-54.

Burch, W. R. Jr. 1971. *Daydreams and Nightmares: A Sociological Essay on the American Environment*. New York: Harper and Row.

Cain, M. 1983. "Fertility As An Adjustment To Risk." *Population and Development Review* 9(4):688-701.

Cain, M. 1985. "On The Relationship Between Landholding and Fertility." *Population Studies* (39):5-15.

Cain, M. 1982. "Perspectives on Family and Fertility in Developing Countries." *Population Studies* 36(2):159-175.

Cain, M. and G. McNicoll. 1988. Population Growth and Agrarian Outcomes. In *Population, Food Supply, and Rural Development*, eds. R. D. Lee, W. B. Arthur, A. C. Kelley, G. Rodgers and T. N. Srinivasan, 101-117. Oxford: Clarendon Press.

Caldwell, J. C. 1977. "The Economic Rationality of High Fertility: An Investigation Illustrated with Nigerian Survey Data." *Population Studies* 31(1):5-27.

Caldwell, J. C. c. 1970. *Family Planning Programs and Offical Policy Decisions in Southern Africa* Canbena, Department of Demography, Australian National University.[Unpublished Paper.]

Caldwell, J. C. 1987. Fertility Control as Innovation: A Report on In-Depth Interviews in Ibadan. In van de Walle, pp. 233-51.

Caldwell, J. C. 1988. "Is the Asian Family Planning Program Model Suited to Africa?" *Studies in Family Planning* 19(1):19-28.

Caldwell, J. C. June 2,1986. "Routes to Low Mortality in Poor Countries." *Population and Development Review* 12(2):171-220.

Caldwell, J. C. September/December 1976. "Toward a Restatement of Demographic Transition Theory." *Population and Development Review* 2(2-3):321-66.

Caldwell, J. C. and P. Caldwell. 1987. "The Cultural Context of High Fertility in Sub-Saharan Africa." *Population and Development Review* 13(3):409-37.

Caldwell, L. K. 1984. *International Environmental Policy: Emergence and Dimensions*. Durham, NC: Duke University Press.

CARDER-Acatora. 1989. *Survey of Households*.

Caribbean Fishery Management Council. 1985. *Fishery Management Plan, Final Environmental Impact Statement, Reeffish Fisheries of Puerto Rico and the U.S. Virgin Islands*.

Carlyle, T. 1843. *Past and Present*. London: Chapman and Hall.

Carlyle, T. 1834. *Sartor Resartus*. London: James Fraser.

Cassen, R. H. 1976. "Population and Development: A Survey." *World Development* (4):785-830.

Centre Pedologique de reconnaissance du Dahomey au 1/200.000 feuille de languiete ORSTOM. 1960. Centre de Cotonou.

Chagnon, N. Is Reproductive Success Equal in Egalitarian Societies. In *Evolutionary Biology and Human Social Behavior: An Anthropological Perspective*, eds. N. Chagnon and W. Irons, North Scituate, MA: Duxbury Pr.

Chagnon, N. 1988. "Life Histories, Blood Revenge and Warfare in A Tribal Population." *Science* 239:985-992.

Chagnon, N. 1982. Sociodemographic Attributes of Nepotism in Tribal Populations: Man the Rule-breaker. In *Current Problems in Sociobiology*, eds. Kings' College Sociobiology Group, Cambridge: Cambridge University Press.

Chandler, T. and G. Fox. 1974. *3000 Years of Urban Growth*. New York: The Academic Press.

Cipolla, C. M. 1974. *The Economic History of World Population*. New York: Penguin Books.

Cipolla, C. M. 1965. *Guns, Sails, and Empires: Technological Innovation in the Early Phases of European Expansion 1400-1700*. New York: Minerva Press.

Clark, J. R., ed. 1985. *Coastal Resources Management: Development Case Studies*. Columbia, SC: Research Planning Institute, Inc. for the National Park Service and USAID. [Coastal Publication No. 3.]

Clark, William. 1987. *Forecasting in the Social and Natural Sciences*. Vienna, Austria: International Institute for Applied Systems Analysis.

Clay, J. W. 1988. *Indigenous Peoples and Tropical Forests: Models of Land Use and Management from Latin America*. Cambridge: Cultural Survival, Inc.

Cleland, J. and J. Hobcraft., eds. 1985. *Reproductive Change in Developing Countries, Insights from the World Fertility Survey*. London: Oxford University Press.

Cleland, J. and C. Wilson. 1987. "Demand Theories of the Fertility Transition: An Iconoclastic View." *Population Studies* 41:5-30.

Cliff, A. D., P. Haggett, J. K. Ord, K. Bassett and R. B. Davies. 1975. *Elements of Spatial Structure: A Quantitative Approach*. Cambridge: Cambridge University Press.

Coale, A. J. and S. C. Watkins. 1986. *The Decline of Fertility in Europe*. Princeton: Princeton University Press.

Coedes, G. 1964. *Angkor: An Introduction*. Hong Kong: Oxford University Press.

Cole, H. S. D. and et al., eds. 1973. *Models of Doom: A Critique of the Limits to Growth*. New York: Universe Press.

Comitas, L. 1973. Occupational Multiplicity in Rural Jamaica. In *Work and Family Life; West Indian Perspectives*, eds. L. Comitas and D. Lowenthal, 157-174. New York: Anchor Press.

Conover, W. J. 1971. *Practical Non-parametric Statistics*. New York: Wiley.

Cook, S. F. and W. Borah. 1971. *Essays in Population History, Volume 1. Mexico and the Caribbean*. Berkeley, CA: University of California Press.

Cordell, J. 1988. Introduction: Sea Tenure. In *A Sea of Small Boats*, ed. J. Cordell, 1-32. Cambridge: Cultural Survival, Inc.

Cordell, J., ed. 1988. *A Sea of Small Boats*. Cambridge: Cultural Survival, Inc.Cultural Survival Report No. 26.

Cordell, J. 1988. Social Marginality and Sea Tenure in Bahia. In *A Sea of Small Boats*, ed. J. Cordell, 125-151. Cambridge: Cultural Survival, Inc.

Corson, W. H., ed. 1990. *The Global Ecology Handbook: What You Can About the Environmental Crisis*. Boston: Beacon Press.

Cottrell, F. 1955. *Energy and Society*. New York: McGraw-Hill.

Cripps, L. 1979. *The Spanish Caribbean: From Columbus to Castro*. Boston: G.K. Hall.

Cruz, W. D. and C. J. Cruz. 1990. "Population Pressure and Deforestatin in the Philippines." *ASEAN Economic Bulletin* 7(2):200-212.

Cummings, R. W. 1989. *Modernizing Asia and the Near East: Agricultural Research in the 1990s* U.S. Agency for International Development, Washington D.C.

Dani, A. A. and G. W. Campbell. 1986. *Sustaining Upland Resources: People's Participation in Watershed Management*. Kathmandu: International Centre for Integrated Mountain Development.

Davis, G. R. Sept. 1990. "Energy for Planet Earth." *Scientific American* 263(3):54-60.

Deevey, E. S. 1960. "The Human Population." *Scientific American* 203:3-9.

deGraff, J. 1986. *The Economics of Coffee*. Wagenigen, the Netherlands: PUDOC.

Demerath, N. J. 1976. *Birth Control and Foreign Policy: The Alternatives to Family Planning*. New York: Harper and Row.

Demographic Database. 1986. *History on Data*. Umea University: Umea.

Denslow, J. S. and C. Padoch., eds. 1988. *People of the Tropical Rainforest*. Berkeley, CA: University of California Press.

Department of Rural Development. 1989. *Monthly Report, March 1989* Ministry of Local Government, Rural and Urban Development, Harare.

Disraeli, B. 1913. *Sybil*. London: Longmans, Green and Co.

Dobyns, H. F. 1976. "Brief Perspective on a Scholarly Transformation:'Widowing' the Virgin Land." *Ethnohistory* (23):95-104.

Dobyns, H. F. 1983. *Their Number Become Thinned: Native American Population Dynamics in Eastern North America*. Knoxville: University of Tennessee Press.

Donner, W. 1987. *Land Use and Environment in Indonesia*. Honolulu: University of Hawaii Press.

Douglas-Hamilton, I. 1988. "Elephant Monitoring and Conservation." *GRID News* 1(3):4-5. [Nairobi, Kenya: UNEP.]

Dove, M. R. 1985. *Swidden Agricultrue in Indonesia: The Subsistence Strategies of the Kalimantan Kantu*. West Germany: Mouton.

Drake, W. D., R. I. Miller and D. A. Schon. 1983. "The Study of Community-Level Nutrition Interventions: An Argument for Reflection-In-Action." *Human Systems Management* 4:82-97.

Drake, W. D., F. D. Zinn and C. Antonakos. 1990. "Determinants of Data Quality for the Center for International Health Information." *USAID Report*.

Dregne, H. E. 1983. *Desertification of Arid Lands*. Chur, Switzerland: Hardwood Academic Publishers.

DuBois, R. and E. L. Towle. 1985. Coral Harvesting and Sand Mining Management Practices. In *Coastal Resources Management: Development Case Studies*, ed. J. R. Clark, pp203-289. Columbia, SC: Research Planning Institute, Inc. for the National Park Service and USAID.

Dyos, H. F. and M. Wolff., eds. 1973. *The Victorian City; Images and Realities*. London: Routledge and Kegan Paul.

Easterlin, R. 1978. The Economics and Sociology of Fertility: A Synthesis. In *Historical Studies of Changing Fertility*, ed. C. Tilly, Princeton: Princeton University Press.

Easterlin, R., G. Alter and G. Condran. 1978. Farms and Farm Families in Old and New Areas: The Northern States in 1860. In *Family and Population in Nineteenth Century America*, eds. T. Hareven and M. Vinovskis, Princeton: Princeton University Press.

The Economist. April 7, 1990. "Now we are ten." *The Economist* 315(7649):48-49.

Ehrlich, P. R. *ZPG Reports*.

Ehrlich, P. R. and A. H. Ehrlich. 1970. *Population, Resources, Environment: Issues in Human Eccology*. San Francisco: W.H. Freeman.

Ehrlich, P. R. and A. H. Ehrlich. 1990. *The Population Explosion*. New York: Simon and Shuster.

Elwell, H. A. 1985. "An Assessment of Soil Erosion in Zimbabwe." *Zimbabwe Science News* 19(3-4):27-31.

Engels, F. 1892. *The Condition of the Working-Class in England in 1844*. London: Allen and Unwin.

Eriksson, I. and J. Rogers. 1978. *Rural Labor and Population Change: Social and Demographic Developments in East-Central Sweden During the Nineteenth Century*. Stockholm: Almqvist and Wiksell International.

ESCAP. 1988. The Geography of Fertility in the ESCAP Region. Asian Population Studies Series, No. 62-K Bangkok: ESCAP Population Division.

Fadiman, J. A. 1982. *An Oral History of Tribal Warfare: The Meru of Mt. Kenya*. Athens, OH: Ohio University Press.

Falkenmark, M. 1990. "Water Scarcity Management and Small Scale Irrigation In Traditional Agriculture." Interagency Preparatory Meeting on Water and Sustainable Agricultural Development.

Farooq, G. M. and D. S. DeGraaff. 1988. *Fertility and Development: An Introduction to Theory, Empirical Research and Policy Issues.*

Farooq, G. M. and G. B. Simons., eds. 1985. *Fertility in Developing Countries: An Economic Perspective on Research and Policy Issues.* New Hampshire: MacMillan Press.

Fearnside, P. M. 1986. *Human Carrying Capacity of the Brazilian Rainforest.* New York: Columbia University Press.

Fearnside, P. M. September 1985. "A Stochastic Model for Estimating Human Carrying Capacity in Brazil's TransAmazon Highway Colonization Area." *Human Ecology* 13:331-69.

Financial Gazette. March 23, 1990. "Mugabe Assures Farmers that Land Acquisition will be Fair."

Financial Gazette. May 18, 1984. "Productivity of Communal Farms Delights Agritex." [Harare.]

Finkle, J. L. November/December 1973. "Too Large for Any Single Country." *Ceres (FAO Review on Development)* 6(6):34-38.

Flandrin, J. L. 1979. *Families in Former Times: Kinship, Household, and Sexuality.* Cambridge: Cambridge University Press. [Translated by Richard Southern.]

Flinn, M. W. 1981. *The European Demographic System 1500-1820.* Baltimore: Johns Hopkins University Press.

Food and Agriculture Organization of the United Nations (FAO). 1987. *1948-1985 World Crop and Livestock Statistics.* Rome: FAO.

Food and Agriculture Organization of the United Nations (FAO). "Developpement des parcs nationaux." *Document de travail no. 7* Les mammiferes du Parc National de la Pendjari. PNUD/FAO/BEN/77/01.

Food and Agriculture Organization of the United Nations (FAO). 1988. *Forest Products World Outlook Projections* FAO, Rome.

Food and Agriculture Organization (FAO). 1988. *The State of Food and Agriculture 1987-88.* Rome: FAO.

Freedman, D. S. and A. Thornton. 1982. "Income and Fertility: The Elusive Relationship." *Demography* 19(1):65-78.

Frenk, J. and et al. 1988. "Health Transition in Middle Income Countries: New Challenges for the Organization of Services." *Health Policy Planning* 4(1):29-39.

Frost, S. 1979. *The Whaling Question.* San Francisco, CA: Friends of the Earth.

Fulkerson, W. et. al. September 1990. "Energy From Fossil Fuels." *Scientific American* 263(3):129-135.

Gassman, N. and M. H. Brenner. 1990. *Personal Communication.*

Gaunitz, S. 1979. "Local history as a Means of Understanding Economic Development." *Economy and History* (22):38-62.

Gaunt, D. 1977. "I slottets Skugga. Om frllse-bnders sociala problem i Borgeby och Lddekoppinge under 1700-talet." *Ale* :15-30.

Gaunt, D. 1983. The Property and Kin Relations of Retired Farmers. In *Family Forms in Historic Europe*, eds. R. Wall, J. Robin and P. Laslett, Cambridge: Cambridge University Press.

Gaunt, D. 1987. "Rural Household Organization and Inheritance in Northern Europe." *Journal of Family History* 12(1-3):121-141.

Georges, E. 1990. *The Making of a Transnational Community: Migration, Development, and Cultural Change in the Dominican Republic.* New York: Columbia University Press.

Gerger, T. and G. Hoppe. 1980. "Education and Society: The Geographer's View." *Acta Universitatis Stockholmensis* (1):1-124.

Giddens, A. 1990. *The Consequences of Modernity.* Stanford,CA: Stanford University Press.

Glantz, M. H. 1987. Drought and Economic Development in Sub-Saharan Africa. In *Drought and Hunger in Africa: Denying Famine a Future*, ed. M. H. Glantz, 37-58. New York: Cambridge University Press.

Gliessman, S. 1984. Resource Management in Traditional Tropical Agroecosystems in Southeast Mexico. In *Agricultural Sustainability in a Changing World Order*, ed. G. Douglass, 191-291. Boulder, CO: Westview Press.

The Global Ecology Handbook. 1990. Boston: Beacon Press.

Goliber, T. November 1989. "Africa's Expanding Populations: Old Problems, New Policies." *Population Buletin* 44(3).

Golley, F. B. 1988. Human Populations From An Ecological Perspective. In *Population and Resources in Western Intellectual Traditions*, eds. M. S. Teitelbaum and J. M. Winter, pp 200. New York: Cambridge University Press.

Gordon, G. 1969. *System Simulation.* Englewood Cliffs, NJ: Prentice-Hall.

Government of Zimbabwe and United Kingdom Government. 1989. *The 1989 Joint GOZ/UK Joint Review of the Resettlement Programme* Department of Rural Development, Harare.[Mimeo.]

Government of Zimbabwe and United Kingdom Government. *The Zimbabwe and United Kingdom Governments' Joint Resettlement Review Mission* Department of Rural Development, Harare.[Mimeo.]

Government of Brazil. *Agricultural and Demographic Censuses.*

Government of Zimbabwe. 1982. *Transitional National Development Plan, 1982/83-1984/85* Government of Zimbabwe, Harare, Zimbabwe.[Two Volumes.]

Government of Zimbabwe, Department of Census and Statistics/Westinghouse, Institute for Resource Development. 1989. *Zimbabwe Demographic and Health Survey, 1988. Preliminary Report* Westinghouse, Columbia, MD.

Gradwohl, J. and R. Greenberg. 1988. *Saving Tropical Forests*. Washington, D.C.: Island Press.

Graedel, T. E. and P. J. Crutzen. Sept. 1989. "The Changing Atmosphere." *Scientific American* 261(3):58-68.

Graham, L. R. 1972. *Science and Philosophy in the Soviet Union*. New York: Knopf.

Grainger, A. 1992. Characterization and Assessment of Desertification Processes. In *Desertified Grasslands: Their Biology and Management*, ed. C. Chapman, 17-33. London: Academic Press.

Grainger, A. 1990a. *Controlling Desertification*. London: Earthscan Publications.

Grainger, A. 1986. *The Future Role of the Tropical Rain Forests in the World Forest Economy*. Oxford Academic Publishers, Oxford. Ph.d thesis, Department of Plant Sciences, University of Oxford.

Grainger, A. 1983. Improving the Monitoring of Deforestation in the Humid Tropics. In *Tropical Rain Forest Ecology and Management*, eds. S. L. Sutton, T. C. Whitmore and A. C. Chadwick, 387-395. Oxford: Blackwell's Scientific Publication.

Grainger, A. 1990b. "Modelling Deforestation in the Humid Tropics." *Deforestation or Development in the Third World?* Scandinavian Forest Economics. Finnish Forest Research Institute, Helsinki.

Grainger, A. 1984. "Quantifying Changes in Forest Cover in the Humid Tropics: Overcoming Current Limitations." *Journal of World Forest Resource Management* 1:3-62.

Grainger, A. 1981. "Reforesting Britain." *The Ecologist* (11):56-81.

Grainger, A. 1980. "The State of the World's Tropical Forest." *The Ecologist* (10):6-54.

Grainger, A. 1993 (In press). *The Tropical Rain Forests and Man*. New York: Columbia University Press.

Green, G. M. and R. W. Sussman. April 13, 1990. "Deforestation History of the Eastern Rain Forests of Madagascar from Satellite Images." *Science* 248:212-215.

Gwynne, M. D. and D. W. Mooneyhan. 1989. The Global Environment Monitoring System and the Need for a Global Resource Database. In

Changing the Global Environment: Perspectives on Human Involvement, eds. Botkin et al, San Diego, CA: Academic Press, Inc.

Habakkuk, H. J. 1955. "Family Structure and Economic Change in Nineteenth-Century Europe." *Journal of Economic History* 15:1-12.

Hafele, W. 1990. "Energy from Nuclear Power." *Scientific American* 263(3):137-144.

Hagerstrand, T. 1952. "The Propagation of Innovation Waves." *Human Geography* 4:3-19. Lund Studies in Geography.

Hajnal, J. 1965. European Marriage Patterns in Perspective. In *Population in History: Essays in Historical Demography*, eds. D. V. Glass and D. E. C. Eversley, Chicago: Aldine Publishing Co.

Hall, K. R. 1985. *Maritime Trade and State Development in Early Southeast Asia*. Honolulu: University of Hawaii Press.

Hall, K. R. and J. K. Whitmore., eds. 1976. *Explorations in Early Southeast Asian History: The Origins of Southeast Asian Statecraft*. Ann Arbor, MI: Center for South and Southeast Asian Studies, University of Michigan.

Hammel, E. A., S. Johansson and C. Gunsberg. Winter 1983. "The Value of Children During Industrialization: Sex Ratios in Childhood in Nineteenth-century America." *Journal of Family History* :400-417.

Hansen, A. M. 1966. *The Process of Planning: A Study of India's Five-Year Plans*. London: Oxford University Press.

Hanson, A. J. 1981. Transmigration and Marginal Land Development. In *Agricultural and Rural Development in Indonesia*, ed. G. E. Hansen, Boulder, Colorado: Westview Press.

Harris, N. 1986. *The End of the Third World*. London: I.B. Tauris and Co.

Harrison, J. 1969. *Owen and the Owenites in Britain and America: The Quest for the New World Moral*. Cambridge, MA: Harvard University Press.

Harrison, P. 1992. *Population and Environment*. London: Penguin Books.

Harrison, P. May 19, 1990. "Too Much Life on Earth?" *New Scientist* 129:28-29.

Hawkes, K. and E. L. Charnov. June 1988. "On Human Fertility: Individual or Group Benefit?" *Current Anthropology* 29:469-471.

Hayes, A. 1990. "Population Quality and Sustainable Development in Indonesia: A Framework for Discussion of Policy Issues." *A Memorandum, Manuscript*.

Hays, S. P. 1959. *Conservation and the Gospel of Efficiency*. Cambridge: Harvard University Press.

Hecht, J. 1988. French Utopian Socialists and the Population Question: Seeking the Future City. In *Population and Resources in Western Intellectual Traditions*, eds. M. S. Teitelbaum and J. M. Winter, 49-73. New York: Cambridge University Press.

Herbst, J. 1989. "Political Impediments to Economic Rationality: Explaining Zimbabwe's Failure to Reform its Public Sector." *Journal of Modern African Studies* 27(1):67-84.

Hermalin, A. L. 1978. "Spatial Analysis of Family Planning Program Effects in Taiwan, 1966-72." *Papers of the East-West Population Institute* 48:37-39.

Heydir, L. 1990. *Inditifikasi Berbagai Aspek Perladangan Liar Di Kawasan Hutan Di Kabupaten Lahat, Sumatra Selatan, Indonesia* Pusat Penelitian Kependudukan Universitas Sriwijaya and The School of Natural Resources, The University of Michigan.

Higgins, G. M. and et. al. 1982. *Potential Population Supporting Capacities of Lands in the Developing World* Food and Agriculture Organization of the United Nations, Rome.[Technical Report FPA/INT/513.]

Hill, J. 1984. "Prestige and Reproductive Success in Man." *Ethology and Sociobiology* 5:77-95.

Hobsbawm, E. and T. Ranger., eds. 1983. *The Invention of Tradition*. Cambridge: Cambridge University Press.

Hollingsworth, T. H. 1957. "A Demographic Study of the British Ducal Families." *Population Studies* 11(1):4-26.

Hughes, A. L. 1986. "Reproductive Success and Occupational Class in Eighteenth-Century Lancashire, England." *Social Biology* 33:109-115.

Hugo, G., T. Hull, V. Hull and G. Jones. 1987. *The Demographic Dimension in Indonesian Development*. Oxford: Oxford University Press.

Hulin, J.-P. 1978. *La Ville et les Ecrivains Anglais 1770-1820*. Universite de Lille III, Lille.

Humphrey, C. R. and F. R. Buttel. 1982. *Environment, Energy, and Society*. Belmont, CA: Wadsworth Publishing Co.

Hurst, P. 1987. "Forest Destruction in Southeast Asia." *The Ecologist* 17:170-4.

Hutterer, K. L., ed. 1977. *Economic Change and Social Interaction in Southeast Asia: Perspectives from Prehistory, History and Ethnography*. Ann Arbor, MI: Center for South and Southeast Asian Studies, University of Michigan.

IBGE. 1980. *Demographic Censuses*. Sao Paulo, Brazil: Government of Brazil.

Indonesia Censuses. 1990. Jakarta, Indonesia: Government of Indonesia.

Inger, G. 1980. *Svensk Rattshistoria*. Liber Laromedel: Lund.

INSP/Fundacion Universo XXI. 1989. *Taller de Trabajo Sobre Losefectos de la Contaminacion Atmosferica en la Salud* (Proceedings), Tepotzotlan, Mexico.

International Energy Agency. 1984. *Energy Prices and Taxes*. Paris: OECD.

International Rice Research Institute (IRRI). 1986. *World Rice Statistics 1985*. Los Banos, Philippines: International Rice Research Institute.

Irons, W. 1979. Cultural and Biological Success. In *Evolutionary Biology and Human Social Behavior: An Anthropological Perspective*, eds. N. A. Chagnon and W. Irons, North Scituate, MA: Duxbury Press.

Ives, J. D. and B. Messerli. 1989. *The Himalayan Dilemma: Reconciling Development and Conservation*. London: Routledge.

Jacobson, H., M. Jacobson, and S. Humphrey. Jan. 1990. "Effects of In Utero Exposure to Polychlorinated Biphenyls and Related Contaminants on Cognitive Functioning in Young Children." *The Journal of Pediatric Medicine*.

Jennings, F. 1975. *The Invasion of America: Indians, Colonialism, and the Cant of Conquest*. Chapel Hill: University of North Carolina Press.

Jenny, H. 1980. *The Soil Resource: Origin and Behavior*. New York: Springer Verlag.

Johnson, N. E. and S. Lean. 1985. "Relative Income, Race, and Fertility." *Population Studies* 39:99-112.

Jorberg, L. 1972. *A History of Prices in Sweden: 1732-1914. 2 volumes*. Lund: CWK Gleerup.

Jorberg, L. 1975. Structural Change and Economic Growth in Nineteenth-Century Sweden. In *Sweden's Development from Poverty to Affluence, 1750-1970*, ed. S. Koblik, 92-135. Minneapolis, MN: University of Minnesota Press.

Kageyama, A. A. and et al. 1987. *O Novo Padrao Agricola Brasileiro: do Complexo Rural aos Complexos Agroindustriais*. Campinas: UNICAMP.

Karlyle, T. 1843. *Past and Present*. London: Chapman and Hall.

Kay, G. 1975. "Population Pressures and Development Prospects in Rhodesia." *Rhodesia Science News* 9(1):7-13.

Kay, G. 1970. *Rhodesia: A Human Geography*. London: University of London Press.

Kay, G. 1980. "Towards A Population Policy for Zimbabwe/Rhodesia." *African Affairs* 79(3-4):95-114.

Keesing's Record of World Events. 1990. 36 (12), p. 37909. London: Longman.

Kempton, W. 1991. "Key Perspectives on Global Environmental Climate Change." *Global Environmental Change* June(183-208).

Kennedy, P. 1987. *The Rise and Fall of Great Powers*. New York: Random House.

Kevles, D. J. 1985. *In the Name of Eugenics: Genetics and the Uses of Human Heredity*. New York: Knopf.

Kingsland, S. 1988. Evolution and Debates Over Human Progress From Darwin to Sociobiology. In *Population and Resources in Western Intellectual Traditions*, eds. M. S. Teitelbaum and J. M. Winter, 167-98. New York: Cambridge University Press.

Kingsley, D. March 1955. "The Origin and Growth of Urbanization in the World." *American Journal of Sociology* 60(5):429-37.

Kingsley, D. 1965. "The Urbanization of the Human Population." *Scientific American* 213(3):41-53.

Klingender, F. D. 1947. *Art and the Industrial Revolution*. London: Noel Carrington.

Klosterman, R. E. 1988. Microcomputer Technology and Regional Planning: Prospects and Challenges for the Developing World. In *Information Systems for Government and Business: Trends, Issues, Challenges*, Nagoya, Japan: The United Nations Centre for Regional Development.

Kneese, A. V. 1988. The Economics of Natural Resources. In *Population and Resources in Western Intellectual Traditions*, eds. M. S. Teitelbaum and J. M. Winter, 281-309. New York: Cambridge University Press.

Knodel, J. 1986. Demographic Transitions in German Villages. In *The Decline of Fertility in Europe*, eds. A. J. Coale and S. Watkins, Princeton: Princeton University Press.

Knodel, J. 1988. *Demographic Behavior in the Past*. Cambridge: Cambridge University Press.

Knodel, J., N. Havanon and W. Sittitrai. 1990. "Family Size and the Education of Children in the Context of Rapid Fertility Decline." *Population Development Review* 16(1):31-62.

Korten, D. C. and R. Klass. 1984. *People-Centered Development: Contributions Toward Theory and Planning Frameworks*. West Hartford, CT: Kumarian Press.

Kurosawa, A. *Dreams* (film).

Lang, J. 1988. *Inside Development in Latin America: A Report from the Dominican Republic, Colombia, and Brazil*. Chapel Hill: University of North Carolina Press.

Lanly, J. P., ed. 1981. *Tropical Forest Resources Assessment Project (GEMS): Tropical Africa, Tropical Asia, Tropical America*. Rome: FAO/UNEP.

Lanly, J. P. 1982. "Tropical Forest Resources." *FAO Forestry Paper No. 30* Published by the FAO in Rome.

Laphan, R. B. and G. B. Simmons., eds. 1987. *Organizing for Effective Family Planning Programs*. Washington, D.C.

Leete, R. 1987. "The Post-Demographic Transition in East and Southeast Asia: Similarities and Contrasts with Europe." *Population Studies* 41.

Lesthlaeghe, R. and C. Wilson. 1986. Modes of Production, Secularization, and the Pace of the Fertility Decline in Western Europe, 1870-1930. In *The Decline of Fertility in Europe: The Revised Proceedings of a Conference on the Princeton European Fertility Project*, eds. A. J. Coale and S. Watkins, Princeton, N.J.: Princeton University Press.

Levin, J. 1960. *The Export Economies: Their Patterns and Development in Historical Perspective*. Cambridge, MA: Harvard University Press.

Lext, G. 1968. "Mantalsskrivningen i Sverige fore 1860." *Meddlanden Frm Ekonomisk-Historiska Institutionen vid Goteborgs Universitet* (13).

Libby, R. T. 1984. Development Strategies and Political Divisions Within the Zimbabwean State. In *The Political Economy of Zimbabwe*, ed. M. G. Schatzberg, New York: Praeger.

Liddle, W. 1990. "The Relative Autonomy of the Third World Politician: Sopeharto and Indonesian Economic Development in Comparative Perspective." Manuscript.

Lieberman, V. 1990. "Secular Trends in Burmese Economic History c. 1350-1830 and Their Implications for State Formation." *Modern Asian Studies* 24(3).

Lieberman, V. 1991. "The Structure of Early Modern Southeast Asian History c. 1350-1830." Manuscript.

Likert, R. 1953. "Public Relations and the Social Sciences," Paper presented at the meeting of the Public Relations Society of America, Inc. Washington D.C., November 1952.

Little, P. D., M. M. Horowitz and A. E. Nyerges, Editors. 1987. *Lands at Risk in the Third World: Local Level Perspectives* Westview Press, Boulder, CO.

Livi-Bacci, M. 1986. Social-group Forerunners of Fertilty Control in Europe. In *The Decline of Fertility in Europe*, eds. A. J. Coale and S. Watkins, Princeton, N.J.: Princeton University Press.

Lo-Johannsson, F. a. I. 1981. *Sveriges Rikes Lag: Gillard och Antagen Parisdagen Ar, 1734, faksimilutgava*. Giulunds: Malmo.

Loeb, E. 1972. *Sumatra: Its History and People*. Kuala Lumpur: Oxford Press.

Lomnicki, A. 1988. *Population Ecology of Individuals* Princeton University Press, Princeton, N.J.

"Love Canals in the Making," *Time Magazine*. 20 May 1991.

Low, B. S. 1989. "Cross-cultural Patterns in the Training of Children: An Evolutionary Perspective." *Journal of Comparative Psychology* (103):311-319.

Low, B. S. 1989. "Occupational Status and Reproductive Behavior in Nineteenth-Century Sweden: Locknevi Parish." *Social Biology* (36):82-101.

Low, B. S. 1990. "Occupational Status, Landownership, and Reproductive Behavior in Nineteenth-Century Sweden: Tuna Parish." *American Anthropologist* (92):457-468.

Low, B. S. *Reproductive Life in Nineteenth Century Sweden: An Evolutionary Perspective on Demographic Phenomena.* In press: Ethology and Sociobiology.

Low, B. S. 1990. "Sex, Power, and Resources: Ecological and Social Correlates of Sex Differences." *International Journal of Contemporary Sociology* (27):49-73.

Low, B. S. and A. L. Clarke. 1991. "Family Patterns in Nineteenth Century Sweden: Impact of Occupational Status and Land Ownership." *Journal of Family History* 16.

MacArthur, R. H. and E. O. Wilson. 1967. *The Theory of Island Biogeography*. Princeton, N.J.: Princeton University Press.

Machlis, G. E. and D. L. Tichnell. 1985. *The State of The World's Parks: An International Assessment for Resource Management, Policy, and Research.* Boulder, CO: Westview Press.

Malgavkar, P. D. 1990. *Quality of Life and Problems of Governance in India.* New Delhi: Center for Policy Research.

Malmstrom, A. 1981. *Successionsratt II.* Uppsala: Iustrus Forlag.

Marr, D. and A. C. Milner., eds. 1986. *Southeast Asia in the Ninth to Fourteenth Centuries*. Singapore: Institute of Southeast Asian Studies.

Marsden, W. 1966. *The History of Sumatra.* Kuala Lumpur: Oxford University Press.

Martine, G. "Changes in Agricultural Production and Rural Migration." The Demography of Inequality in Latin America, Gainesville, FL, February 1988.

Martine, G. August 1990. "Fases e Faces da Modernizacao Agricola." *Rivista de Planejamento e Politicas Publicas* 1(3):23-77.

Martine, G. and A. R. Arias. 1987. "A Evolucao do Emprego no Campo." *Revista Brasileira de Estudos de Populacao, Sao Paulo* 4(2):39-84.

Maxwell, J., ed. 1982. *The Malay-Islamic World of Sumatra: Studies in Polities and Culture*. Melbourne: Monash University.

McCay, B. J. 1980. "Appropriate Technologies, Fisheries, and Resource Management." *APPROTECH* 2(4):7-12.

McCay, B. J. 1981. "Development Issues in Fisheries as Agrarian Systems." *Culture and Agriculture* 11:1-8.

McCormick, J. 1989. *Reclaiming Paradise: The Global Environmental Movement*. Bloomington, IN: Indiana University Press.

McEvedy, C. and R. Jones. 1978. *Atlas of World Population History*. New York: Penguin.

McNeely, J. A. 1988. *Economics and Biological Diversity: Developing and Using Economic Incentives to Conserve Biological Resources*. Gland: IUCN.

McNeill, W. 1976. *Plagues and Peoples*. New York: Doubleday.

McNeill, W. 1990. *Population and Politics Since 1750*. Charlottesville: University of Virginia Press.

McNeill, W. 1982. *The Pursuit of Power: Technology, Armed Force, and Society Since 1000 AD*. Chicago: University of Chicago Press.

McNeill, W. 1963. *The Rise of the West*. Chicago: University of Chicago Press.

Meadows, D. L. and D. H. Meadows., eds. 1973. *Toward Global Equilibrium: Collected Papers*. Cambridge, MA: Wright Allen Press.

Meadows, D. H., et. al. 1972. *The Limits To Growth: A Report for the Club of Rome's Project on the Predicament of Mankind*. New York: Universe Books.

Meganck, R. A. and J. M. Goebel. 1979. "Shifting Cultivation: Problems for Parks in Latin America." *Parks* 4(2):4-8.

Meldrum, A. March-April 1990. "Mugabe's Folly." *Africa Report* 35:54-57.

Mendels, F. F. 1981. *Industrialization and Population Pressure in Eighteenth-century Flanders*. New York: Arno Press.

Merrick, T. July 1986. "Population Presures in Latin America." *Population Bulletin* 41(3).

Merton, R. K. 1968. *Social Theory and Social Structure*. New York: The Free Press.

Mok, S. T. "Sustainable Management and Development of Tropical Forest in ASEAN." ASEAN Seminar on The Management of Tropical Forests for Sustainable Development, January 24, 25, 1990.

Morris, W. 1888. *A Dream of John Ball and a King's Lesson*. London: Reeves and Turner.

Morris, W. 1890. *News from Nowhere*. Boston: Roberts Bros.

Mosher, L. 1986. At Sea in the Caribbean? In *Bordering on Trouble: Resources and Politics in Latin America*, eds. A. Maguire and J. W. Brown, 235-269. Bethesda, MD: Adler and Adler.

Mugabe, R. G. "Address by His Excellency the President of The Republic of Zimbabwe." International Forum on Population in the 21st Century, Amsterdam, November 6-9, 1989. 281. Munslow, B. Jan. 1985. "Prospects for the Socialist Transition of Agriculture in Zimbabwe." *World Development* 13(1):41-58.

Muttalib, J. A. 1973. *The History and Society of South Sumatra 1800-1920: Publications in New York Libraries* Southern Asian Institute, Columbia University, New York, NY.

Myers, N. 1980. *Conversion of Tropical Moist Forests: Report Prepared by Myers for Committee on Research Priorities in Tropical Biology of the National Research Council*. Washington, D.C.: Academy of Sciences.

Myers, N. 1984. *The Primary Source: Tropical Forests and Our Future*. New York: W.W. Norton.

Myers, N. 1991. "Population and the Environment: Issues, Prospects and Policies. Draft Paper for UNFPA."

Myint, N. 1990. "Recent Levels and Trends of Fertility and Mortality in the Union of Myanmar." *Manuscript, UNFPA/Institute of Economics Seminar on Population and Development, Yangon 21-22 August.*

Myrdal, G. 1968. *Asian Drama*. New York: The 20th Century Fund.

Nahn, V. Q. and R. Hannenberg. 1990. "The 1988 Demographic Survey of Viet Nam." *Asia-Pacific Population Journal* 4(3):3-14.

Naning, M. I. and et al. 1988. *Evaluasi Dampak Pengembangam Perkebunan Rakyat Terhamp Potensi Sumberdays Air di Kecamatan Jarai Kabupten Lahat* Indonesia, Pusat Penelitian Universitas Sriwijaya.

National Commission on Floods. 1980. *Ministry of Energy and Irrigation Report* Government of India Press, New Delhi.[2 volumes.]

Ness, G. D. and H. Ando. 1984. *The Land is Shrinking: Population Planning in Asia*. Baltimore: Johns Hopkins University Press.

Ness, G. D. 1967. *Bureaucracy and Rural Development in Malaysia: A Study of Complex Organizations in Stimulation Economic Development in New States*. Berkeley: University of California Press.

Ness, G. D. 1988. *Program Performance II: Family Planning Programs Performance in Asia, 1960-1985.*. Ann Arbor, MI: Population Planning and International Health.

Ness, G. D. and S. R. Brechin. forthcoming. "An Ecological Perspective on World Systems Theory."

Ness, G. D. and J. Finkle. 1985. *Managing Delivery Systems: Identifying Leverage Points for Improving Family Planning Program Performance*. Ann Arbor: University of Michigan Department of Population Planning and International Health.

Ness, G. D., J. T. Johnson and S. Bernstein. 1984. Programme Performance: The Assessment of Asian Family Planning Programmes. In *Population Research Leads*, Bangkok: The Population Division, ESCAP.

Ness, G. D. and W. Stahl. 1 January 1977. "Western Imperialist Armies in Asia." *Comparative Studies in Society and History* XIX(1).

New York Times. November 25, 1989. "Uncertain Lies Land of the Farmer as Winds of Change Hit Zimbabwe." *The New York Times*. :4.

Norberg, A. and S. Akerman. 1973. Migration and the Building of Families: Studies on the Rise of the Lumber Industry in Sweden. In *Aristocrats,*

Farmers, Proletarians. Essays in Swedish Demographic History, eds. K. Agren and et. al., Stockholm: Esselte Studiuns.

Norberg, A. and M. Rolen. June 1977. Migration and Marriage: Some Empirical Results from Tuna Parish 1865-1894. In *Reports from the Symposium Time, Space, and Man in Essays on Microdemography*, eds. J. Sundin and E. Soderlund, Stockholm: Almqvist and WIksell.

Norman, H. and H. Runlom. 1988. *Transatlantic Connections: Nordic Migration to the New World after 1800*. Norway: Norwegian University Press.

Notestein, F. 1982. "Demography in the United States: A Partial Account of the Development of the Field." *Population and Development Review* 8:462.

Notestein, F. 1945. Population-The Long View. In *Food for the World*, ed. T. W. Schultz, Chicago: University of Chicago Press.

Nystuen, J. and F. D. Zinn. forthcoming. "Computer-aided Management Advice for Loan Programs Run by Indonesian Women."

ODA (Overseas Development Administration). 1991. "Population Environment and Development: An Issues Paper for the Third UNCED Prepatory Committee."

OECD. 1991. *The State of the Environment*. Paris: OECD.

Omran, A. R. March 1977. "A Century of Epidemiologic Transition in the United States." *Preventive Medicine* 6(1):30-51.

Ostergren, R. 1988. *A Community Transplanted: The Trans-Atlantic Experience of a Swedish Immigrant Settlement in the Upper Middle West 1835-1915*. Madison: University of Wisconsin Press.

Owen, N. G.,ed. 1987. *Death and Disease in Southeast Asia: Explorations in Social, Medical and Demgraphic History*. Oxford: Oxford University Press.

Owen, N. G. July 1971. "The Industry of Mainland Southeast Asia, 1850-1914." *Journal of the Siam Society* 59:78-142.

Owen, N. G. 1987. Measuring Mortality in the Nineteenth Century Philippines. In *Death and Disease in Southeast Asia*, ed. N. Owen, Oxford: Oxford University Press.

Palmer, R. H. 1977. *Land and Racial Domination in Rhodesia*. Berkeley/Los Angeles: University of California Press.

Palmore, J. A. and R. W. Gardner. 1983. *Measuring Mortality, Fertility, and Natural Increase: A Self-teaching Guide to Elementary Measures*. Honolulu: East-West Population Institute.

Palo, M., G. Mery and J. Salmi. 1987. "Deforestation in the Tropics: Pilot Scenarios Based on Quantitative Analyses." *Deforestation or Development in the Third World* Editors M. Palo and J. Salmi. Finnish Forest Research Institute, Helsinki.

Parry, J. H. 1974. *The Discovery of the Seas*. New York: Dial Press.

Parry, J. H. 1959. *The Establishment of the European Hegemony: 1415-1715*. New York: Harper and Row.

Parry, J. H. 1949. *Europe and a Wider World 1415-1715*. London: Hutchinson's University Library.

Partridge, W. L. Spring 1984. "The Humid Tropics Cattle Ranching Complex: Cases from Panama Reviewed." *Human Organization* 43(1):76-80.

Peck, T. 1984. The World Perspective. In *Forest Policy: A Contribution to Resource Development*, ed. F. Hummel, 21-66. The Hague: Nijhoff/Junk.

Peluso, N. L. and M. Poffenberger. 1989. "Social Forestry in Java: Reorienting Management Systems." *Human Organization* 48(4):333-343.

Pernia, E. M. 1980. *A Note on Migration and Fertility* University of Philippines, School of Economics, Quezon City.

Perrow, C. 1984. *Normal Accidents: Living with High-Risk Technologies*. New York: Basic Books.

Pianka, E. R. 1970. "On r- and K-selection." *American Naturalist* (106):581-588.

Pick, D. 1989. *Faces of Degeneration: A European Disorder of 1848 - c1918*. Cambridge: Cambridge University Press.

Pinkerton, E., ed. 1989. *Co-operative Management of Local Fisheries: New Directions for Improving Management and Community Development*. Vancouver, Canada: University of British Columbia Press.

Piotrow, P. T. 1973. *World Population Crisis: The United States' Response*. New York: Praeger.

Planning Commission. 1990. *Approach to the Eighth Five Year Plan*. New Delhi: Planning Commission.

Plucknett, D. H. and N. J. H. Smith. Jan. 1986. "Sustaining Agricultural Yields: As Productivity Rises, Maintenance Research is Needed to Uphold the Gains." *BioScience* 36:40-45.

Plumwood, V. and R. Routley. 1982. "World Rainforest Destruction-The Social Factors." *The Ecologist* (12):4-22.

Poffenberger, M. 1990. The Evolution of Forest Management Systems in Southeast Asia. In *Keepers of the Forest: Land Management Alternatives in Southeast Asia*, ed. M. Poffenberger, West Hartford, CT.: Kumarian Press.

Poffenberger, M. 1990. Facilitating Change in Forestry Bureaucracies. In *Keepers of the Forest: Land Management Alternatives in Southeast Asia*, ed. M. Poffenberger, West Hartford, CT.: Kumarian Press.

Poffenberger, M., ed. 1990. *Keepers of the Forest: Land Management Alternatives in Southeast Asia*. West Hartford, CT.: Kumarian Press.

Pogiliano, C. 1984. "Scienze e stirpe: eugenica in Italia (1912-1939)." *Pessado e presente* :61-79.

Polanyi, K. 1957. *The Great Transformation*. Boston: Beacon Press.

Population Reference Bureau (PRB). 1991. *World Environment Data Sheet*. Washington, D.C.: PRB.

Postel, S. Nov/Dec 1988. "Global View of a Tropical Disaster." *American Forests* 94:25-29.

Potter, L. October 1988. "Eating the Forests in Gulps and Nibbles: Concessionaires, Transmigrants and Free Loggers in South Kalimantan." *Inside Indonesia* :19-21.

Pray, C. E. May 1983. "Private Agricultural Research in Asia." *Food Policy* :131-140.

Pullen, J. R. 1981. "Malthus's Theological Ideas and Their Influence on His Principle of Population." *History of Political Economy* (13).

Purdy, B. A. Sept. 1988. "American Indians after A.D. 1492: A Case Study of Forced Culture Change." *American Anthropologist* 90(3):640-655.

Pursell, C., ed. 1973. *From Conservation to Ecology: The Development of Environemental Concern*. New York: Thomas Crowell Co.

Ranger, T. O. 1985. *Peasant Consciousness and Guerilla War in Zimbabwe: A Comparative Study*. London: James Currey.

Ranger, T. O. 1967. *Revolt in Southern Rhodesia, 1896-97: A Study in African Resistance*. Evanston: Northwestern University Press.

Rapport, D. J. Autumn 1989. "What Constitutes Ecosystem Health?" *Perspectives in Biology and Medicine* 33(1).

Reid, A. 1988. *Southeast Asia in the Age of Commerce 1450-1680*. New Haven: Yale University Press.

Reid, W. V., J. N. Barnes and B. Blackwelder. 1988. *Bankrolling Success: A Portfolio of Sustainable Development Projects*. Washington, D.C.: Environmental Protection Institute, National Wildlife Federation.

Reilly, J. and R. Bucklin. 1989. "Climate Change and Agriculture." *World Agriculture Situation and Outlook Report* USDA/ARS WAS-55, Washington, D.C.[pp43-46.]

Repetto, R. C. 1988. *The Forest for The Trees? Government Policies and the Misuse of Forest Resources*. Washington, D.C.: World Resources Institute.

Repetto, R. and M. Gillis., eds. 1988. *Public Policies and the Misuse of Forest Resources*. New York: Cambridge University Press.

Repetto, R. C., W. Magrath, M. Wells, C. Beer and F. Rossini. 1989. *Wasting Assets: Natural Resources in National Income Accounts*. Washington, D.C.: World Resources Institute.

Republic of Indonesia. *Act No. 4, 1982 on Basic Regulation on Enviornmental Management*.

Republic of Indonesia and Ministry of Forestry. *Forest in Indonesia, It's Potential and Management*.

Richardson, G. P. and A. L. Pugh. 1981. *Introduction to System Dynamics Modeling with DYNAMO.* Cambridge: Massachusettes Institute of Technology Press.

Robben, A. C. G.M. 1985. "Sea Tenure and Conservation of Coral Reef Resources in Brazil." *Cultural Survival Quarterly* 9(1):45-47.

Roberts, P. C. 1978. *Limits to Growth Revistited: Modelling Large Systems.* New York: Halsted Press.

Rogers, E. 1969. *Diffusion of Innovations.* New York: Free Press of Glencoe.

Rogoff, M. H. and S. L. Rawlins. December 1987. "Food Security: A Technological Alternative: Biotechnology can Convert Biomass into a Stable Food Supply." *BioScience* 37:800-807.

Rolen, M. 1979. *A Forest Community in Transition: Studies in Population Development, Migration and Social Change in Revsund 1820-1977.* Uppsala: University of Uppsala. [English Summary.]

Romsan, A. "The Future Role of Public Participation in Environment and Decision Making Process: A Case Study of Indonesian Transmigration Sites in the Province of South Sumatra." [Unpublished Master's Thesis, Dalhousie University, Nova Scotia, Canada.]

Rosenberg, N. J. 1986. Climate, Technology, Climate Change and Policy: The Long Run. In *The Future of the North American Granery: Politics, Economics and Resource Contraints in North American Agriculture*, ed. C. F. Runge, Ames, IA: Iowa State University Press.

Rosenblat, A. 1976. The Population of Hispaniola at the Time of Columbus. In *The Native Population of the Americas in 1492*, ed. W. Denevan, Madison: University of Wisconsin Press.

Rowe, J. S. 1984. *Understanding Forest Landscapes: What You Conceive is What You Get.* Vancouver, B.C.: xxx. [The Leslie H. Schaffer Lectureship in Forest Science.]

Rowe, J. S. Fall 1989. "What on Earth is Environment?" *Trumpeter* 6(4):123-126.

Le Roy Ladurie, E. 1971. *Times of Feast, Times of Famine, A History of Climate Since the Year 1000.* New York: Doubleday.

Rubino, M. C. and R. W. Stoffle. 1989. Caribbean Mithrax Crab Mariculture and Traditional Seafood Distribution. In *Proceedings of the Thirty-Ninth Annual Gulf and Caribbean Fisheries Institute (Hamilton, Bermuda, 1986)*, eds. G. T. Waugh and M. H. Goodwin, 134-145. Charleston, SC: Gulf and Caribbean Fisheries Institute.

Rubino, M. C. and R. W. Stoffle. Winter 1990. "Who Will Control The Blue Revolution?: Economic and Social Feasibility of Caribbean Crab Mariculture." *Human Organization* 49:386-94.

Rudel, T. K. 1989. "Population, Development, and Tropical Deforestation: A Cross-National Study." *Rural Sociology* 54(3):327-338.

Ruegg, E. F. and et al. 1987. Impactos dos agrotoxicos sobre o ambiente e a saude. In *Os Impactos Sociais da Modernizacao Agricola*, eds. G. Martine and R. Garcia, Sao Paulo: Editora Caetes/Hucitec.

Ruttan, V. W., ed. forthcoming, 1990. St. Paul, MN: University of Minnesota Department of Agriculture and Applied Economics. Health Constraints on Agricultural Development.

Ruttan, V. W., ed. December 1989. *Biological and Technical Constraints on Crop and Animal Productivity: Report on a Dialogue*. St. Paul, MN: University of Minnesota Department of Agricultural and Applied Economics.

Ruttan, V. W., ed. forthcoming, 1990. *Resource and Environmental constraints on Sustainable Growth in Agricultural Production*. St. Paul: University of Minnesota Department of Agricultural and Applied Economics.

Ruttan, V. W. 1986. "Toward a Global Agricultural Research System: A Personal View." *Research Policy*.

Ruttan, V. W. and Y. Hayami. 1985. *Agricultural Development: An International Perspective 2nd Ed.*. Baltimore: The Johns Hopkins University Press. [pp.41-72, 255-328.]

Ruttan, V. W. and Y. Hayami. 1987. Population Growth and Agricultural Productivity. In *Population Growth and Economic Development: Issues and Evidence*, eds. D. G. Johnson and R. D. Lee, Madison: University of Wisconsin Press.

Sadik, D. N. 1989. *Safeguarding the Future*. New York.

Sadlie, M. and T. K. Wie. August 1990. *Industry, Technology and Natural Resource* Congress of Indonesian Economist Association, Bandung, Indonesia.

Salas, R. M. 1979. "International Population Asabwe's Agricultural Policies." *The Journal of Modern African Studies* 27(1):85-107.

Salim, E. "Sustainable Development." Congress of Indonesian Economist Association, Bandung, Indonesia, August.

Sanchez, P. A. 1976. *Properties and Mangement of Soils in the Tropics*. New York: John Wiley and Sons.

Schon, D. A., W. D. Drake and R. I. Miller. 1984. Social Experimentation as Reflection-in-Action. In *Knowledge: Creation, Diffusion, Utilization*, 5-36.

Schultz, T. P. 1988. Economic Demography and Develoment: New Direction in an Old Field. In *The State of Development Economics: Progress and Perspectives*, eds. G. Ranis and T. P. Schultz, 416-51. Oxford: Basil Blackwell.

Seager, J., ed. 1990. *The State of the Earth Atlas*. Simon and Schuster, Inc.

Secrett, C. 1986. "The Environmental Impact of Transmigration." *Ecologist* 16:77-88.

Seddon, D. 1987. *Nepal: A State of Poverty*. Delhi: Vicas Publishing House.

Selznick, P. 1949. *TVA and the Grass Roots: A Study of Politics and Organizations*. Berkeley: The University of California Press.

Shils, E. A. 1962. *Political Development in the New States*. Gravenhage: Mouton and Co.

Simmons, I. G. 1989. *Changing the Face of the Earth: Culture, Environment and History*. Oxford: Basil Blackwell.

Simon, J. 1981. *The Ultimate Resource*. Princeton, N.J.: Princeton University Press.

Sinopse Preliminar do Censo Agropecuario. 1985.

Skalnes, T. March 1989. "Group Interests and the State: An Explanation of Zimbabwe's Agricultural Policies." *The Journal of Modern African Studies* 27(1):85-107.

Smith, G. 1968. *Dickens, Money, and Society*. Berkeley, CA: University of California Press.

Smith, M. E. 1977. Introduction. In *Those Who Live From The Sea: A Study In Martime Anthropology*, ed. M. E. Smith, 1-22. St. Paul, MN: West Publishing Company.

Smith, N. 1980. "Anthrosols and Human Carrying Capacity in Amazonia." *Annals of the Association of American Geographers* 70:553-566.

Sommer, A. 1976. "Attempt at an Assessment of the World's Tropical Rain Forests." *Unasylva* 28(112-113):5-25.

South China University. 1990. Memorandum.

Spengler, J. J. 1945. "Malthus's Total Population Theory: A Restatement and Reappraisal." *Canadian Journal of Economics and Political Science* ii.

Sriwijaya University. 1990. *Fact Sheets on Population in South Sumatra Province* Population Study Center, Sriwijaya University, South Sumatra Indonesia, Palembang.

Sriwijaya Post. August 22, 1990b. "Harus Pindah, 500 KK Perambah Hutan Lindung."

Sriwijaya Post. August 21, 1990a. "Kebun Kopi Dibakar, Pelakunya Ditangkap."

Stargardt, J. 1986. Hydraulic Works and South East Asian Politics. In *Southeast Asia in the Ninth to Fourteenth Centuries*, eds. D. G. Marr and A. C. Milner, 23-48. Pasir Panjang, Singapore: Institute of Southeast Asian Studies.

1985. *State of India's Environment 1984-85 The Second Citizens Report* Center for Science and Environment, New Delhi.

Steinberg, D. I. 1981. *Burma's Road Toward Development: Growth and Ideology Under Military Rule*. Boulder, CO: Westview Press.

Steinberg, D. I. 1982. *Burma, A Socialist Nation of Southeast Asia*. Boulder,CO: Westview Press.

Steinberg, D. J. and D. P., eds. Chandler. 1987. *In Search of Southeast Asia: A Modern History*. Honolulu: University of Hawaii Press.

Stern, F. 1976. Capitalism and the Cultural Historian. In *From Parnassus. Essays in Honor of Jacques Barzun*, eds. D. B. Wiener and W. K. Keylor, New York: Harper and Row.

Stinchcombe, A. September 1961. "Agricultural Enterprise and Rural Class Relations." *American Journal of Sociology* LXII(2):165-76.

Stinchcombe, A. L. 1974. *Creating Efficient Industrial Administrations*. New York: Academic Press.

Stocking, M. A. 1986. *Effect of Soil Erosion on Soil Nutrients in Zimbabwe* FAO, Rome.

Stoffle, R. W. 1986. *Caribbean Fishermen Farmers: A Social Assessment of Smithsonian King Crab Mariculture* Institute for Social Research, Ann Arbor, MI.

Stoffle, R. W., D. B. Halmo and B. W. Stoffle. 1990. Inappropriate Management of an Appropriate Technology: A Restudy of Mithrax Crab Mariculture in the Dominican Republic. In *Sociocultural Aspects of Small-Scale Fisheries in Developing Countries*, eds. J. Poggie and R. Pollnac, Kingston, RI: International Center for Marine Resource Development, University of Rhode Island.

Stoffle, R. W., D. B. Halmo and B. W. Stoffle. 1990. *Un Analisis del Proyecto Maricultura de Centolla en la Comunidad de buen Hombre, Republica Domincana* School of Natural Resources, University of Michigan, Ann Arbor, MI.[Report prepared for the Department of Fishery Resources, Office of the Secretary of Agriculture, Government of the Dominican Republic.]

Stoffle, R. W., M. C. Rubino and D. L. Rasch. 1988. "Fishermen and Crab Mariculture in the Caribbean." *Practicing Anthropology* 10(1):10-11,14.

Stokes, E. 1957. *The English Utilitarians and India*. Oxford: Clarendon Press.

Stoneman, C. and L. Cliffe. 1988. *Zimbabwe: Politics, Economics and Society*. New York: Pinter Publishers.

Studenski, P. 1958. *The Income of Nations; Theory, Measurement and Analysis: Past and Present; A Study in Applied Economics and Statistics*. New York: New York University Press.

Sumarwoto, O. and I. Sumarwoto. "Forest and Global Environmental Problems." ASEAN Seminar on the Management of Tropical Forest for Sustainable Development, January 24, 25, 1990.

Sumitro, D. "The Emerging Indonesian Economy in the Context of Global Development." The Indonesia Forum, Jakarta, July 10, 1990.

Sundin, J. and L. G. Tedebrand. 1981. Mortality and Morbidity in Swedish Iron Foundries 1750-1875. In *Tradition and Transition: Studies in microdemography and social change*, eds. A. Brandstrom and J. Sundin, Umea: Demographic Database, University of Umea.

Sundin, J. 1976. "Theft and penury in Sweden 1830-1920: A comparative study at the county level." *Scandinavian Journal of History* (1):265-292.

Suparlan, P. and H. Suigit. 1980. Culture and Fertility: The Case of Indonesia. Research Notes and Discussion Paper No. 18 Singapore: Institute of Southeast Asian Studies.

Surapaty, S. C., E. Roflin and L. Heydir. *Studi Dinamika Kependudukan-Lingkungan Hidup diKapubupaten Lahat, Sumatra Selatan, Indonesia* Pusat Penelitan Kependudukan Universitas Sriwijaya.

Sylvester, C. 1986. "Zimbabwe's 1985 Elections: Search for National Mythology." *Journal of Modern African Studies* 24(1):229-255.

Symonds, R. and M. Carder. 1973. *The United Nations and the Population Question, 1945-1970*. New York: McGraw-Hill Book Company.

Taller de trabajo sobre losefectos de la contaminacion atmosferica en la salud, Tepotzotalan, Mexico, 1989.

Tchabi, V. 1986. *Etude Preliminaire sur l'Ecologie et les Ressources Pastoreales de la Zone d'Exploitation Controlee du Qibier (Z.E.C.Q.) de la Pendjari au Benin.*

Teitelbaum, M. S. 1975. "The Relevance of Demographic Trasition Theory for Developing Countries." *Science* (188):420-5.

Teitelbaum, M. S. and J. M. Winter. 1985. *The Fear of Population Decline*, Chapter 3. Orlando: Academic Press.

Teitelbaum, M. S. and J. M. Winter, eds. 1988. *Population and Resources in Western Intellectual Traditions*. New York: Cambridge University Press.

Tempo. 1990. "Api dan Operasi Lestari." September 15.

Thornton, R. D. 1987. *American Indian Holocaust and Survival: A Population History Since 1492*. Norman, OK: University of Oklahoma Press.

Thrupp, L. A. Fall 1989. "Politics of the Sustainable Development Crusade: From Elite Protectionism to Social Justice in Third World Resource Issues." *Environment, Technology and Society* 58:1-7.

Tilly, C. 1978. The Historical Study of Vital Processes. In *Historical Studies of Changing Fertility*, ed. C. Tilly, 1-55. Princeton: Princeton University Press.

Tisdell, C. 1988. "Sustainable Development: Differing Perspective ofEcologists and Economists, and Relevance to LDCs." *World Development* 16:373-384.

Trenbath, B. R. 1989. Mathematical Models of Shifting Cultivation. In *Mineral Nutrients in Tropical Forest and Savanna Ecosystems*, ed. J. Proctor, 353-369. Oxford: Blackwell Scientific Publications.

Turke, P. W. 1989. "Evolution and the demand for children." *Population and Development Review* 15(1):61-90.

UNEP. 1982. *GEMS: Estimating Human Exposure to Air Pollutants*. Tunbridge Wells, England: Penshurst Press, Ltd.

UNEP. 1983. *GEMS/Water Data Evaluation Report*. Tunbridge Wells, England: Penshurst Press, Ltd.

UNEP 1987. *GEMS: Global Pollution and Health*. Tunbridge Wells, England: Penshurst Press, Ltd.

UNEP. 1985. *GRID*. Tunbridge Wells, England: Penshurst Press Ltd.

United Nations. 1973. *The Determinants and Consequences of Population Trends*, 631-32. New York: United Nations.

United Nations. 1988. *World Demographic Estimates and Projections 1950-2025*. New York: United Nations.

United Nations. 1988. *World Population Prospects 1988*. New York: The United Nations.

United Nations Department of International Economic and Social Affairs. 1973. *The Determinants and Consequences of Population Trends*. New York: United Nations.

United Nations Department of International Economic and Social Affairs. 1982. *Levels and Trends of Mortality Since 1950*. New York: United Nations.

United Nations Department of International Economic and Social Affairs. 1984. *Population and Vital Statistics Report 1984*. New York: United Nations.

United Nations Department of International Economic and Social Affairs. 1989. *Prospects of World Urbanization 1988*. New York: United Nations.

United Nations Department of International Economic and Social Affairs. 1989. *World Population Prospects 1990*. New York: United Nations.

United Nations Fund for Population Activities (UNFPA). 1989. *Global Population Assistance Report*. New York: UNFPA.

United Nations Fund for Population Activities (UNFPA). 1985. *Population Perspectives: Statements by World Leaders*. New York: UNFPA.

United States Agency for International Development (USAID). 1983. *AID Population Sector Strategy* Agency for International Development, Washington, D.C.[Population Sector Council (March).]

United States Agency For International Development (USAID). 1986. "Implementing AID Privitazation Objectives." [Policy Determination. Agency for International Development. Washington, D.C.(June).]

United States Agency For International Development (USAID). 1982. *Population Assistance: Policy Paper* USAID, Washington, D.C.[Bureau For Program and Policy Coordination (September).]

United States Bureau of the Census. 1961. *Historical Statistics of the United States*. Washington D.C.: U.S. Department of Commerce.

Uphoff, N. T. 1986. *Local Institutional Development: An Analytical Source Book with Cases*. West Hartford, CT: Kumarian Press.

Urbina, M. 1989. "Fertility and Health in Mexcio." *Sal Pub Mex* 32(2):168-176.

Utterstrom, G. 1955. "Climatic Fluctuations and Population Problems in Early Modern History." *Scandinavian Economic History Review* (XX).

Utterstrom, G. 1957. *Jordbrukets arbetare den Svenska arbetarklassens historia*, Stockholm.

Vatkiotis, M. January 12, 1989. "Tug-of-war Over Trees: Indonesia Juggles Timber Money and Conservation." *Far Eastern Economic Review* 143(41).

Vayda, A. P. and A. Sahur. 1985. "Forest Clearing and Pepper Farming by Bugis Migrants in East Kalimantan." *Indonesia* 39:93-110.

Vazquez de Espinosa, A. 1942. Compendium and Description of the West Indies, Book Two. Washington, D.C.: Smithsonian Institution.

Vogt, W. April-June 1946. "Mexico's National Parks." *National Parks Magazine* :13-16.

Voland, E. 1990. "Differential Reproductive Success within the Krummhnorn Population (Germany, 18th and 19th Centuries)." *Behav. Ecol. and Sociobiol.* 26:65-72.

Walsh, J. September 12, 1986. "Famine Early Warning Closer to Reality." *Science* 233(News and Comment):1145-1147.

Walsh, J. January 15, 1988. "Famine Early Warning System Wins Its Spurs." *Science* 239(News and Comment):249-250.

Warwick, D. P. 1982. *Bitter Pills: Population Policies and their Implementation in Eight Developing Countries*. New York: Cambridge University Press.

Wattal, P. K. 1917, 1958. *The Population Problem in India: A Census Study*. London: Bennet, Coleman & Co.

Weeks, J. December 1988. "The Demography of IslamicNations." *Population Bulletin* 43(4).

Weiner, D., S. Moyo, B. Munslow and P. O'Keefe. 1985. "Land Use and Agricultural Productivity in Zimbabwe." *Journal of Modern African Studies* 23(2):251-85.

Weins, J. A. 1989. "Spatial scaling in ecology." *Functional Ecology* (3):385-397.

West, P. C. and S. R. Brechin. 1991b. National Parks, Protected Areas, and Resident Peoples: A Comparative Assessment and Integration. *Resident Peoples and National Parks: Social Dilemmas and Strategies in International Conservation*, eds. P. C. West and S. R. Brechin, Tucson: University of Arizona Press.

West, P. C. and S. R. Brechin., eds. 1991a. *Resident Peoples and National Parks: Social Dilemmas and Strategies in International Conservation*. Tucson: University of Arizona Press.

Wetterberg, G. B. 1974. *The History and Status of South American National Parks and an Evaluation of Selected Management Options*. Phd. Dissertaion, University of Washington.

Whitlow, J. R. 1980. Deforestation in Zimbabwe: Problems and Prospects. Supplement to *Zambezia*, Harare: University of Zimbabwe.

Whitlow, J. R. 1980. "Environmental Constraints and Population Pressure in the Tribal Areas of Zimbabwe." *Zimbabwe Agricultural Journal* 77(4):173-81.

Whitlow, J. R. 1988. *Land Degradation in Zimbabwe: A Geographical Study* National Resources Board, Harare.

Whitlow, J. R. 1980. "Land Use, Population Pressures, and Rock Outcrops in the Tribal Areas of Zimbabwe." *Zimbabwe Agricultural Journal* 77(1):3-11.

Whitlow, J. R. 1988. "Soil Conservation History in Zimbabwe." *Journal of Soil and Water Conservation* 3(4):299-303.

Whitlow, J. R. and B. Cambell. 1988. "Factors Influencing Erosion in Zimbabwe: A Statistical Analysis." *Journal of Environmental Management*.

Whitten, A. J. October 1987. "Indonesia's Transmigration Program and Its Role in the Loss of Tropical Rain Forests." *Conservation Biology* 1(3):239-256.

Wilken, G. C. 1987. *Good Farmers: Traditional Agriculture and Resource Management in Mexico and Central America*. Berkeley: University of California Press.

Wilken, G. C. 1972. "Microclimate Management by Traditional Farmers." *Geographical Review* 62(4):544-560.

Williams, G. C. 1966. *Adaptation and Natural Selection: A Critique of Some Current Evolutionary Thought*. Princeton, N.J.: Princeton University Press.

Williams, M. 1989. *Americans and Their Forests: A Historical Geography*. Cambridge: Cambridge University Press.

Williams, R. 1973. *The Country and the City*. New York: Oxford University Press.

Willigan, J. D. and K. A. Lynch. 1982. *Sources and Methods of Historical Demography*. New York: Academic Press.

Winch, D. 1987. *Malthus*. Oxford: Oxford University Press.

Winter, J. M. 1988. Socialism, Social Democracy and Population Questions in Western Europe: 1870-1950. In *Population and Resources in Western Intellectual Traditions*, eds. M. S. Teitelbaum and J. M. Winter, New York: Cambridge University Press.

Wirosuhardjo, K. and S. Yosephine. March 1990. "Population Mobility and Urbanization towards the year 2000." Seminar on Economic Prospect, University of Indonesia.

Wohlin, N. 1912. *Den Svenska Jordstyckningspolitiken i de 18: de och 19: de Arhundradena*. Stockholm: Kungl. Boktryckeriet P. A. Norstedt and Soner.

Wolf, E. R. 1966. *Peasants*. Englewood Cliffs, NJ: Prentice Hall.

Wolfson, M. 1983. *Profiles in Population Assistance: A Comparative Review of the Principal Donor Agencies*. Paris: OECD Development Center.

Working Group on Population and Economic Development. 1986. *Population Growth and Economic Development: Some Policy Questions*. Washington, D.C.: Committee on Population of the National Research Council.

World Bank. 1981. *Accelerated Development in Sub-Saharan Africa: An Agenda for Action*. Washington, D.C.: World Bank.

World Bank. 1989. *Striking a Balance: The Environmental Challenge of Development*. Washington, D.C.: The World Bank.

World Bank. 1989. *Sub-Saharan Africa: From Crisis to Sustainable Growth*. Washington, D.C.: The World Bank.

World Bank. 1982. *World Development Report 1982*. Washington, D.C.: Oxford University Press.

World Bank. 1984. *World Development Report, 1984*. New York: Oxford University Press.

World Commission on Environment and Development. 1987. *Food 2000: Global Policies for Sustainable Agriculture*. London: Zed Books.

World Commission on Environment and Development. 1987. *Our Common Future*. New York: Oxford University Press.

World Resources Institute (WRI). 1988. *World Resources 1988-1989*. New York: Basic Books.

The World Resources Institute. 1990. *World Resources 1990-91*. New York: Oxford University Press.

Worldwatch Institute. 1988. *State of the World 1988 (Annual)*. New York: Norton.

Worldwatch Institute 1991. *State of the World 1991*. New York: Norton.

Worster, D. 1987. "The Vulnerable Earth: Towards a Planetary History." *Environmental Review* 11(2):87-103.

Wrigley, E. A. 1988. The Limits to Growth: Malthus and the Classical Economists. In *Population and Resources in Western Intellectual Traditions*, eds. M. S. Teitelbaum and J. M. Winter, 30-48. New York: Cambridge University Press.

Wrigley, E. A. and A. Schofield. 1981. *Population History of England 1541-1871: A Reconsideration*. Cambridge: Harvard University Press.

Zachariah, K. C. and M. T. Vu. 1988. *World Population Projections, 1987-88* The John Hopkins University Press, Baltimore/London.

Zelizer, V. A. 1985. *Pricing the Priceless Child: The Changing Social Value of Children*. New York: Basic Books, Inc.

Zimbabwe National Family Planning Council. 1988. *Establishment of a Secretariat for Population Policy Development* ZNFPC (November), Harare.

Zinyama, L. and R. Whitlow. 1986. Changing Patterns of Population Distribution in Zimbabwe. In *GeoJournal*, 365-84. Dordrecht and Boston: Reidel Publishing Company.

Zola, E. 1899. *Recondite*. Paris: E. Pasquelle.

Contributing Authors

Valentin Agbo
Professor
National University of Benin
Benin

Steven R. Brechin
Lecturer
Sociology

Research Fellow
Center for Energy and Environmental Studies
Princeton University

C. Gaye Burpee
Professor
Crop and Soil Sciences
Michigan State University

Alice Clarke
Project Fellow
Behavioral Ecology
School of Natural Resources & Environment
The University of Michigan

William D. Drake
Professor
Resource Planning and Conservation
School of Natural Resources & Environment

Population Planning and International Health
School of Public Health
Urban, Technological, and Environmental Planning
College of Architecture & Urban Planning
The University of Michigan

Alan Grainger
Professor
School of Geography
University of Leeds
England

David B. Halmo
Research Associate
School of Natural Resources
University of Arizona

Laurel Heydir
Research Associate
Sriwijaya University
Indonesia

John Hough
Senior Advisor
World Wildlife Fund
Madagascar

Bobbi S. Low
Professor
Resource Ecology
School of Natural Resources & Environment
The University of Michigan

George Martine
President
Institute for the Study of Society, Population, and Nature
Brasilia, Brazil

Alison McIntosh
Professor
Population Planning and International Health
School of Public Health
The University of Michigan

Gayl D. Ness
Professor
Sociology
College of Literature, Science, and Arts

Population Planning and International Health
School of Public Health
The University of Michigan

Eddy Roflin
Research Associate
Sriwijaya University
Indonesia

Vernon Ruttan
Professor
Agriculture and Applied Economics
University of Minnesota

Nestor Sokpon
Professor
National University of Benin
Benin

Brent W. Stoffle
Research Associate
Ohio Wesleyan University
Delaware, Ohio

Richard W. Stoffle
Professor
Bureau of Applied Research in Anthropology
University of Arizona

Surya Chandra Surapaty
Professor and Director
Population Research Center
Sriwijaya University
Indonesia

Michael Teitelbaum
Program Officer
Alfred P. Sloan Foundation
New York, New York

Patrick C. West
Professor
Outdoor Recreation
School of Natural Resources & Environment
The University of Michigan

Andrew L. Williams
Project Fellow
Anthropology
The University of Michigan

Jay Winter
Lecturer, History
Pembroke College
Cambridge, England

Kartomo Wirosuhardjo
Director, Research Institute
University of Indonesia
Indonesia

Frank D. Zinn
Professor
Urban Planning
Michigan State University

Index

Abortion, 3
Academic specialization, 20, 361
Acquired immune deficiency syndrome (AIDS), 12, 66-67, 69
Adaptionist strategies, 63
Aerial photography, 360
Africa, 357, 388, 398, *see also* specific countries
  agricultural transition in, 316, 317
  agriculture in, 57
  AIDS in, 67
  birth rates in, 310
  calorie intake per capita in, 66
  carrying capacity of earth in, 42
  cereal output in, 51
  deforestation in, 75, 87, 96, 97
  demographic transition in, 45, 309, 310, 340
  droughts in, 363
  epidemiological transition in, 313
  famine in, 363
  fertility in, 52, 312
  food consumption in, 95
  food production in, 53
  forest area trends in, 98
  forestry transition in, 322, 323
  future of, 52-53
  mortality rates in, 312
  nightmare scenario for, 52-53
  population of, 36
  population growth in, 52-53
  tuberculosis in, 67
  water resources in, 53
Agbo, Valentin, 283
Aggression, 29
Agricultural transition, 315-317, 318, 341, 397, 399, 400, *see also* Agriculture
  in Brazil, 167-186
    Conservative Modernization Era in, 168-175
    Crisis Period in, 176-179
    efficiency of, 181, 183
    Super-Harvests Era in, 179-180, 182
Agriculture, 57-70, 305, *see also* Agricultural transition; Food production
  advent of, 8
  in Asia, 51, 57
  biological constraints on, 58-61
  climate change effects on, 100
  conservation farming methods in, 134
  demands placed on, 63-64
  in Dominican Republic, 263-265, 266-267
  emergence of, 34
  environmental constraints on, 62-65
  environmental effects of modernization of, 172
  environmentally compatibile, 65
  external factors affecting, 266-267
  governmental policies on, 65, 181, 387-388
  health constraints on development of, 65-70
  in Indonesia, 159-161
  local factors affecting, 266-267
  mixed crop, 264
  modeling of systems in, 100
  permanent, 13, 78
  plantation, 117, 118
  price supports and subsidies in, 65, 181
  research in, 58, 60, 61
  resource-based, 57
  resource constraints on, 61-65
  science-based, 57
  shifting, 13, 71, 73, 78, 80, 291-294
  in Southeast Asia, 113, 117, 118
  in South Sumatra, Indonesia, 249
  in Tamougou, Northern Benin, 287, 288, 289-294
  technological constraints on, 58-61
AIDS, *see* Acquired immunodeficiency syndrome
Air pollution, 324, 338
Aldrin, 175
Animal feed efficiency, 60
Animal nutrition, 60
Animal vs. human populations, 28-29
Anthropology, 394
Asia, *see also* specific countries
  agricultural transition in, 317, 318
  agriculture in, 51, 57
  birth rates in, 312
  carrying capacity of earth in, 42
  deforestation in, 75, 87, 97
  demographic transition in, 45
  economic development of, 49-52
  food consumption in, 95
  food production in, 49
  forest area trends in, 98
  forestry transition in, 321
  future of, 49-52
  mortality rates in, 312

Asia (*continued*)
population of, 36
population growth in, 49
Southeast, *see* Southeast Asia
Atmospheric chemical changes, 33
Atmospheric pollution, 66, 67, 324, 338

Baby booms, 27
Bangladesh, 68, 341, 391
Beaches, 268-269
Behavioral ecology, 197
Behavior laws, 19
Benin, 331, 332, 382, 400
Northern, *see* Tamougou, Northern Benin
BHC, 175
Biochemistry, 60
Biological diversity, 226
Biological factors, 17, 19, 26
Biologists, 196
Biology, 23, 26, 58, 60
Biomass burning, 62, 63
Biomass production, 63, 64
Biotechnology, 58, 60, 61
Birth control, *see* Contraception; Family planning
Birth order, 198, 211, 213
Birth rates, *see also* Fertility
in Africa, 310
in Asia, 311
demographic transition and, 44, 309
high, 34-35
history of, 5-6
in Southeast Asia, 118, 310
in Zimbabwe, 147
BKKBN, *see* National Family Planning Coordinating Board, Indonesia
Black fly, 68
Botswana, 52
Boundaries of ecosystems, 386-387, 393, 397-402
Brazil, 167-186, 388, 399, 401, 404
Conservative Modernization Era in, 168-175
Crisis Period in, 176-179
deforestation in, 87
food supply and distribution in, 185
mechanization in, 180
Super-Harvests Era in, 179-180, 182
urbanization in, 184, 185
Brazilian Amazonia, 72, 89, 175
Brechin, Steven R., 225, 357, 377
Buen Hombre, Dominican Republic, 381
agriculture in, 263-265, 266-267
beaches in, 268-269
coral reef exozones in, 269-273
demographics of, 275
description of, 259-262
efficiency of effort in, 276
fishing in, 273-276, 279
forests in, 266-267
lagoons in, 268-269
mangroves in, 268-269
marine ecozones in, 267-268
resources in, 262-263, 267-268
soil in, 262-263
terrestrial ecozones in, 262-263
Bureaucracy, 338
Bureaucratically portable technology, 47
Burma, 115, 116
demographic transition in, 400
fertility in, 126, 127, 128, 398
mortality rates in, 120, 121
political systems in, 132
population of, 111
rice output in, 122, 123, 124, 125
Burpee, C. Gaye, 253

Calorie intake per capita, 66
Canada, 334, 401
Capitalism, 119
Carbon dioxide, 62, 63, 64
Carrying capacity of earth, 42
Cato Institute, 28
Cattle ranching, 73
Causation, 29
Cereal output, 38, 50, 51, 70
CFCs, *see* Chlorofluorocarbons
CGIAR, *see* Consultative Group on International Agricultureal Research
Chad, 363
Chaos theory, 398
Chernobyl, 27
Child Spacing and Family Planning Council, Zimbabwe, *see* Zimbabwe National Family Planning Council
Child survival rates, 68
China, 51, 315, 322, 326-327, 390, 405
Chlordano, 175
Chlorofluorocarbons (CFCs), 62
Ciguatera, 272
Clarke, Alice, 195
Clearance, 74
Climate, 190, 359
agriculture and, 100
changes in, 62, 65, 100, 226
deforestation and, 253
ecosystems and, 100
industrialization and, 62
models of, 305
prediction of, 305
in Sweden, 203
Coffee production cycle, 240-241, 246
Communities, 5, 6, 7

Complexity, 305-306, 380-385, 392-396
Comprehensive planning, 360-369
Congenital malformations, 67
Conservation, 63, 82, 134
Conservation management, 246, 248-249
Conservatism, 25, 28
Constrained ecosystems, 284, 288, 294-295
Consultative Group on International Agricultural Research (CGIAR), 61, 68
Contraception, 47, 48, 128-129, *see also* Family planning
Coral reef exozones, 269-273
Cornacopianism, 28
Correlation analysis, 83
Costa Rica, 390
Cost-benefit analysis, 3
Cross-sectional analysis, 83, 86, 92
Cuba, 92, 96
Czechoslovakia, 326

DDT, 175
Decadence, 23
Decentralization, 15
Deforestation, 12, 14, 33, 62, 71-101, 341, 393, 397, 399
  annual rates of, 85, 92, 93
  in Brazil, 72, 89, 175
  causes of, 72, 73, 74-82
    in South Sumatra, Indonesia, 226, 227, 231-243
  cross-sectional analysis of, 83, 86, 92
  current rates of, 83-84
  defined, 73, 74
  dynamic analysis of, 88-89
  economic factors in, 242-243, 250
  empirical studies of, 100
  equilibrium and, 81-82
  estimation of rates of, 73
  future research on, 100-101
  future trends in, 79-81
  global climate and, 253
  governmental policies on, 79, 227, 247-251
  history of, 87, 91
  international cross-sectional analysis of, 86, 92
  intra-national variation in, 88-89
  logging and, 73, 74, 226, 227, 393
  long-term trends in rates of, 84
  modeling of, 73-82, 92-96, 305
    national, 86-88, 89
    quantification in, 82-83
    scenario design for, 94-95
    simulations in, 95-96
    structure for, 92-94
  monitoring of, 100, 248
  physical factors in, 78-79
  population growth and, 72-73
    in South Sumatra, Indonesia, 227, 231-239
  prevention of, 90, 91
  rates of, 95, 96
    annual, 85, 92, 93
    current, 83-84
    estimation of, 73
    forest area per capita vs., 89-90, 91, 92
    long-term trends in, 84
    regional trends in, 97
  regional trends in rates of, 97
  research on, 100-101
  socioeconomic factors in, 77, 82
  in South Sumatra, Indonesia, 226-227
    causes of, 227, 231-243
    economic factors in, 242-243
  as spatial phenomenon, 78-79, 100
  static analysis of, 88
  trends in, 84, 97
Degenerative disease, 34, 313, 353, *see also* Epidemilogical transition
Demand for children, 224
Demand patterns, 70
Demand side, 48
Demographic transition, 5, 19, 34, 309-312, 353, 399
  in Africa, 45, 310, 312, 340
  in Asia, 45
  basis for, 72
  birth rates and, 308
  in Brazil, 184
  child survival rates and, 68
  in Europe, 45
  fertility and, 215-216
  history of, 44-48
  ideas behind, 309
  in Latin America, 45
  modified theory of, 73
  mortality rates and, 215-216, 308
  in North America, 309, 310, 340
  present, 44-48
  in Southeast Asia, 309, 310, 400
  in Sweden, 195, 215-216
  in Thailand, 400
  theory of, 12, 24, 26, 73
  timing of, 340
  trajectory of, 312, 340
  variants of, 44
Demography, 11-12, 24-28, 196, 197, 217
  biological factors and, 26
  change in, 5
  deforestation and, 73
  of Dominican Republic, 256-269, 275

Demography (*continued*)
fertility and, 220
goals in, 3
of Hispaniola, 256-259
population and, 377
population from perspective of, 196, 377
variables in, 2
Denmark, 325
Dependency ratio, 66
Desertification, 62, 71, 72, 99, 100
Determinism, 17, 24
Diarrheal disease, 66, 298
Diarrheal Research Center, Bangladesh, 68
Dieldrin, 175
Dietary deficiencies, 68
Disequilibriums, 109, 284, *see also* specific types
Domestication of animals, 34
Dominican Republic, 253-282, 381
agriculture in, 263-265, 266-267
beaches in, 268-269
Buen Hombre Community in, *see* Buen Hombre, Dominican Republic
coral reef exozones in, 269-273
deforestation in, 96
demographics of, 275
demography of, 256-269
droughts in, 260, 266
ecological assessment studies of, 256
efficiency of effort in, 276
fishing in, 273-276, 279
forest area per capita in, 92
forests in, 266-267
historic demography of, 256-269
lagoons in, 268-269
mangroves in, 268-269
marine ecozones in, 267-268
multiple transitions in, 400
population of, 257
resources in, 262-263, 267-268
seafood in, 277-279
social assessment studies of, 256
soil in, 262-263
terrestrial ecozones in, 262-263
weather in, 271
Drake, William D., 305, 377
Droughts, 1, 260, 266, 330, 363
Dynamic analysis of deforestation, 88-89
Dysfunctional disequilibriums, 284

East Asia, 57, *see also* specific countries
East Java, 132
Ecologists, 6, 7, 17, 20
Ecology, 9, 28-30
behavioral, 197
complexity in, 29
of Dominican Republic, 256
economics vs., 29-30
of family, 217-223
fertility and, 220
as "subversive science," 30
of Tamougou, Northern Benin, 283
Economic development, 13, 27, 128
in Asia, 49-52
deforestation and, 226
expansion of, 22
fertility and, 72, 73
governmental policies on, 49, 388
land use changes and, 77
limits to, 28
mortality rates and, 72
output per capita and, 49
in Southeast Asia, 112
Economics, 3, 17, 24-28, 72, 197, 283, 399
deforestation and, 242-243, 250
ecology vs., 29-30
Keynesian, 24, 27
population growth and, 72
in Tamougou, Northern Benin, 283, 297
in Zimbabwe, 142
Economic stagflation, 27
Economists, 2, 20
Ecosphere, 6, 7
Ecosystems, 6, 20, 110, 254
boundaries of, 386-387, 393, 397-402
climate change effects on, 100
constrained, 284, 288, 294-295
culturally defined, 399
defined, 399
health of, 9, 13, 402-403
landscape, 6
linguistically defined, 399
local, 28
multiple, 400-402
theory of, 386, 387
Ecotesseras, 6
Educational transition, 337-338
EEC, *see* European Economic Community
Electronic monitoring, 360
El Salvador, 92, 96
Empirical studies, 100
Employment, 146-148, 162-164, 202
Energy conservation, 63
Energy consumption, 51, 54, 332
Energy efficiency, 62, 333, 391
Energy prices, 64, 305
Energy production, 387-388

Energy transformations, 33, 38-43, *see also* Fossil fuels; Sails; specific types
Energy transition, 338
Energy use revolutions, 38
Engineering technology, 64
Enteric fermentation, 62
Environment, 105-106, 189-191, 383
  abstract view of, 383
  artificial categories of, 386
  boundaries of, 386
  conceptualization of, 2
  defined, 6-7, 12
  family and, 196-198
  global persepctive on, 12-13
  governmental policies on, 3
  heredity vs., 12, 197
  interdependence of genes and, 196
  language style and, 20-21
  local monitoring of, *see* Local monitoring
  thought styles concerning, 17-20
  Western theory of, 12, 21-24
Environmental degradation, 29, 33, *see also* specific types
Environmental determinism, 24
Environmental experimentation, 29
Environmentalism, 28-30, 73, 226
Environmentally compatibile agriculture, 65
Environmental protection, 1
Environmental research, 254
Epidemics, 12, 66, 308, *see also* specific types
Epidemiological transition, 29, 312-313, 340, 342, 353, 400
Epidemiology, 25
Equilibrium, 81-82
Erosion, 62, 64, 69, 357, 358
  in Brazil, 175
  causes of, 133
  prevention of, 357
  in Zimbabwe, 133, 134-137
Ethiopia, 363
Eugenics, 20, 23-24
Europe, 395, 405, *see also* specific countries
  carrying capacity of earth in, 42
  demographic transition in, 45
  ecological disasters in, 405
  energy consumption in, 54
  environmental disasters in, 391
  epidemiological transition in, 312
  fertility in, 398
  future of, 53-54
  gasoline prices in, 334
  industrialization of, 21, 54
  population of, 36
  population growth in, 53-54
  toxicity transition in, 325-326
  urbanization transition in, 330
European Economic Community (EEC), 182
Explorers, 39, 42
Exponential functions, 349, 350
Extinctions of species, 33, 34, 64

Family ecology, 217-223
Family planning, 48, 72, *see also* Contraception; Fertility
  in Asia, 51
  in Indonesia, 156, 158
  in Southeast Asia, 128
  in South Sumatra, Indonesia, 249
  village-based, 156
  in Zimbabwe, 140-142
Famine, 1, 288, 308, 363
Famine Early Warning System (FEWS), 363, 364
FAO, *see* United Nations Food and Agriculture Organization
Farmland, defined, 94
Fascism, 26
Feedback, 29, 342, 343, 402, 404, 405
Feigenbaum's graphic analysis, 396
Fermentation, 62
Fertility, 1, 11, 12, 196, 197, 391, 394, 398, *see also* Birth rates; Family planning
  in 1940s and 50s, 27
  in Africa, 52, 311
  biologistic explanations of decline in, 25, 26
  contraception and, 47
  cultural factors in, 48
  current, 47, 48
  demographic approach to, 220
  demographic transition and, 46, 47, 48, 215-216
  ecological approach to, 220
  economic development and, 72, 73
  in Europe, 398
  governmental policies on, 387-388
  high, 22
  in Indonesia, 155, 398
  low rates of, 27
  migration and, 214
  mortality rates and, 34
  resources and, 198, 204-211
  in Southeast Asia, 109, 125-130, 132, 398
  in South Sumatra, Indonesia, 249
  in Sweden, 201, 202, 211, 214
    in demographic transition, 215-216
    investment and, 217-223

Fertility (*continued*)
in Sweden (*continued*)
resources and, 204-211
in Zimbabwe, 138, 140-142
Fertilizer, 58, 60, 65, 69, 72, 305
in Asia, 51
in Brazil, 168, 172, 179
FEWS, *see* Famine Early Warning System
Fishing, 273-276, 279, 295
Flooding, 65, 341, 358
Food and Agriculture Organization, *see* United Nations Food and Agriculture Organization
Food consumption, 95
Food costs, 208
Food demand, 58
Food distribution, 185
Food prices, 80
Food production, 20, 22, 49, 51, 53, 290, *see also* Agriculture
Food supply, 185
Food-system perspective, 65
Ford Foundation, 61, 121, 153, 155
Forest area per capita, 89-90, 91, 92, 99
Foresters, 73
Forestry transition, 316, 318-323, 353
Forests
conservation of, 82
in Dominican Republic, 266-267
external factors affecting, 266-267
governmental policies on, 387-388
growth rates for, 91
in Indonesia, 161
local factors affecting, 266-267
loss of, *see* Deforestation
population density and, 84-86, 87
protection of, 82
regional trends in area of, 98
Fossil fuels, 62-63, 332
Fossil fuel transition, 33, 42-43, 109, 332-334, 338
France, 325, 334
Functional characteristics, 352-354
Fungicides, 172

Gasoline prices, 332, 334
GEMS, *see* Global Environmental Monitoring System
Genetic engineering, 58
Genetics, 20, 23
Geographers, 2
Geographic information systems (GIS), 101, 361, 362, 363, 370, 373
Geography, 394
Germany, 325, 332, 334
Ghana, 92, 96
GIS, *see* Geographic information systems
Global change, 33, 64
Global environment, *see* Ecosystems
Global Environmental Monitoring System (GEMS), 362-363
Global perspective, 4, 11-15
Global warming, 1, 33, 62, 63
Gompertz function, 350, 351
Governmental policies, 387-391, 394, 399, *see also* specific types
on agriculture, 65, 181, 387-388
conflicting, 359
on deforestation, 79, 227, 247-251
on economic development, 49, 388
on energy production, 387-388
on environment, 3
on fertility, 387-388
on forests, 387-388
on health, 387-388
on land use, 75
on mortality rates, 387-388
on population, 3, 49, 139, 142-143, 195
on resource management, 63
in South Sumatra, Indonesia, 247-251
in Tamougou, Northern Benin, 283
on trade, 387-388
transition theory and, 344-348
in Zimbabwe, 139, 142-143, 148-150
Grainger, Alan, 71
Graphic analysis, 396
Great Lakes, 346
Greenhouse gases, 52, 62, *see also* specific types
Greenhouse model of Earth, 1
Green movement, 28
Green party, 28
Green Revolution, 51, 168, 172
GRID project, 363
Groundwater pollution, 62, 64, 324
Growth hormones, 60
Guinea, 96

Halmo, David B., 253
Health, 25, 358, 394
agricultural development and, 65-70
epidemiological transition and, 312
governmental policies on, 387-388
improvements in, 29, 72
in Mexico, 339
in Tamougou, Northern Benin, 297-298
Heavy metals, 67
Hepatitis, 326
Heptachlorex, 175
Herbicides, 179
Heredity, 12, 196, 197

Heritage Foundation, 28
Heydir, Laurel, 225
High-yielding crops, 72, 134, 168
High-yielding variety (HYV) seeds, 168, 172
Hispaniola, 256-259
Hong Kong, 51, 391, 401
Hough, John, 283
HPV, *see* High-yielding variety
Human perfectibility, 22
Hungary, 325

IARCs, *see* International agricultural research centers
IGBP, *see* International Geosphere-Biosphere Programs
IIASA, *see* International Institute for Applied Systems Analysis
Incentive compatible institutions, 63
Income, 58, 63, 70
India, 51, 163, 330, 331, 390, 391, 401
Individual decision models, 220
Indochina, 111
Indonesia, 153-165, 390
  agriculture in, 159-161
  employment in, 162-164
  family planning in, 156, 158
  fertility in, 129, 155, 398
  forests in, 161
  local monitoring in, 369-372
  Ministry of Population and Environment in, 364
  mortality rates in, 120, 121
  political systems in, 132
  population of, 111, 154-159
  Population-Environment Monitoring Systems in, 369-372
  population growth in, 51, 154-159
  rice output in, 122, 123, 124
  South Sumatra in, *see* South Sumatra, Indonesia
  urbanization transition in, 162-164, 331
Industrialization, 21, 42, 341
  in China, 51
  climate and, 62
  demographic transition and, 48
  in Europe, 54
  in North America, 54
  population growth and, 43-44
Industrial pollution, 226
Industrial revolution, 26
Infant mortality, 65, 68, 132, 216, 361
Infectious disease, 29, 34, 66, 68, 117, 312, 313, 353, *see also* Epidemilogical transition; specific types
Insecticides, 168, 172, 175, 179
Insects, 62
Institutions, 387-391, 402-405, *see also* specific types
Integrated monitoring, 360-364, 366-369
Intercropping, 264
International agricultural research centers (IARCs), 67
International cross-sectional analysis, 86, 92
International donor community, 67
International Geosphere-Biosphere Programs (IGBP), 64
International Institute for Applied Systems Analysis (IIASA), 89
International Planned Parenthood Federation (IPPF), 153, 155
International Rice Research Institute (IRRI), 121, 122, 124, 160
International Union for the Scientific Study of Population (IUSSP), 26
Interventions, 387-391, *see also* specific types
Investment, 217-223, 342
IPPF, *see* International Planned Parenthood Federation
IRRI, *see* International Rice Research Institute
Irrigation, 51, 71
Italy, 334
IUSSP, *see* International Union for the Scientific Study of Population
Ivory Coast, 96

Japan, 115, 401
  energy consumption in, 51, 54, 332
  energy efficiency in, 391
  gasoline prices in, 334
  population growth in, 51
  toxicity transition in, 326
Java, 113

Kampuchea, 116
  economic development in, 128
  fertility in, 126, 127, 128
  internal turmoil in, 128
  mortality rates in, 120, 121
  rice output in, 122, 123
Kenya, 52
Keynesian economics, 24, 27
Knowledge, 29, 64
Kwashiorkor, 288

Labor resources, 359
Lagoons, 268-269
Lahat District, *see* South Sumatra, Indonesia

Laissez-faire thought, 28, 119
Land capability classification studies, 100
Land cover, 359
Land degradation, 99
Land extension, 314
Land fragmentation, 226
Land productivity, 314-315
Land reform, 143-148
Landscapes, 6
Landslides, 358
Land tenure in Zimbabwe, 134-137
Land use, *see also* specific types
  empirical studies of, 71
  intensity of, 70
  local planning for, 365
  morphology of, 75-77
  national, 75-78
  planning for, 365
  population and transitions in, 72-73
  in Sweden, 202, 209, 211
  systems model of, 76
  technology and, 65
  theoretical studies of, 71
  transformation of, 64
  transitions in, 71, 72-73, 96, 133
    economic development and, 77
    population and, 82-92
    population growth and, 77
    socioeconomic factors in, 77, 82
  in Zimbabwe, 133, 143-150
Language styles, 20-21
Laos, 116, 120, 121, 122, 123, 126, 127, 128
Latin America, *see also* specific countries
  agriculture in, 57
  deforestation in, 75, 87, 97
  demographic transition in, 45
  food consumption in, 95
  forest area trends in, 98
  forestry transition in, 323
  population of, 36
  population growth in, 52
LDCs, *see* Lesser-developed countries
Lesser-developed countries (LDCs), 195, 196, 217, 223, 224, *see also* specific countries
Libertarianism, 28
Life expectancy, 65, 67, 68, 361
Lignocellulose, 60, 65
Local ecosystems, 28
Local expertise, 357-358
Local monitoring, 302-303
  adaptability of, 368
  attributes of, 366-368
  for comprehensive planning, 360-369
  flexibility of, 368
  in Indonesia, 369-372
  integrated, 360-364, 366-369
  methodology in, 368
  presentation in, 368
  technlogy in, 368
Local pollution, 324, 337
Locknevi Parish, 199, 214, *see also* Sweden
  aggregate analysis of, 199
  demographic transition in, 216
  fertility in, 204-211, 216, 217
  food costs in, 208
  individual analysis in, 201-202
  land ownership in, 209, 211
  migration in, 206, 212-213
  population growth in, 200
  resources in
    fertility and, 204-211
    migration and, 212-213
    mortality rates and, 213-214
  response to costs in, 205-210
Logging, 73, 74, 226, 227, 393
Logistical function, 349-351
Low, Bobbie S., 195

Macro systems, 4, 28, 388, 389
Maintenance research, 58
Malaria, 66, 118
Malaya, 116
Malaysia, 388
  deforestation in, 87
  fertility in, 127, 129, 130, 398
  political systems in, 132
  population of, 111
  population growth in, 158
  rice output in, 122, 123
Mali, 363
Malnutrition, 288
Mammoths, 34
Manageability, 305-306
Mangroves, 268-269
Marasmus, 288
Marine ecozones, 267-268
Market economy, 119
Market proximity, 80
Markets, 359
Martime, George, 167
Marxism, 18, 19, 20, 22, 23, 24, 28, 30
Maternal mortality, 132
Mathematical functions, 349-355, 396, *see also* specific types
Mathematics, 25
Mauritania, 363
McIntosh, Alison, 133
Mechanical technology, 64
Mechanization, 168, 180
Metals, 67
Methane, 62, 64

Metropolization, *see* Urbanization transition
Mexico, 313, 390, 401
  epidemiological transition in, 341-342
  health in, 339
  Population-Environment Monitoring Systems in, 372
  urbanization transition in, 331, 332
Micro systems, 4, 28, 388, 389
Middle East, 305, 323, *see also* specific countries
Middle Eastern War, 27
Migration, 197
  fertility and, 214
  resources and, 212-213
  rural to urban, 330
  in South Sumatra, Indonesia, 237-239
  in Sweden, 206, 212-213
Ministry of Population and Environment, Indonesia, 364
Mirex, 175
Mixed crop agriculture, 264
Models, *see also* specific types
  advances in, 305
  of agricultural systems, 100
  of climate, 305
  of deforestation, *see* under Deforestation
  greenhouse, 1
  individual decision, 220
  of land use, 76
  mathematical, 349-355, 396
  Population, Organization, Environment, and Technology, 380
  sector-spanning, 305
  systems, 76
Molecular biology, 58, 60, 61
Monitoring, 9, *see also* specific types
  of deforestation, 100, 248
  electronic, 360
  integrated, 360-364, 366-369
  local, *see* Local monitoring
  of population environment dynamics, 302-303
  resource, 64
  of soil erosion, 69
Monitoring technologies, *see also* specific types
Monocultures, 175
Mortality rates, 29, 68
  in Africa, 312
  in Asia, 312
  demographic transition and, 44, 46, 215-216, 309
  economic development and, 72
  epidemiological transition and, 312, 313
  governmental policies on, 387-388
  high, 34-35
  history of, 5-6
  in infants, 65, 68, 132, 216, 361
  maternal, 132
  pollution and, 67
  rapid decline in, 27
  resources and, 197, 213-214
  in Southeast Asia, 120-121, 132
  in Sweden, 215-216
  technology limiting, 47
  in United States, 314
Mozambique, 363
Multiple causation, 29
Multiple ecosystems, 400-402
Multiple transitions, 393-394, 400

National Erosion Survey, Zimbabwe, 134, 136, 137
National Family Planning Coordinating Board, Indonesia (BKKBN), 156, 157
National parks, 82
National Urban Development Strategy Project (NUDSP), Indonesia, 163
Nation states, 4
Natural gas, 42
Natural law, 30
Natural resources, 29, 73, 227, 394, *see also* Resources; specific types
Nature-nuture controversy, 12, 197
Negative reference thinking, 19
Neo-Malthusian perspective, 72
Nepal, 324
Ness, Gayl D., 33, 357, 377
Netherlands, 317
Newly industrialized countries (NICs), 154, 401, *see also* specific countries
New right, 28
NGOs, *see* Non-governmental Organizations
NICs, *see* Newly industrialized countries
Niger, 363
Nigeria, 92, 96
Nitrate, 69
Nitrous oxides, 62, 325
Non-governmental Organizations (NGOs), 153, 155
North America, 399, *see also* specific countries
  agricultural transition in, 318, 319
  demographic transition in, 309, 310, 340
  forestry transition in, 322
  future of, 53-54
  industrialization of, 21, 54

North America (*continued*)
population of, 36, 311
population growth in, 53-54
urbanization transition in, 331
North American Trade Agreement, 401
Northern Benin, *see* Tamougou, Northern Benin
North Sumatra, 129
Nuclear wastes, 325
NUDSP, *see* National Urban Development Strategy Project
Nutritional deficiencies, 65

Oceania, 36
Oil, 42
Oil spills, 1
One-factor determinists, 17
Organization, 2-3, 7, 12
Organo-chlorides, 175
Output per capita, 49
Overcrowding, 27
Overpopulation, 29
Ozone layer, 20, 33

Parasitic diseases, 66, 68, *see also* specific types
Pardigms, 377-380, *see also* specific types
Parks, 82
Pastoralism, 18, 71
Pasture, 94
Pathogens, 62
PCBs, *see* Polychlorinated biphenyls
PEMS, *see* Population-Environment Monitoring Systems
Per-capita income, 58
Permanent agriculture, 13, 78
Permanent pasture, 94
Permeability, 401
Pesticides, 51, 60, 62, 65, 72, 168, 172, 179, *see also* specific types
Pest management technology, 65
Petroleum, 306
Philippines, 1116
deforestation in, 96
demographic transition in, 400
fertility in, 126, 127, 128, 129, 398
forest area per capita in, 92
homelessness in, 163
independence of, 119
mortality rates in, 121
population of, 111
population growth in, 158
rice output in, 122, 123
Photography, 360
Plagues, 66
Plantation agriculture, 117, 118
Plant nutrients, 62
Poaching, 297
POET, *see* Population, Organization, Environment, and Technology model
Poland, 326
Political environmentalism, 28
Political institutions, 390-391, 403-405, *see also* Governmental policies
Pollution, *see also* specific types
air, 324, 338
atmospheric, 66, 67, 324, 338
groundwater, 62, 64, 324
industrial, 226
local, 324, 338
soil, 67, 175
surface water, 324, 338
toxicity transition and, 324
water, 62, 64, 67, 324, 338
Polychlorinated biphenyls (PCBs), 272, 347
Population, 104-105, 188, 382-383
abstract view of, 382
of Africa, 36
animal vs. human, 28-29
biologist view of, 196
conceptualization of, 2
control of, 3, 249
decline in, 11
defined, 5-6, 377
deforestation and, 72-73
demographic view of, 196, 377
density of, *see* Population density
distribution of, 100, 138
of Dominican Republic, 257
family and, 196-198
future of, 49-54
global perspective on, 11-12
governmental policies on, 3, 49, 139, 142-143, 195
growth of, *see* Population growth
health of, 12
homogenization of, 399
of Indonesia, 154-159
land use transition and, 72-73, 82-92
language styles and, 20-21
local monitoring of, *see* Local monitoring
maldistribution of in Zimbabwe, 137-138
of North America, 310
priorities in, 17-20
quality of, 11
slow net increases in, 33
of Southeast Asia, 111
of South Sumatra, Indonesia, 249

stable, 2
of Tamougou, Northern Benin, 284, 287
thought styles concerning, 17-20
total, 36, 309, 311, 344
Western theory of, 12, 21-24
world total, 36
of Zimbabwe, 137-138, 139, 142-143
Population, Organization, Environment, and Technology (POET) model, 380
Population density, 89
forest cover vs., 84-86, 87
soil degradation and, 133
in South Sumatra, Indonesia, 233-237, 246
in Zimbabwe, 133, 138
Population-Environment Monitoring Systems (PEMS), 9, 302, 303, 361, 369, 402
in Indonesia, 369-372
in Mexico, 372
multiple transitions and, 394
transition evidence and, 393, 394
in Zimbabwe, 372-373
Population growth, 1, 14, 17, 344, 399
in Africa, 52-53
agricultural production and, 314
in Asia, 49
deforestation and, 72-73, 227
in South Sumatra, Indonesia, 227, 231-239
dietary deficiencies and, 68
economic conditions and, 72
environmental effects of, 359
in Europe, 53-54
expectations of unlimited, 27
exponential, 33, 111
geometric, 20
history of, 5, 34-37
in Indonesia, 154-159
industrialization and, 43-44
land use changes and, 77
in Latin America, 52
limits on, 19, 28
in North America, 53-54
output and, 38
in Philippines, 158
productivity and, 38
protein cycles and, 294-295
rapid, 8, 13, 29, 33
rate of, 3
recent, 33
restraints on, 12
in Southeast Asia, *see* under Southeast Asia
in South Sumatra, Indonesia, 231-239
in Sweden, 200
in Tamougou, Northern Benin, 288, 294-295
as threat of disaster, 22
urbanization transition and, 43-44
worldwide concern about, 195
in Zimbabwe, 137-139
Power function, 350, 351
Power imbalances, 3-4
Price supports and subsidies, 65, 181
Productive transition, 400
Productivity, 12, 13, 37, 49, 58, 99, 182, 314
Protein cycles, 294-295
Proximate variables methods, 220
Public health, 25, 29, 394, *see also* Health
Public policies, *see* Governmental policies
Pulmonary malfunction, 67

QUICKMAP, 370

Radiative forcing, 62, 64
Rain forests, 1, 30, 71
destruction of, *see* Deforestation
Recycling, 336
Reductionism, 17
Reforestation, 249, 295
Regional transitions, 393-394
Relocation, 250-251
Remote sensing technology, 253, 360
Research
agenda for, 405
agricultural, 58, 60, 61
on deforestation, 100-101
on disease, 68
environmental, 254
on environmentally compatibile agriculture, 65
health related, 67
maintenance, 58
transfer of, 60
on transitions, 391-405
Resource-based agriculture, 57
Resources, 17, *see also* specific types
control of, 382
in Dominican Republic, 262-263, 267-268
exploitation of, 73, 110
fertility and, 197, 198, 204-211
labor, 359
management of, 63, 251
measurement of, 223
migration and, 197, 212-213
monitoring of, 64
mortality rates and, 197, 213-214
natural, 29, 73, 227, 394

Resources (*continued*)
richness of, 217
scarcity of, 29
in South Sumatra, Indonesia, 227, 251
in Sweden
decline in, 215
fertility and, 204-211
investment and, 217-223
migration and, 212-213
mortality rates and, 213-214
technology for expansion of, 29
water, 51, 53, 64, 68
wood, 291-294
Respiratory disease, 66
Rhodesia, *see* Zimbabwe
Rice prices, 58
Risk analysis, 345-347, 355
Risk assessment, 346
Risk factors, 9
Risk reduction, 355
Rockefeller Foundation, 61, 121, 153
Roflin, Eddy, 225
Rural environment, 18, 19, 21
Rural-urban balance, 21
Rural-urban transition, *see* Urbanization transition

Sails, 33, 38-42, 109
Salinization, 62, 64, 65
Sanitary conditions, 297-298
Satellite imagery, 360
Scale, 339-340, 341, 392-396
Science, 20, 29
Science-based agriculture, 57
Scientific laws of behavior, 19
Selection, 197
Shifting agriculture, 13, 71, 73, 78, 80
in Tamougou, Northern Benin, 291-294
Sierra Leone, 96
Singapore, 391, 401
fertility in, 126, 127, 398
mortality rates in, 121
political systems in, 132
population growth in, 51, 158
Slash and burn farmers, 226
Smallpox, 117
Smallpox vaccination, 118
Small-scale societies, 4
Social class, 382
Social Darwinism, 20
Social disorder, 29
Societal vulnerability to transitions, 341-343
Socioeconomic factors in land use transitions, 77, 82
Sociology, 2, 217, 379
Soil, 359
in Brazil, 175
degradation of, 133
in Dominican Republic, 262-263
erosion of, *see* Erosion
fertility of, 21, 72, 80, 263, 357
loss of, 62, 64, 175, 357
pollution of, 67, 175
population density and degradation of, 133
in Zimbabwe, 133
Sokpon, Nestor, 283
Solid waste, 337
Southeast Asia, 109-132, 154, *see also* Asia; specific countries
agricultural transition in, 400
agriculture in, 57, 113, 117, 118
birth rates in, 118, 309
colonial period in, 115-118
contraceptives in, 128-129
deforestation in, 87
demographic transition in, 309, 310, 400
description of, 110-112
economic growth in, 112
family planning in, 128
fertility in, 109, 125-130, 132, 398
forestry transition in, 321
history of, 113-118
infant mortality in, 132
maternal mortality in, 132
military ventures of, 114
modern period in, 118-130
fertility in, 125-130
mortality rates in, 120-121
rice production in, 121-125
mortality rates in, 120-121, 132
political power in, 113, 119
political systems in, 130-132
population of, 111
population growth in, 117, 131, 154, 158
agricultural growth and, 113
average annual rate of, 110
economic growth and, 112
exponential, 111
rapid, 109
slow, 114
trade and, 113
rice output in, 122, 123, 131
technology in, 109, 130-132
Southern Rhodesia, *see* Zimbabwe
South Korea, 51, 390, 401
South Sumatra, Indonesia, 225-251
agriculture in, 249
coffee production cycle in, 240-241, 246

conservation management in, 246, 248-249
deforestation in, 226-227
causes of, 227, 231-243
economic factors in, 242-243
description of, 227-231
fertility in, 249
government policies in, 247-251
history of, 241
methods of study of, 227-231
migration in, 237-239
multiple transitions in, 400
natural resources in, 227
population control in, 249
population density in, 233-237, 246
population growth in, 231-239
protected areas in, 226-227, 229, 230, 232, 241-242
reforestation in, 249
relocation in, 250-251
resources in, 251
Soviet Union, 314, 323, 325, 391, 401, 405
Specialization, 20, 361
Species destruction, 33, 34, 64
Sri Lanka, 51, 391
Stable population theory, 2
Stagflation, 27
Static analysis of deforestation, 88
Stoffle, Brent W., 253
Stoffle, Richard W., 253
Stress disorders, 29
Stresses, 9, *see also* specific types
Sudan, 363
Sulfur dioxide, 325
Sumatra
North, 129
South, *see* South Sumatra
West, 120, 121, 129
Super-Harvests Era in Brazil, 179-180, 182
Supply side, 48
Surapaty, Surya Chandra, 225
Surface water pollution, 324, 337
Sweden, 397
climate in, 203
commodity costs in, 203-204
cost of living in, 203, 215
demographic transition in, 195, 215-216
employment in, 202
fertility in, 201, 202
in demographic transition, 215-216
investment and, 217-223
migration and, 214
resources and, 204-211
food costs in, 208
infant mortality in, 216
investment in, 217-223
land ownership in, 202, 209, 211
Locknevi Parish in, *see* Locnevi Parish
migration in, 206, 212-213
mortality rates in, 215-216
Nineteenth Century, 198-204
population growth in, 200
resources in
decline in, 215
fertility and, 204-211
investment and, 217-223
migration and, 212-213
mortality rates and, 213-214
statistical analysis in, 204
toxicity transition in, 325
Tuna Parish in, *see* Tuna Parish
urbanization transition in, 330, 331
weather in, 203
Symbolic communication, 7
Systems models, 76

Taiwan, 51, 390, 401
Tamougou, Northern Benin, 283-300
agriculture in, 287, 288, 289-294
ecology of, 283
economics of, 283, 297
famine in, 288
fishing in, 295
food production in, 290
geography of, 283
governmental policies in, 283
health in, 297-298
malnutrition in, 288
people of, 289
poaching in, 297
population of, 284, 287
population growth in, 288, 294-295
reforestation in, 295
sanitary conditions in, 297-298
shifting agriculture in, 291-294
wood resources in, 291-294
Technology, 2, 7, 12, 399
advances in, 25
in agriculture, 58-61
bureaucratically portable, 47
death limiting, 47
effective application of, 29
engineering, 64
in local monitoring, 368
mechanical, 64
rapid expansion of, 128
role of, 29-30
in Southeast Asia, 109, 130-132
transfer of, 60
transitions in, 333, 335-336, 338

Technology (*continued*)
 in Zimbabwe, 142
Teitelbaum, Michael, 17
Terrestrial ecozones, 262-263
Thailand, 115, 116, 390
 deforestation in, 96
 demographic transition in, 400
 exports of, 124-125
 fertility in, 126, 127, 128, 398
 forest area per capita in, 92
 mortality rates in, 121
 political systems in, 132
 population of, 111
 population growth in, 51, 158
 rice output in, 122, 123, 124-125
Thought styles, 17-20
Three Mile Island, 27
Tightly linked population-environment relationship, 283, 284
Timber volume, 91
Tonkin, 115
Tourism, 226
Toxicity transition, 324-329, 337, 353
 societal vulnerability to, 341
 timing of, 340
Toxic waste, 325
Toxin release, 33, *see also* specific types
Trade, 113, 387-388, 401
Trade route transformations, 39, 40, 41
Transitions, 8, *see also* specific types
 agricultural, 314, 315, 317, 341, 397, 399, 400
 analytical properties of, 343-344
 to bureaucracy, 338
 causes of, 394
 characteristics of, 339-343
 complexity of, 392-396
 defined, 306, 392
 deforestation, *see* Deforestation
 demographic, *see* Demographic transition
 differences in, 307
 dynamics of, 307-308
 educational, 336-337
 energy, 337
 epidemiological, 29, 312-313, 340, 342, 353, 400
 evidence of, 393-394
 family of, 5, 8, 306, 307, 339, 344, 385, 400
 forestry, 316, 318- 323, 353
 fossil fuel, 33, 42-43, 109, 332-334, 338
 functional characteristics of, 352-354
 functions and, 349-355
 future research on, 391-405
 hypothesizing the shape of, 351-352
 interconnectedness of, 342, 385, 395, 396, 402
 in land use, *see* under Land use
 linkages between, 395-396
 mathematical description of, 396
 mathematical functions and, 349-355
 multiple, 393-394, 400
 natural, 8
 productive, 400
 regional, 393-394
 research on, 391-405
 resource-based to science-based agriculture, 57
 rural-urban, *see* Urbanization transition
 scale and, 339, 341, 392-396
 shape of, 351-352
 similarities in, 307
 societal opportunities during, 343
 societal vulnerability to, 341-343
 taxonomy of, 347, 392
 technological, 333, 335-336, 337
 theory of, 301-302, 338-348, 396
  defined, 348
  development of, 347-348
  governmental policy implications of, 344-347
  mathematical functions and, 349-355
  risk analysis and, 345-347, 355
  steps in development of, 347-348
 timing of, 307, 340
 toxicity, *see* Toxicity transition
 trajectory of, 312, 338, 339, 340, 343, 355
 urbanization, *see* Urbanization transition
 volatility of, 393
Transportation, 25, 336, *see also* specific types
Tribal Trust Lands (TTLs), 135, 136
Tropical forests, 13
 destruction of, *see* Deforestation
TTLs, *see* Tribal Trust Lands
Tuberculosis, 66, 67
Tuna Parish, 199, 214, *see also* Sweden
 aggregate analysis of, 199
 demographic transition in, 215-216
 fertility in, 204-211, 215-216
 food costs in, 208
 individual analysis in, 201-202
 land ownership in, 209, 211
 migration in, 206, 212-213
 population growth in, 200
 resources in
  fertility and, 204-211
  migration and, 212-213

mortality rates and, 213-214
response to costs in, 205-210
Tunisia, 390

Uganda, 96
Underproduction, 27
UNEP, *see* United Nations Environment Program
UNESCO, 363
UNFAO, *see* United Nations Food and Agriculture Organization
UNFPA, *see* United Nations Population Fund
United Kingdom, 333, 334
United Nations, 36, 154, 158
United Nations Environment Program (UNEP), 154, 362-363
United Nations Food and Agriculture Organization (UNFAO), 73, 89, 94, 226, 365
United Nations International Conference on Population, 139
United Nations Population Fund (UNFPA), 139, 141, 153, 155
United States, 314, 401
energy consumption in, 51, 333
energy efficiency in, 391
gasoline prices in, 334
mortality rates in, 314
technological transition in, 335
toxicity transition in, 327, 330
urbanization transition in, 330
wildlife protection in, 405
United States Agency for International Development (USAID), 1, 363, 364
University of Minessota, 61
University of the Philippines, 122
Urban environment, 18, 21, 23, 30, *see also* Urbanization transition
Urbanization transition, 8, 18, 19, 330-331, 332
in Benin, 332
in Brazil, 184, 185
defined, 330
deforestation and, 81
demographic transition and, 48
in India, 332
in Indonesia, 162-164, 332
in Mexico, 332
population growth and, 43-44
in Sweden, 332
in United States, 332
Urban-rural balance, 21
USAID, *see* United States Agency for International Development
Utopianism, 18, 19, 20, 22, 28
Vietnam, 113, 116
deforestation in, 96
fertility in, 126, 127, 128, 129
forest area per capita in, 92
internal turmoil in, 128
mortality rates in, 120, 121
rice output in, 122, 123
Vietnam War, 27
Village-based family planning programs, 156
Violence, 29

Wall Street Journal, 28
Wastes, 324, 337
Water availability, 65
Water demand, 65
Water-logging, 62
Water pollution, 62, 64, 67, 324, 338
Water resource development projects, 68
Water resources, 51, 53, 64
Watershed protection, 64
Water use efficiency, 65
Weather, 203, 271
Weeds, 62
West, Patrick, 283
Western population and environment theory, 12, 21-24
West Java, 129, 132
West Sumatra, 120, 121, 129
Wheat prices, 58
WHO, *see* World Health Organization
Wildlife protection, 405
Williams, Andrew L., 253
Winter, Jay, 17
Wirosuhardjo, Kartomo, 153
WMO, 363
Wood resources, 291-294
World Bank, 403
World Health Organization (WHO), 67, 155, 363
World Resource Institute (WRI), 226
World systems, 20
World War II, 27
WRI, *see* World Resource Institute

Xylenes, 330

Zimbabwe, 52, 133-151, 388, 390, 404
age structure in, 147
economy of, 142
employment in, 146-148
erosion in, 134-137
family planning in, 140-142
fertility in, 138, 140-142
governmental policies in, 139, 142-143, 148-150
land reform in, 143-148
land tenure in, 134-137

Zimbabwe (*continued*)
land use in, 143-150
multiple transitions in, 400
National Erosion Survey in, 134, 136, 137
non-agricultural employment in, 146-148
population of, 137-138, 139, 142-143
Population-Environment Monitoring Systems, 372-373
population growth in, 137-139
resettlement in, 143-148
technology in, 142
Zimbabwe National Family Planning Council (ZNFPC), 140, 141
Zimbabwe Reproductve Health Survey (ZRHS), 141
Zinn, Frank D., 357
ZNFPC, *see* Zimbabwe National Family Planning Council
ZRHS, *see* Zimbabwe Reproductive Health Survey